The Big Picture Approach to Learning Anatomy and Physiology

Dr. Bruce Forciea

Copyright ©, 2011 Bruce Forciea. All rights reserved.

Scientific information continuously progresses and changes. The author has done substantial work in order to ensure the information presented in this book is accurate, up to date, and within acceptable standards at the time of publication. The author is not responsible for errors or omissions, or for consequences from the application of the information contained in this book and makes no warranty, expressed or implied, with regard to the contents of this book.

Library of Congress Cataloging-in-Publication Data

Forciea, Bruce

>The Big Picture Approach to Learning Anatomy and Physiology, First edition.
>Includes index

>ISBN#: 978-1-257-82032-0

About the Author

Bruce Forciea is a full-time science instructor at Moraine Park Technical College. He primarily teaches anatomy and physiology. Besides developing courses, teaching, and dabbling in digital media, he enjoys playing guitar and writing music. Dr. Forciea is trained as a chiropractor and attended Parker College of Chiropractic. He ran a full-time practice for more than a dozen years and treated thousands of patients before accepting a teaching position. He has also presented his various projects at regional, national and international conferences.

Acknowledgements

Many thanks to Wikimedia Commons for making good anatomical images available, and to my students for helping to edit this book with their suggestions and comments.

For students...

In hopes of easing their struggles.

Contents

Introduction	7
Chapter 1 Body Basics	10
Chapter 2 Taking the Creepiness out of Cells	24
Chapter 3 Monstrous Metabolism	42
Chapter 4 Don't be Terrified of Tissues	52
Chapter 5 Gimme Some Skin Man	60
Chapter 6 Bones, Bumps and Holes	68
Chapter 7 Managing Muscles	98
Chapter 8 What's a Joint Like this Doing in a Girl Like You?	142
Chapter 9 Don't Get Nervous About the Nervous System	154
Chapter 10 Making Sense of the Sensory System	188
Chapter 11 Learning the Hell out of Hormones	204
Chapter 12 I Want to Drink Your Blood But I First Want to Know What's In It	218
Chapter 13 Emphatic over the Lymphatic System	228
Chapter 14 Don't be Immune to the Immune System	234
Chapter 15 Gettin the Blood to All the Right Places	240
Chapter 16 In With the Good Air, Out With the Bad	260
Chapter 17 There's A Lot to Making Pee Pee	276
Chapter 18 You Already Know About Fluids and Electrolytes, How About Some Acid-Base Balance?	294
Chapter 19 Taking the Indigestion out of the Digestive System	304
Chapter 20 Fun with Reproduction	326
Answers to Review Questions	341
Appendix One	347
Index	353
Image Credits	368

Introduction

The Big Picture Approach to Learning Anatomy and Physiology

You wouldn't think that learning anatomy and physiology would be so difficult, especially considering that you've lived in your own human body your whole life. As simple as your body may seem, in reality it is a very complex thing and it becomes even more complex when you really get into the details about how it's put together and how all the pieces work. There are literally thousands of anatomical parts and many complex systems in the human body. Learning about it can be a daunting task for even the savviest student.

This book presents the concept that learning anatomy and physiology can be easy if you learn the big picture about a concept before learning the details (what I will call the nitty-gritty). The problem is that traditional textbooks *begin* with all of the nitty gritty details causing many students to get lost before they even have a chance to *see* the big picture.

This book will definitely help you to see the big picture first. In fact this is the way I want you to learn:

1. See the Big Picture
2. Understand the Big Picture
3. Learn the nitty-gritty details

I will present each section by giving you the Big Picture concept. You really need to work to understand this first. Once you do you will begin to fill in the details. I will present a good deal of details but not as many as a traditional 1000 plus page textbook, so you may still need to consult your text. What you will get is an *understanding* of what you need to know in a traditional two-semester course in college anatomy and physiology.

The key to learning anatomy and physiology is to understand how the body is put together. That's what the big picture approach helps you to do. Once you understand this you can fill in the details. You learn the general concepts first then the specific details. The problem is that many students get into memorizing parts without understanding how they are put together.

I realized this back in my student days when I took a rigorous gross anatomy class in professional school. This class included about fifteen hours of cadaver dissection per week and went on for two semesters. We were responsible for knowing everything we dissected including not only the major muscles, bones, nerves and organs, but even the more obscure things such as the blood and nerve supply to muscles and anatomical spaces.

I would go back to my small apartment and put together lists of anatomical structures and try to memorize them by brute force. This was an exhausting way to learn and I did not by any means see the big picture. About a third of the way through the course I picked up my old A&P book, the one that I used early on in my undergraduate studies. I went through the chapters and made my own notes summarizing *how* the body was put together. This helped me immensely. For example, once I understood *how* the arm was put together I could more easily learn the names of all of the parts.

I would always begin my anatomy study time by going over the big picture first, then examining the details. I carried this technique with me throughout my studies and still use it to this day. Now a good portion of what I do is to get students to see the big picture.

This book is organized according to systems much like a traditional college level anatomy and physiology textbook. In fact you can and should use this book with your text. This book won't cover all of the concepts and details that a traditional 1000+ page text will but it will help you to understand the big picture enough so that you can better understand the details in your text.

I personally use this book in each of my college level anatomy and physiology courses. Students find the book much easier to read than the traditional texts and say that it helps them to learn the concepts much faster. Many go through the courses without opening a traditional text.

The important thing is to approach your studies with an open mind and don't get discouraged if you don't understand something right away. Sometimes concepts need to sink in. Also, the study of anatomy and physiology builds on information learned in earlier sections. You may need to frequently review earlier concepts. That's okay, I still review concepts every semester too.

I sincerely wish you the best success in your study of anatomy and physiology and I am grateful to be of any help in the process. I hope you enjoy this book as much as I have putting it together.

Please visit our support site for videos, powerpoints and much more:

www.learnanatomyphysiology.com

www.bruceforciea.com

Chapter 1

Body Basics

Chapter 1

Body Basics

Before we delve into the nitty gritty of the human body it's a good idea to get some basics under our belts. In this chapter we will learn some basic concepts and terminology that apply to anatomy and physiology. It shouldn't be too painful and remember to keep the big picture in mind.

First of all let's talk about what the words *anatomy* and *physiology* mean. Anatomy basically refers to the body parts (structure) while physiology refers to how they work (function). How parts are constructed goes together with how they work.

Homeowhatsthis…??

One important concept in A&P is the concept of homeostasis or the ability of an organism to maintain a range of values of something (equilibrium). The human body does this by way of feedback mechanisms. Let's use an analogy to learn about feedback mechanisms.

A good example of this is the thermostat. The thermostat monitors the temperature and feeds this information back to the furnace or air conditioner. For example if the thermostat is set at 70 degrees and the room becomes cooler the thermostat will kick in and turn on the heat. Likewise if the temperature rises above 70 degrees the heat turns off and the air conditioner kicks in to cool the room.

This is an example of what is known as negative feedback. We can think of this in terms of stimulus and response. In negative feedback the stimulus and response are opposite. In the case of the room getting cold the stimulus is "room colder" and the response is "heat up room" so the stimulus and response are opposite. The same goes for the stimulus "room hotter" and the response "cool room." Stimulus and response are opposite and again we have negative feedback.

Now let's say that I installed a thermostat in my house. I am not an electrician by any means. So true to form I wired up the darned thing backwards. Now when the room gets warmer the heat turns on instead of the air conditioner. The result is to make the room hotter and hotter. This is an example of positive feedback. In positive feedback the stimulus and response are the same. Our stimulus "room getting hotter" results in the response "make room even hotter." You can see where things get magnified with positive feedback.

Fortunately there are few systems in the body that use positive feedback. Most use negative feedback. An example of negative feedback is the regulation of blood sugar or glucose. If blood glucose gets too high the body responds by secreting a hormone to bring it back down. Likewise if blood glucose gets too low the body responds by secreting a different hormone to bring it back up. Examples of positive feedback are the contractions that occur during labor and delivery. The contractions begin and become more and more intense as labor progresses.

The Big Picture: How the Body is Organized

The body consists of solid parts and hollow parts. The solid parts are the arms and legs while the hollow parts are in the head, back, chest and stomach.

There is a big hollow part in the front and a smaller hollow part in the back. Hey, the organs have to go somewhere! The front of the body is known as the ventral part and the back of the body is known as the dorsal part (think of a dorsal fin on a shark). So the hollow part in the front is called the ventral cavity. The hollow part in the back is called the dorsal cavity.

Now to add a bit more detail. The upper part of the ventral cavity is in the chest and is called the thoracic cavity. The lower part of the ventral cavity is the stomach and called the abdominopelvic cavity (think abdomen and pelvis). The two parts are separated by the diaphragm. There's gotta be a hollow part in your head to contain your brain and that's called the cranial cavity. There's also a hollow part that contains the spinal cord called the spinal cavity (fig. 1.2).

The arms and legs are not hollow (at least they should not be) and are called the upper and lower extremities.

Freeze! It's Anatomical Position Time

We all know how movable the body is. This is a good thing for us and a not so good thing for anatomists who are trying to speak a common language to describe all of those parts. We have to have a reference point and we do. It's called the anatomical position.

Figure 1.3 shows us what anatomical position looks like. The body is facing forward with the arms at the sides and feet facing forward. This is the reference position that allows us to tell where all those parts are.

Anatomical Terms (fig. 1.9, 1.10)

What? You don't know your cephalon from your coccyx?

The study of anatomy goes way back in human history and so do the terms. Many have Latin roots and are hard to pronounce. Anatomy is like a whole new language so the time spent learning terms is time well spent that will help you to learn anatomy. The best way to do this is to begin with a few terms then use them in sentences. Practice, practice, practice. Hey, you'll sound smarter too!

Some terms relate to position. We use positional terminology to specify locations of anatomical structures. The positional terms usually go in pairs. For example superior and inferior go together. Superior means above and inferior means below.

We could write a statement saying that the head is superior to the chest, or to be more specific—the cephalon is superior to the thorax.

AND...

The reverse would also be true:

The thorax is inferior to the cephalon.

Here are some other terms:

Anterior means toward the front while posterior means toward the back.

Example: *the sternum is anterior to the heart...*

Or...

The heart is posterior to the sternum.

Medial means toward the midline of the body while lateral means away from the midline.

Example: *the ears are lateral to the nose*

Or...

The nose is medial to the ears.

Proximal means toward the trunk of the body while distal means away from the trunk.

Proximal and distal are usually used when describing structures in the extremities.

Example: *the elbow is proximal to the wrist.*

Or...

The wrist is distal to the elbow.

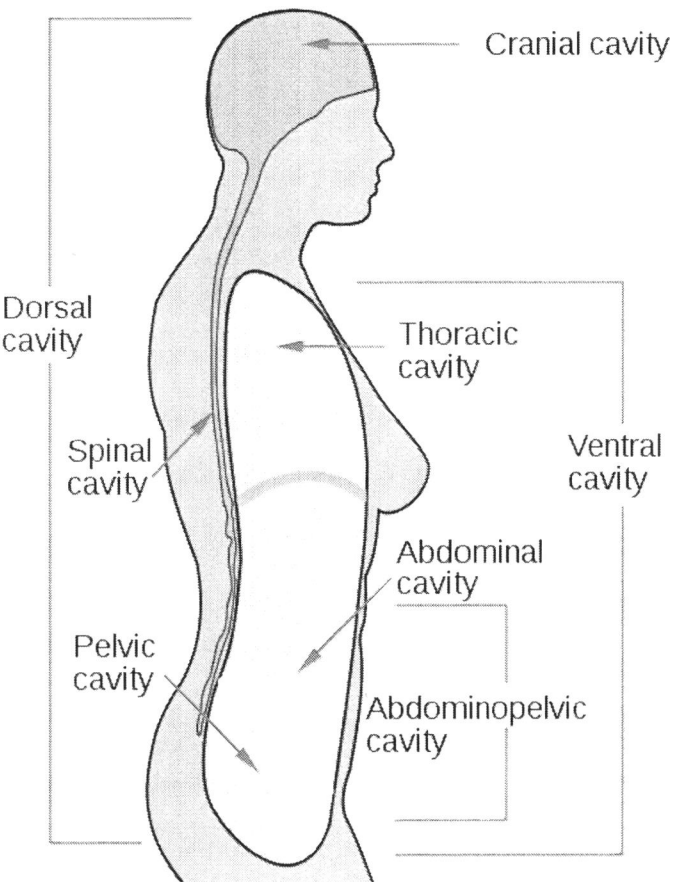

Figure 1.2 Body Cavities

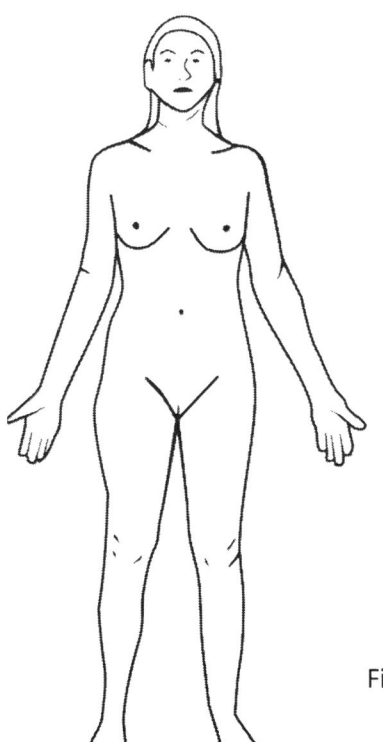

Figure 1.3. Anatomical Position.

Table 1.1

Common Term	Anatomical Term	Region
Foot	Pes	Pedal
Shin	Crus	Crural
Calf	Sura	Sural
Front of knee	Patella (knee cap)	Patellar
Back of knee	Popliteus	Popliteal
Thigh	Femorus	Femoral
Groin	Inguina	Inguinal
Butt	Buttock	Gluteal
Stomach	Abdomen	Abdominal
Low Back	Lumbus	Lumbar
Chest and Middle back	Thorax	Thoracic
Lateral chest	Pectorus	Pectoral
Middle chest	Sternum	Sternal
Neck	Cervicis	Cervical
Chin	Mentum	Mental
Head	Cephalon	Cephalic
Shoulder	Acromion	Acromial
Arm	Brachium	Brachial
Elbow (front)	Antecubitus	Antecubital
Elbow (back)	Olecranon	Olecranal
Wrist	Carpus	Carpal
Hand	Manus	Manual
Forearm	Antebrachium	Antebrachial

Superficial means toward the surface while deep means under the surface.

Example: *the skin is superficial to the stomach.*

Or...

The stomach is deep to the skin.

Ipsilateral means on the same side while contralateral means on the opposite side.

Example : *the right shoulder and elbow are ipsilateral.*

Or...

The right shoulder and left elbow are contralateral.

The Big Picture: Anatomical Planes

You can slice the body up 3 different ways.

Slicing and Dicing

Anatomists like to slice up the body into sections to really see how it's put together. Also, health care providers use CT scans and MRIs to look at slices of the body in order to detect problems. We can slice the body up in a number of ways using just a few basic planes.

Body Planes (figs. 1.6, 1.7)

The sagittal plane divides body into right and left portions.

A mid-sagittal plane runs right down the middle of the body (dividing it into right and left portions).

A parasagittal plane runs laterally to the mid-sagittal plane on either side.

The transverse plane divides body into superior and inferior portions.

The coronal plane divides body into anterior and posterior portions.

The oblique plane divides the body at an angle.

Since the abdomen contains a lot of organs it is often the source of pain or illness. Often health care providers need to describe the location of a particular abdominal problem. In these cases two methods of dividing the abdomen can be used. In one method the abdomen is divided into nine sections much like tic-tac-toe. The other method is a bit simpler in that the abdomen is divided into four sections.

Four planes are needed in order to divide the abdomen into nine equal sections. There are two parasagittal planes (sometimes called lateral lines) and two transverse planes. The superior transverse plane is called the transpyloric plane and the inferior plane is called the transtubercular plane. The center of the nine regions is the umbilicus. The three superior regions are the epigastric and right and left hypochondriac. The middle regions are the umbilical and right and left lumbar. The lower regions are the hypogastric and right and left inguinal (fig. 1.5).

The other method of dividing the abdominal area consists of using a transverse and mid-sagittal plane intersecting at the umbilicus. This results in 4 quadrants including the right and left upper quadrants and right and left lower quadrants (fig. 1.4).

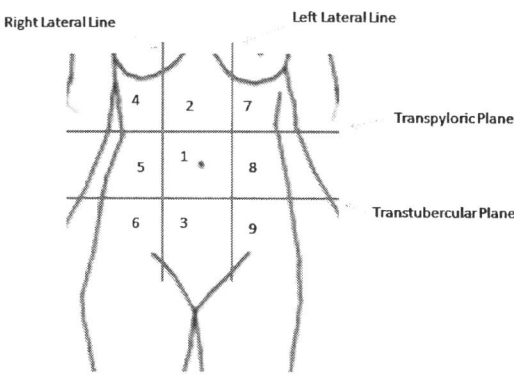

Figure 1.5. Planes dividing the abdominal region into 9 areas.

1. Umbilical
2. Epigastric
3. Hypogastric
4. Right hypochondriac
5. Right lumbar
6. Right Iliac
7. Left hypochondriac
8. Left lumbar
9. Left iliac

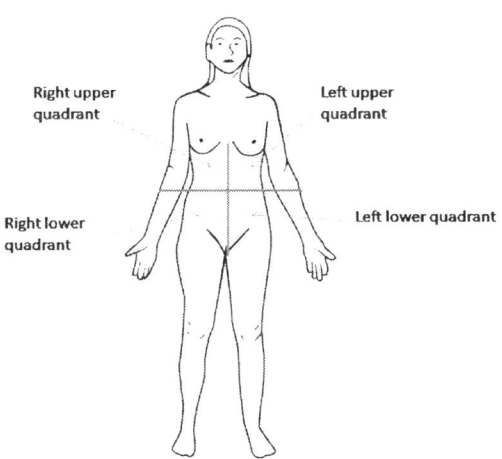

Figure 1.4 Abdominal quadrants

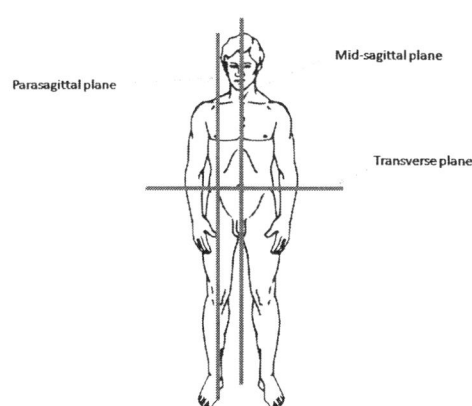

Figure 1.6. Anatomical planes

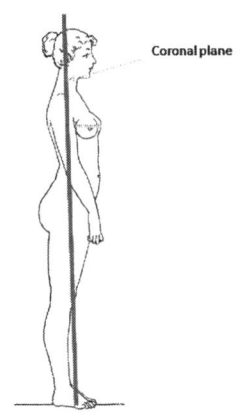

Figure 1.7. Coronal plane

Overview of Body Systems

Let's look at an overview of all of the body systems. We will be covering these in more detail in subsequent chapters but take a look at the big picture.

These include:

Integumentary (skin)	Skeletal
Muscular	Nervous
Endocrine	Lymphatic
Digestive	Respiratory
Urinary	Reproductive

The integumentary system consists of the hair, skin, nails, sweat glands, and sebaceous glands. Its function is protection and regulation of body temperature. The integumentary system also supports sensory receptors that send information to the nervous system.

The skeletal system consists of the bones, ligaments, and cartilage. It provides protection and support and produces red blood cells. It also stores chemical salts.

The muscular system produces movement, helps to maintain posture, and produces heat.

The nervous system consists of the brain, spinal cord, and receptors. It receives sensory information, detects changes and responds by stimulating muscles and glands.

The endocrine system is a series of glands that secrete hormones. The endocrine system contains many feedback systems to help maintain homeostasis. The glands include:

Pituitary	Thyroid	Parathyroid
Adrenal	Pancreas	Ovaries
Testes	Pineal	Thymus
Hypothalamus		

The cardiovascular system includes the heart, arteries, capillaries and veins. The function of the cardiovascular system is to transport blood.

The lymphatic system includes the lymph vessels, lymph nodes, thymus and spleen. The function of the lymphatic system is to return fluid to blood as well as transport some absorbed food molecules and defend against infection.

The respiratory consists of the nasal cavity, lungs, pharynx, larynx, trachea, and bronchi. The respiratory system supplies the body with oxygen and eliminates carbon dioxide.

The digestive system consists of a large tube Known as the alimentary canal and includes:

Mouth	Tongue	Teeth
Salivary glands		Pharynx
Esophagus	Liver	Gallbladder
Pancreas	Intestines	

The function of the digestive system is to receive, break-down, and absorb food. It also eliminates wastes.

The urinary system includes:

Kidneys   Ureters   Urinary bladder
Urethra

The function of the urinary system is to remove wastes, maintain water and electrolyte balance, and store and transport urine.

The male reproductive system includes:
Scrotum   Epididymis
Vasa deferens   Testes
Seminal vesicles   Prostate
Bulbourethral glands   Urethra
Penis

The female reproductive system includes:
Ovaries   Uterine tubes   Uterus
Vagina   Clitoris   Vulva

The function of the reproductive systems is to pass genetic information down to future generations.

Putting it all Together

You might want to start seeing the body with x-ray vision. It will help to understand where all of those parts are by studying a transparent view of the body. Figure 1.8 will help you to see the organs through the skin.

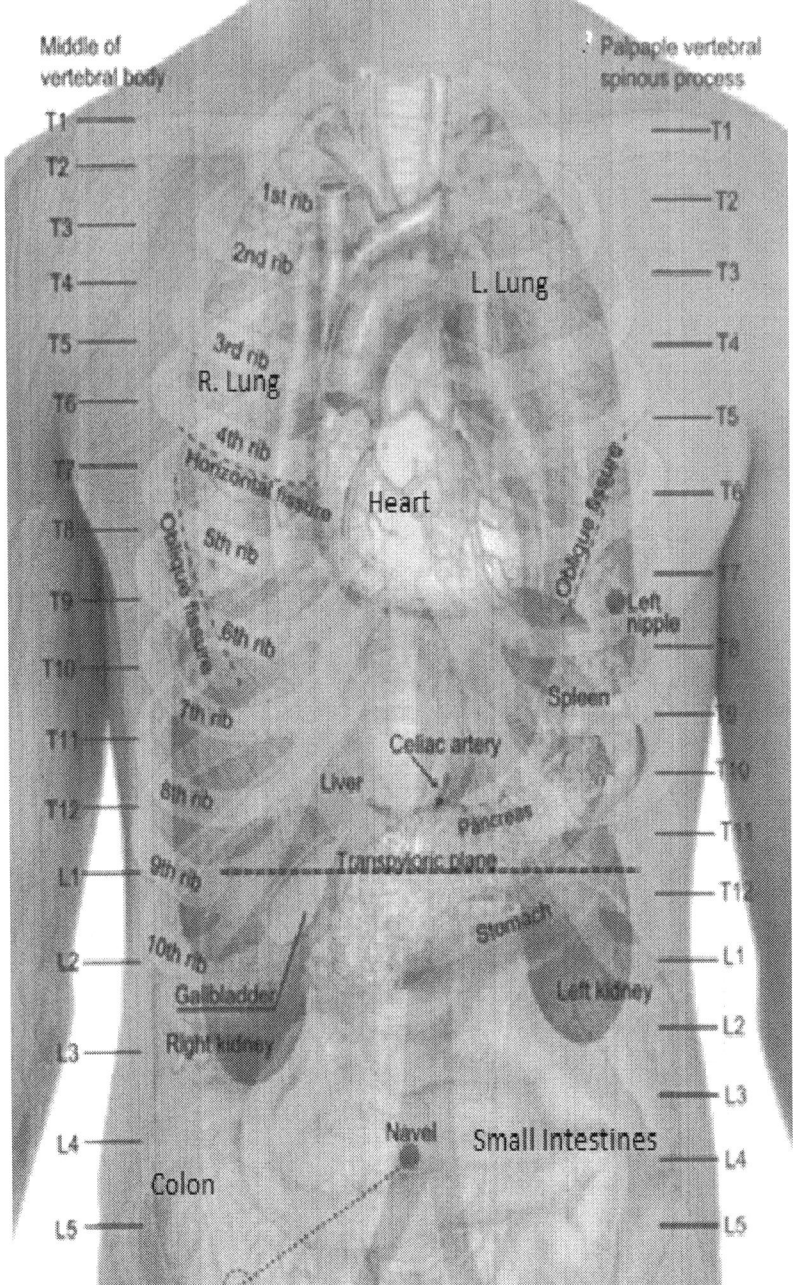

Fig. 1.8. See-through version of the body.

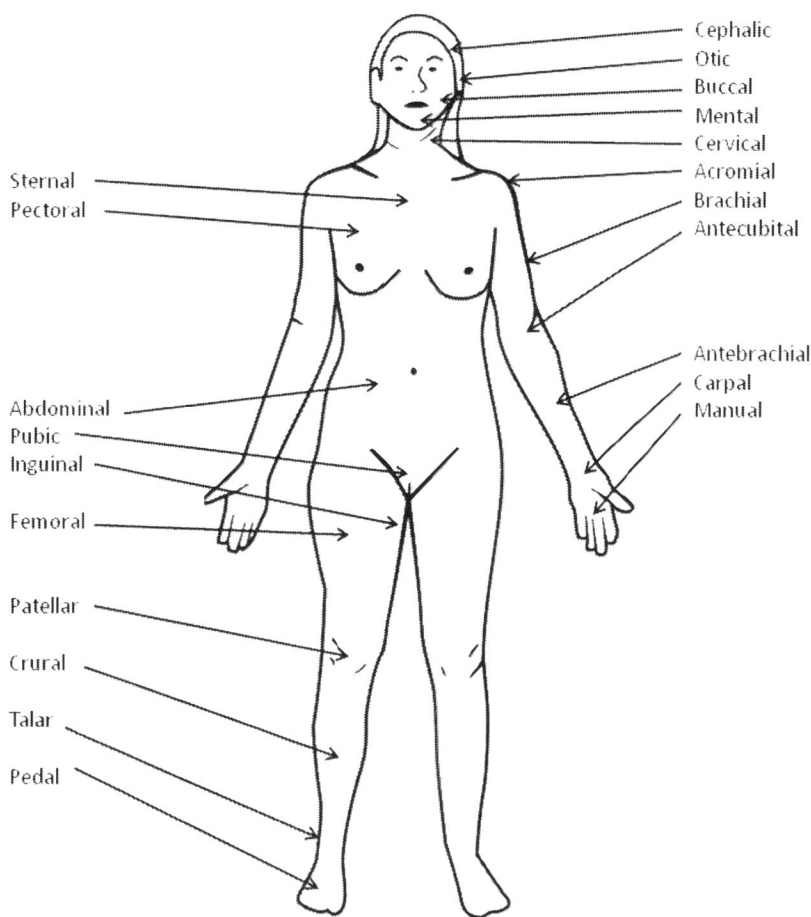

Figure 1.9. Anatomical regions, anterior view.

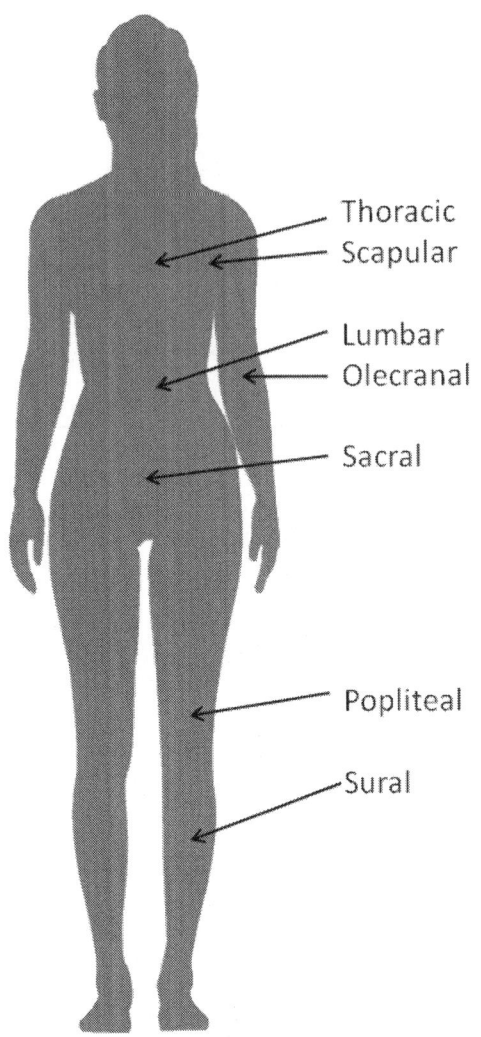

Figure 1.10 Anatomical regions, posterior view.

Review Questions Chapter 1

1. The study of physiology involves which of the following:

   a. Structure of the body
   b. Function of the body
   c. The position of the body
   d. All of the above

2. Homeostasis incorporates the use of _____.
   a. Muscles
   b. Feedback systems
   c. Equilibrium
   d. Movement

3. Which of the following is an example of negative feedback:
   a. Thermostat
   b. Sound system feedback
   c. Stimulus and response are the same
   d. An increase in secretion of a substance causing a subsequent increase in the same substance

4. The _____ level of complexity is greater than the _____ level of complexity:
   a. Organ, molecule
   b. Atom, molecule
   c. Organ system, organism
   d. Molecule, cell

5. Which body system secretes hormones:
   a. Skeletal
   b. Muscular
   c. Integument
   d. Endocrine

6. Nails are part of which body system:
   a. Endocrine
   b. Muscular
   c. Skeletal
   d. Integument

7. Joints are part of which body system:
   a. Muscular
   b. Integument
   c. Nervous
   d. Skeletal

8. In the 9 region abdominal divisions the _____ plane is superior to the _____ plane:
   a. Transpyloric, transtubercular
   b. Sagittal, transverse
   c. Mid-sagittal, coronal
   d. Transtubercular, transpyloric

9. The _____ region is inferior to the _____ region:
   a. Umbilical, hypogastric
   b. Right inguinal, right hypochondriac
   c. Epigastric, umbilical
   d. Left hypochondriac, right hypochondriac

10. In the 4 quadrant abdominal regions 2 planes intersect at the _____
    a. Stomach
    b. Diaphragm
    c. Umbilicus
    d. Bladder

11. The _____ plane divides the body into anterior and posterior sections:
    a. Sagittal
    b. Coronal
    c. Transverse
    d. Oblique

12. The head is called the _____:
    a. Cervical
    b. Cephalix
    c. Cephalon
    d. Cranish

13. The anterior part of the leg below the knee is called the _____ region:
    a. Femoral
    b. Popliteal
    c. Crural
    d. Sural

14. The forearm is called the _____
    a. antebrachium
    b. brachium
    c. cubital
    d. olecranal

15. The highest part of the shoulder is called the _____ region:
    a. Pectoral
    b. Acromial
    c. Thoracic
    d. Cervical

16. This positional term means "in front of"
    a. Medial
    b. Anterior
    c. Distal
    d. Superior

17. This positional term means "on the same side as"
a. Proximal
b. Ipsilateral
c. Anterior
d. Medial

18. This positional term is typically used for describing the extremities:
a. Anterior
b. Superficial
c. Medial
d. Proximal

Chapter 2

Taking the Creepiness Out of Cells

Chapter 2

Taking the Creepiness Out of Cells

Cells, cells, cells. Yup there are billions of these little guys (or gals) in your body. Some are simple little buggers with just a few parts while others are pretty complicated. Generally the more complicated ones are like people; they have brains to speak of. This tiny cell brain is called the nucleus and it works a lot like your brain telling your body what to do. Your brain can tell you to pick up your feet so you don't trip on a step or it can tell your heart to beat faster when you are frightened (like just before an A&P exam). In order for it to do this there must be muscles to move your bones and a heart to beat faster. In your body these things are called organs. In cells they're called organelles (fig. 2.1).

The Big Picture: Cells

Cells are like tiny versions of our bodies. They even have little organs called organelles (how cute is that!).

So the cell is like a wee version of your body. It has a skin (called the cell membrane), organs (organelles), and a brain (the nucleus). Like a lot of bodies, cells have a bit of fat in them as well. In fact this fat, or lipid, is in the cell membrane. Also, like your body, the cell has a good amount of fluid inside. This fluid is called the cytoplasm. The organelles are suspended in the cytoplasm.

Cells also have a skeleton called the cytoskeleton. The cell skeleton does not contain bones but instead contains rod-like shaped proteins called protein filaments. Some of these rods are hollow and act as kind of a cell circulatory system.

Like that mass of grey matter between your ears, the nucleus also has a large store of information. The information isn't in grey matter though. It's locked up in a molecule called DNA or (take a deep breath) deoxyribonucleic acid. DNA contains lots of info and is locked up safely in the nucleus like your brain is locked up in your cranium.

The Big Picture: DNA, One Crazy Twisted Ladder

DNA consists of right-angle pieces called nucleotides connected together and twisted into a double helix.

Let's build us one crazy twisted ladder. Our ladder consists of two sides and a series of rungs. The problem is that we only have right angle pieces to build the ladder. How can this be done?

Well we could connect the right angle pieces together like in fig. 2.2. This would give us one side of the ladder. We could then build the other side and connect the two sides together like in fig. 2.2. We now have a complete ladder.

This is much like how DNA is constructed. Each right angle piece is called a nucleotide. The nucleotide consists of three parts we will call S, B, and P. The 3 parts go together as illustrated in figure 2.3. In reality the "S" is called a deoxyribose sugar, the "P" is called a phosphate and the "B" is a nitrogen containing base.

There are four different types of "B" pieces that fit together in a specific way. We can call these BA, BG, BC, and BT. BA fits into BT and BG fits into BC. That's the only way they fit together. In reality "BA" represents the base adenine, "BG" represents guanine, "BC" represents cytosine, and "BT" represents thymine.

Once we construct our ladder we can now perform the crazy twisted part. Let's take our ladder and twist it so it looks like figure 2.4.

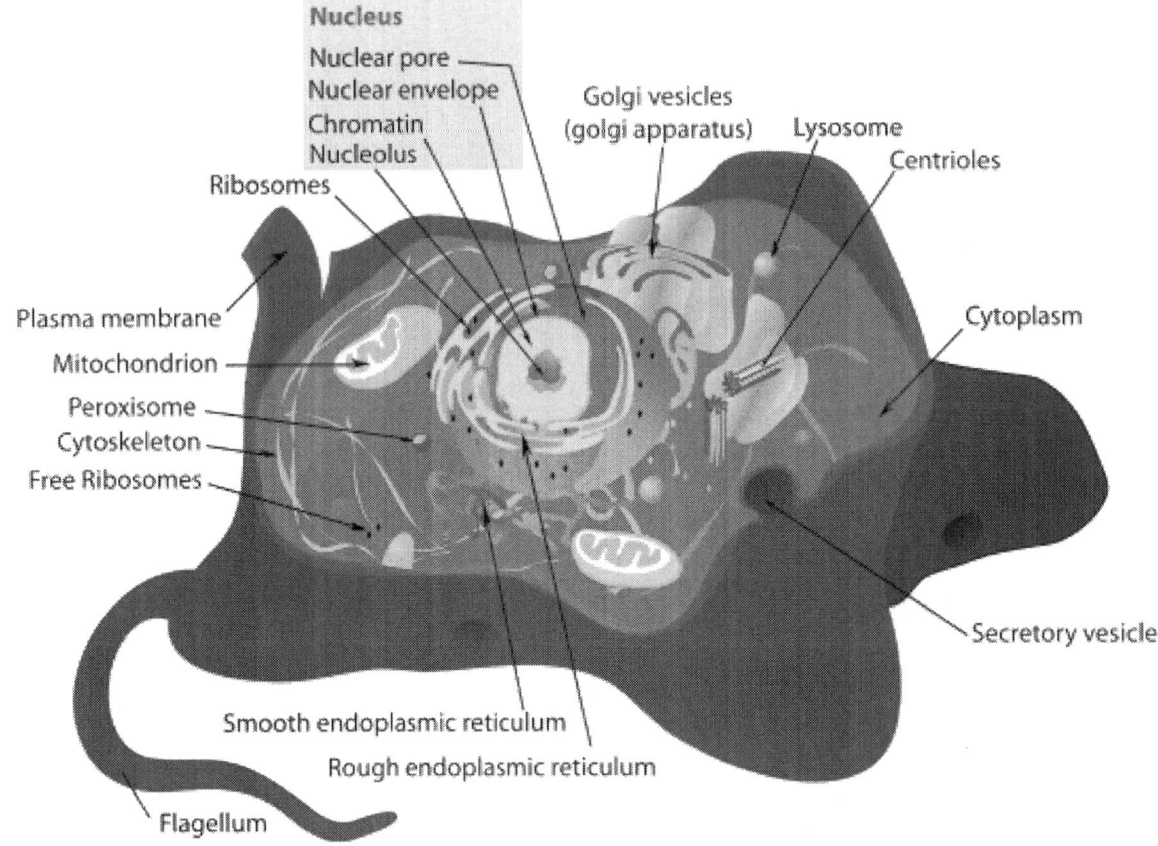

Figure 2.1. The cell contains a variety of organelles.

Taking the Creepiness Out of Cells 27

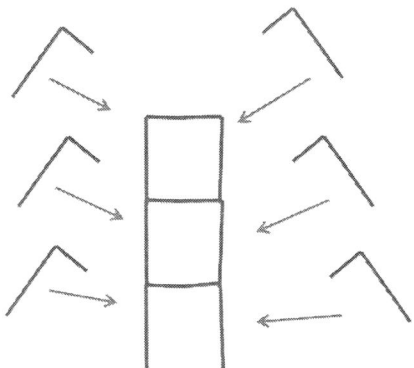

Fig. 2.2. DNA is constructed with right-angle-like pieces.

Fig 2.3. DNA is constructed from right-angle pieces called nucleotides. Each nucleotide has a 5-carbon sugar, a phosphate group and one of 4 nitrogen bases.

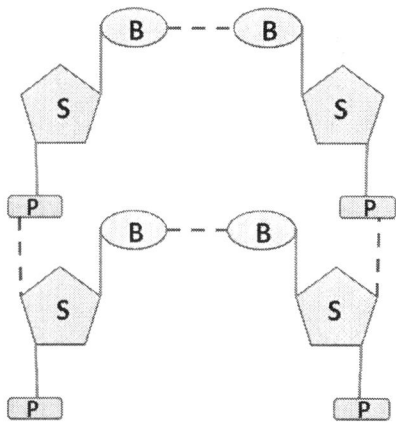

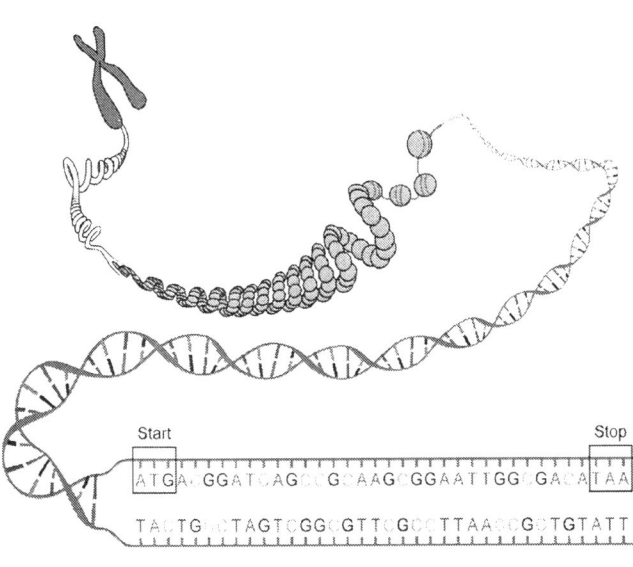

Figure 2.4. DNA is coiled into chromosomes. DNA contains information in its base pairs.

The Big Picture: Transcription and Translation

Information moves from the nucleus to the other parts of the cell by way of DNA, RNA and proteins.

The ACME Toy Rocket Company

Remember that the "brain" of the cell is in the nucleus. Well, information needed to control the cell is "locked up" in the nucleus. The cell has to have a way to transport this information out of the nucleus to other parts of the cell or to other cells. Fortunately, the cell has a system for moving information out of the nucleus. This system consists of two processes called transcription and translation.

Let's visit the ACME Toy Rocket Company to learn about transcription and translation (figs.2.5, 2.6). The boss is kinda high strung. His name is Dwight Nathan Anderson (DNA for short). Dwight is very paranoid about trade secrets and is always afraid that someone is going to steal his secret method for building toy rockets. To help protect his valuable information Dwight keeps it in his head. He is a walking library of toy rocket information.

Myron N. Ackley (mRNA) is Dwight's faithful assistant. He has been with the company for as long as anyone can remember. He is always available when Dwight needs something. He also understands Dwight's special shorthand.

Terry N. Applegate (tRNA) is the chief rocket assembler. She diligently works at her bench and loves her work at ACME. She is so fond of her job that she feels right at home at her bench with pictures of her family and especially her pet pooch Rocko (ribosome). In fact she actually named her work bench after Rocko.

Okay, so here's how toy rockets get built at ACME. Dwight gets an order for a toy rocket. Each rocket is custom built and includes a cone, midsection and tail. Also, each of the three parts can be a different color. So the information Dwight accesses includes the part and the color for each of the three parts.

Since Dwight is so paranoid he has a secret code that only he and Myron know. For example, let's say an order just came in for a custom toy rocket. The customer wants a red cone, a blue midsection and a green tail. Dwight's code then is ATT for the red cone, CGC for the blue midsection and CAT for the green tail.

Because of the secrecy, Myron also has a code that corresponds to Dwight's. For every "A" of Dwight's code Myron writes a "U." Likewise for every "T" Myron writes an "A." The same sort of thing happens with "G" and "C." Dwight tells Myron "G" and he writes down a "C." It's kind of an opposite thing.

So for Dwight's code of ATT Myron writes down UAA. In other words Myron has *transcribed* the information from Dwight.

Now Myron has to take his message out of Dwight's office (nucleus) over to Terry the chief assembler. He gives the message to Terry who understands what each three letter code means. She reads the first code "UAA" and grabs a red cone. She read the next code "GCG" and grabs a blue midsection. She then reads the last code "GUA" and grabs a green tail piece. Terry has translated the message from Myron and assembled the rocket just as the customer ordered. All the work was done on Terry's desk (Rocko the ribosome).

Now the rocket needs to be shipped. It needs to go over to section G (golgi apparatus) for shipping. Section G then packs up the rocket in a neat package (vesicle) and sends it on its way out of the company. It exits the company by way of exocytosis.

Think of lots of Terrys sitting at desks in the assembly department. The assembly department is like the organelle called the endoplasmic reticulum studded with ribosome desks. This is where toy rockets (proteins and polypeptides) are assembled.

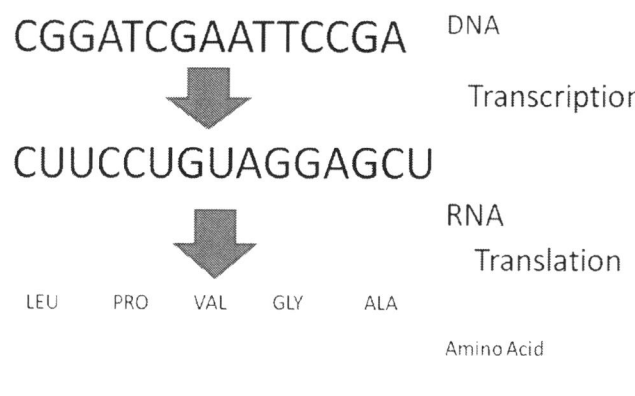

Figure 2.5. Information flows from DNA (transcription) to RNA (translation) to Protein. Each 3-base sequence codes for an amino acid.

Understanding how the ACME Toy Rocket Company works should help you to understand transcription and translation.

During transcription the information encoded by the sequence of bases on DNA (Dwight) needs to get out to the protein making machinery at the ribosome (Terry's desk). The first step in translation is to expose the information on DNA. A special enzyme called RNA polymerase helps to unwind the DNA to expose the bases.

Next a single stranded messenger RNA (mRNA) (Myron) molecule "reads" the sequence of bases. The first 3-base code is a start code followed by codes for various amino acids. Some amino acids have more than one code so there is some redundancy in the coding. The 3-base code on mRNA is known as the codon.

The mRNA (Myron) then takes its message out of the nucleus by moving through a nuclear pore and delivers it to the protein making machinery known as the ribosome (Terry's desk).

Translation occurs at the ribosome. Ribosomes contain large and small subunits. The messenger RNA (mRNA)(Myron) carries the information for making the protein to the ribosome. There is a binding site for mRNA on the small ribosomal subunit. The transfer RNA (tRNA) (Terry) attaches to a binding site on the large ribosomal subunit.

Transfer RNA (Terry) reads the information on the mRNA (Myron) and assembles the protein (toy rocket) accordingly. The 3 base code on the tRNA is known as the anticodon. There are 20 mRNA molecules, one for each of the 20 amino acids. On the other end of the tRNA is an amino acid binding site.

The protein (toy rocket) is assembled one amino acid at a time. The process stops when a stop code is reached. The completed protein then detaches from the tRNA (Terry) and the ribosome (Terry's desk) splits into its 2 subunits. Sometimes another ribosome attaches to the same strand of mRNA and transcribes it. When several ribosomes attach to one strand of mRNA we call this a polyribosome.

And there we have it!

Energy

Big Picture: Energy

Much of the energy in the body comes from a special energy molecule called ATP. ATP is like a little ball of energy that can be used to power many of the body's systems.

Cells make energy to keep us alive. Our bodies need energy to keep our hearts beating and our minds thinking. The energy comes from a little molecule called ATP (adenosine triphosphate).

These molecules are produced in an organelle called the mighty mitochondrion. Think of the mitochondrion as an energy factory that makes little energy balls (ATPs). It assembles the energy balls by taking an ADP (adenosine diphosphate) and adding another P (phosphate). ATP can store energy (add a P to ADP) or release energy (break a P off of ATP).

Hair and Tails

Not all cells are bald. Some have hair called cilia. Cilia are protein filaments that work to move substances across the surface of cells. The hair moves in waves pushing stuff along the surface of a group of cells toward a destination. For example the cells lining your respiratory system move mucous and debris toward your esophagus. So every time you swallow you are swallowing some mucous from your respiratory system. The debris gets broken down by the acid and enzymes in your stomach.

Some cells have tails called flagella. The human sperm cell has a long flagellum that moves the cell along. The big difference between cilia and flagella is that cilia move substances while a flagellum moves the cell. Cells can have lots of hair (cilia) but only one flagellum (tail).

Cell Sex

The Big Picture: Mitosis

Remember the acronym "IPMAT" for interphase, prophase, metaphase, anaphase and telophase.

Interphase = cell rests and gets ready to divide.

Prophase = nuclear membrane disappears, chromosomes and mitotic spindle forms.

Metaphase = chromosomes all line up in the center of the cell.

Anaphase = chromosomes pulled apart.

Telophase = nuclear membrane reappears, chromosomes uncoil, mitotic spindle disassembles and there are now 2 cells.

Cells are lonely creatures when it comes to sex. In fact they "do it" all by themselves in a process known as mitosis. There is nothing exciting about the sex life of a cell. It just rests and then splits into two cells. Cells need to rest before the big event in a rest phase called interphase. During this "getting ready" time the cell will reproduce its organelles and DNA. When it's ready, the cell enters the next phase called prophase.

During prophase strands of genetic material in the nucleus coil up and become chromosomes (fig. 2.6). These babies have to divide into two identical groups so both of the new baby cells have the same number of chromosomes. Something's gotta pull the chromosomes apart and that's the job of the mitotic spindle. Since the chromosomes are in the nucleus the membrane surrounding the nucleus begins to disappear.

Think of pulling apart a wishbone with another person. You both hold the wishbone with one hand and pull until the wishbone separates. In the mitotic spindle you and your partner represent structures called centrosomes while your arms and hands represent microfilaments that contact the chromosomes (wishbone). The only difference is that the chromosomes evenly divide. There is no "winner" with the larger piece.

In prophase the cell hasn't pulled apart the chromosomes yet but is getting ready to. In order to get the chromosomes to divide they need to line up in a nice straight line. This is what happens in the next phase called metaphase (fig. 2.7). In metaphase you still haven't pulled the wishbone apart but are still getting ready to.

The pulling apart happens in the next phase called anaphase (fig. 2.8). In anaphase the chromosomes have divided into two equal sets. Snap—the wishbone has broken.

The final phase is called telophase (fig. 2.9). Here the cell cleaves into 2 baby cells and both complete the division process by reforming the cell membrane and nuclear membrane. The chromosomes uncoil and the mitotic spindle breaks up.

More Cell Sex

Mitosis isn't the only way cells divide in the body. If they did then every time we humans reproduce we'd have more chromosomes. Chromosomes are nice to have but too many could be a big problem. So cells must have a way to divide so that there is half the number of chromosomes in each new cell. That way mommy human has half and daddy human has half so baby human has a full set (goo goo).

This other method of cell division is called meiosis. You probably guessed where this occurs. Yup it happens in the gonads (aka "nads"). The goal of the cells that eventually become our precious offspring is to produce half the number of chromosomes. So what happens to the other half? Well in males there are just two sperm cells with half the number of chromosomes. So the goal in males is to make millions and millions of sperm cells. In the female the goal is to make one nice big egg cell. So females drop off half of the chromosomes in what is called a polar body. The polar body just disintegrates.

We'll talk more about meiosis in the last chapter on everybody's favorite system, the reproductive system. So you'll have to wait until then to find out more about meiosis.

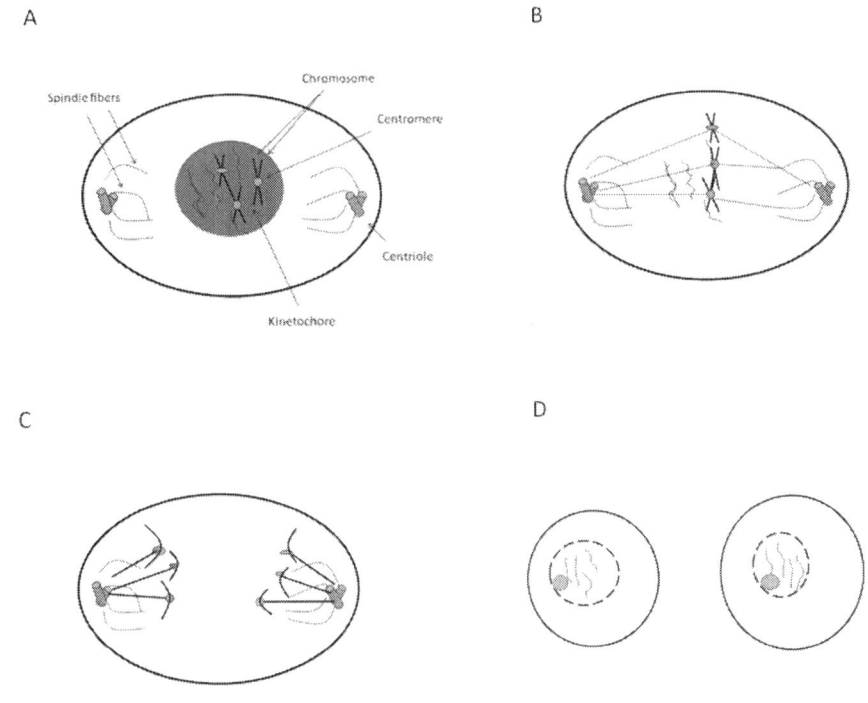

Figure 2.6. Mitosis. A. Prophase. B. Metaphase. C. Anaphase. D. Telophase.

The Big Picture: Getting Stuff in and Out of Cells

Cellular transport is like going to a party.

Cells need to transport stuff across their fatty cell membranes. One thing to remember is that fat dissolves in fat. Think of how water dries up on your skin but oil is absorbed. In other words things that dissolve in fat are called lipid soluble substances. Lipid soluble substances can move right through the cell membrane because lipid dissolves in lipid. There are a number of ways that stuff gets in and out of cells. These go by the names of diffusion, facilitated diffusion, active transport and osmosis.

To help you to get the big picture let's go to a party.

Let's say that I am going to an office Christmas party. I start hanging out in the main room and after awhile find myself getting pushed around because of the crowd gathering there. People start to realize this and begin to move into other areas of the building.

We could say that everyone moved from a higher concentration of people (in the main room) to a lower concentration of people (other rooms). They did this until there were about the same number of people in each room (equilibrium). The people diffused out of the room. This is how diffusion works. Substances move from higher to lower concentrations until reaching equilibrium. Since lipid soluble substances can move right across a cell membrane then they move by diffusion.

Diffusion = Substances move from higher to lower concentration until reaching equilibrium.

Okay, so I solve this problem by grabbing the bag of chips I brought and showing it at the kitchen door. They let me in and I escape the crowd. Here is an example of facilitated diffusion. In cells, non-lipid soluble substances move through protein channels in the cell membrane. They still move from area of higher to lower concentration but they move *through a protein channel*. In the party I moved from an area of more people to less people through the kitchen door (protein channel).

Faciliated Diffusion = Similar to diffusion but substances move through a protein channel.

I hand over the chips and find that the kitchen people are boring. Even though there are less people (less concentration of people) I decide that the talk about Oprah and Dr. Phil was not my thing so I try to get back to the main room where there is a conversation about the local football team going on.

Even though there are a lot more people in the main room who are trying to get into the kitchen I want to go there. Since I didn't eat before I came I eat a few chips (energy) so I can go against the crowd to get out. You could say this is like active transport. It requires energy in the form of ATP to move substances against a gradient.

Active Transport uses energy to move substances against gradients.

Okay, so I made it back into the main room and start talking about how I would run the football team. I am still hungry and start to eat lots of salty peanuts and chips. I start to get real thirsty so I seek out something to drink. I find some of my favorite beer and have a drink (or several). I could say this is like osmosis in that water follows salt. The more salt I ingest, the more water I require.

In fact in osmosis water moves across a semipermeable membrane (a membrane that only allows water to pass through) toward an area of higher concentration of solute.

Osmosis = movement of water though a semipermeable membrane toward an area of higher solute concentration.

So now my bladder is full and I need to seek out the men's room. I find the men's room and pass through the door. Only men are allowed in the men's room, so only men are allowed to pass through the door. This is another example of facilitated diffusion. Substances move across cell membranes through a protein channel. The channel only allows certain substances to move.

After about an hour I decide that it's time to leave. Because I am such a loudmouthed lush I have to open both the double doors at the entrance to let me and my date out. Just as I leave someone else enters (as they utter "what's their problem"). You could say that this is an example of an antiporter. As I move down my concentration gradient it allows someone else to move in the opposite direction. They use my energy to do so. Also as I leave a few people follow behind me. Again I provided the energy to open the door and they just came along for the ride. This is an example of a symporter. One substance moves down its gradient while others tag along for the ride.

Finally, I get home. I am feeling no pain and my cell phone goes off. I realize that I am a surgeon (this is an imaginary story) and am on call for my practice. I have to sober up (once again) to go in for an emergency consult (I like to live on the edge). I decide to brew a pot of coffee (very strong coffee) to help my situation. I put lots of water in to speed up the process. This is an example of filtration. The more water the higher the pressure the more filtrate is produced.

I sober up and get into work and am thankful that the patient did not have appendicitis---just gas from Mexican food (not that there is anything wrong with that). I go home to sleep it off—grateful to have dodged another lawsuit.

Nitty-Gritty Substance Transport in Cells

Now that we've seen the big picture with regard to transport systems, let's spend a little time on the nitty-gritty.

Remember that the cell membrane is made up of phospholipids. Since the membrane is composed of phospholipids, then lipid soluble substances can move across the membrane. Remember some examples of lipid soluble substances include oxygen, carbon dioxide and steroids.

But what pushes or pulls a substance across the membrane?

Diffusion

Diffusion is the movement of substances from an area of higher concentration toward an area of lower concentration until reaching equilibrium. The force that drives diffusion comes from differences in concentration called concentration gradients.

The actual mechanism behind diffusion is quite complex and has to do with the second law of thermodynamics. This law states that in any given system there must be an increase in entropy. To explain this in simpler terms, substances tend to move from an organized state (concentrated state) to a more disorganized state (less concentrated state).

The process of diffusion can be illustrated by making a glass of Kool Aide. When the powder first hits the water it is in higher concentration than its surrounding fluid. The powder will dissolve and then begin to distribute evenly throughout the glass of water. The powder is said to move from an area of higher concentration to lower concentration until it is equally distributed throughout the glass.

Another example is with an aerosol spray. Let's say I stood in front of the class and sprayed a room freshener into the air. The particles would eventually distribute evenly throughout the room so that even the students in the back of the room could smell it.

Diffusion tends to happen with lipid soluble substances because they can move across the phospholipid cell membrane. Examples of lipid soluble substances include oxygen, carbon dioxide and steroids.

Other examples of lipid soluble substances include alcohols, fatty acids and lipid soluble drugs.

Substances can diffuse at different rates. The rate of diffusion depends on a number of factors. These include:

- Molecule size—smaller molecules diffuse faster than larger molecules.
- Size of concentration gradient—the larger the difference in concentration the faster substances will diffuse.
- Temperature—because diffusion relies on the movement of molecules, higher temperatures will cause more movement and speed up diffusion. For example, substances will diffuse faster at body temperature than room temperature.
- Distance—the shorter the distance, the faster substances will diffuse.
- Electrical forces—Cells generally carry a negative charge on the inside. Negative charges will attract positive electrolytes and repel negative ones. This can speed up or slow down the rate of diffusion.

Facilitated Diffusion

Now we know how lipid soluble substances pass through cell membranes powered by diffusion but how do non-lipid soluble substances get in and out of the cell?

The answer has to do with what is known as facilitated diffusion (fig. 3.11).

Cell membranes contain proteins. Some of these proteins go all the way through the membrane and act as channels for specific substances. Examples of substances that move by facilitated diffusion include sodium, potassium, and chloride.

The force that moves substances in facilitative diffusion is the same as diffusion. That is, the difference in concentration or concentration gradient. Again, substances still move from areas of higher to lower concentration but this time they move through a protein channel.

Substances moving in and out of cells by facilitated diffusion must bind to receptors on the protein channel. Once they bind, the protein changes its shape allowing the substance in or out of the cell.

Since proteins only have so many receptors, once the receptors become saturated there cannot be movement of any additional substances. Therefore in some cases a higher concentration gradient will not move substances at a faster rate (unlike diffusion). The rate of diffusion then partially depends on the saturation of the receptors on the protein channel.

One example of a substance transported into the cell via facilitated diffusion is glucose. Glucose is used by cells to make ATP, an important energy molecule in the body. Muscle cells require glucose to make ATP for muscle contraction. In order for glucose to move into a muscle cell it not only needs to connect to a receptor on the protein channel but another hormone called insulin also needs to connect to a special insulin receptor on the protein.

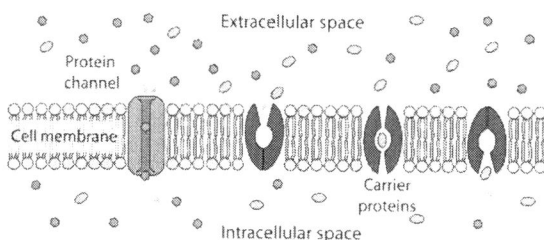

Figure 2.7. Carrier proteins allow non-lipid soluble substances to move in and out of cells in facilitated diffusion.

In some cases the insulin receptors become resistant to insulin causing blood glucose levels to rise. This occurs in what is known as insulin resistant diabetes. We will learn more about diabetes in a later chapter.

Osmosis

Water moves within the human body across a variety of membranes. The membranes are called semipermeable because they only allow water to move across them, not solute. The movement of water across a semipermeable membrane has a special name; osmosis.

Water moves like every other substance in our universe, from an area of higher concentration to lower concentration. However, we typically do not talk about concentration in terms of water. We usually talk about concentration in terms of solute.

So you could think of osmosis in two ways:

1. Water moves across a semipermeable membrane from a higher area of concentration of water to a lower concentration of water.
2. Water moves across a semipermeable membrane from an area of lower

concentration of solute to an area of higher concentration of solute.

One simple way to remember osmosis is the phrase "water follows salt" to mean that water always moves toward an area of higher concentration of solute. Here is a simple osmosis experiment (fig. 3.12-3.19).

Isotonic/Hypotonic/Hypertonic Solutions

Remember that the force exhibited by osmosis is called osmotic pressure. This pressure is related to the solute concentration of the solution. In chemistry we describe concentration in terms of osmolarity. However, in physiology when we are concerned with concentration with regard to cells we use the term tonicity. Tonicity then is related to the human cell whereby osmolarity is the number of osmoles per liter.

Osmolarity depends on the number of particles of solute. For example one mole of glucose in water would equate to one osmole since glucose remains as one molecule in water. However, one mole of sodium chloride would equate to two osmoles because sodium chloride dissociates in water to form two particles.

If a solution has the same osmolarity as body fluids we say the solution is isotonic. The human body's osmolarity is close to .30 osmoles or 300 milliosmoles.

If a solution is less concentrated than body fluids we say the solution is hypotonic. And if the solution is more concentrated than body fluids we say the solution is hypertonic. Tonicity is important when it comes to introducing solutions to the human body. Let's see why.

Here is another short experiment regarding

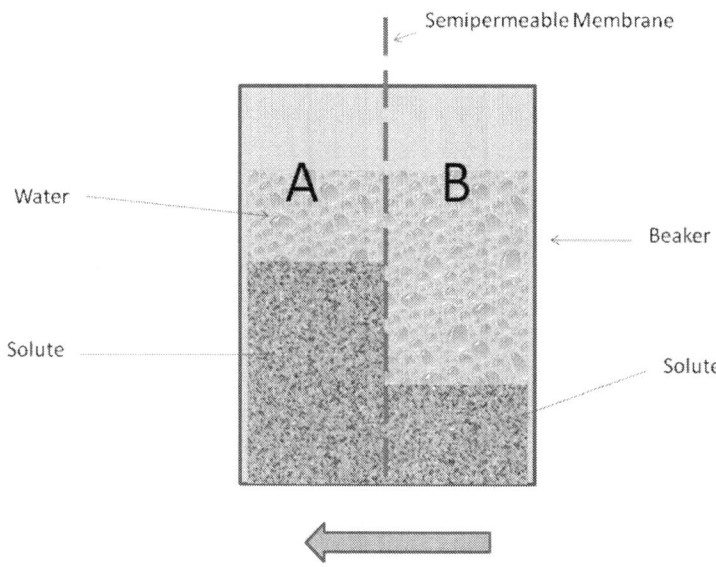

Fig. 2.8 Osmosis demonstration.

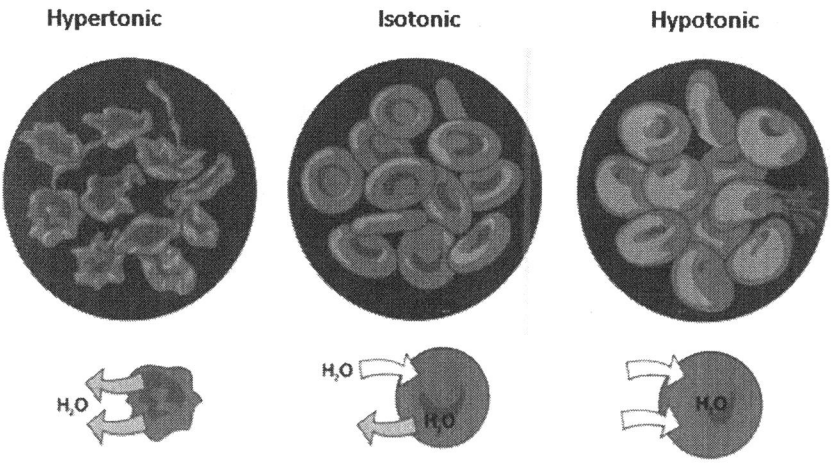

Figure 2.9. Osmosis experiment.

In the first image on the left a red blood cell is placed in a hypertonic solution. Since the solution is more concentrated than the red blood cell, the cell shrivels up or crenates. In the middle picture the red blood cells are placed in an isotonic solution. Since the tonicity is equal nothing happens to the cell. In the final picture on the right the cells are placed in a hypotonic solution. Since there is more concentration inside the cells water flows in and the cells swell (and can burst).

Filtration

Sometimes cells arrange themselves in thin layers and substances can move between the cells. These layers or membranes work the same way as filters. Filters sort substances based on size. Smaller substances move through the spaces and larger substances do not. Think of a coffee filter. The filter has very small holes that only allow the water to move through. The grounds are too large to fit through the holes.

The force that drives filtration is fluid pressure. This pressure is also known as hydrostatic pressure. In order to move substances through a filter they must move from an area of higher pressure to lower pressure.

There are many examples of filters in the body. These include the capillaries and kidneys.

Active Transport

So far we have seen how substances move down their respective concentration gradients in diffusion and facilitative diffusion. But what if a substance needs to be moved *against* its concentration gradient?

In active transport substances are moved against their concentration gradients by carrier proteins. However, there is an energy cost to be paid for this action. So the carrier proteins use ATP as an energy source.

An example of an active transport protein is the sodium potassium pump (fig. 2.10). Normally there is more sodium outside of the cell than in so sodium would move from outside to in.

Also, there is usually more potassium inside the cell than out, so potassium would follow its concentration gradient and move out of the cell.

However, we want to move these molecules against their concentration gradients. So this can be done but energy must be used to do so. Energy is used by the pump in the form of ATP.

The sodium potassium pump is vital to the human body and works to maintain and establish the concentration gradients that keep us alive.

Other Transport Mechanisms

Other ways that substances can move in and out of cells include cotransport, exocytosis and endocytosis (fig. 2.11).

In endocytosis substances enter cells via vesicles. There are three types of endocytosis. They include phagocytosis, pinocytosis and receptor-mediated endocytosis. All involve the cell membrane wrapping around and engulfing a vesicle.

In pinocytosis a cell can take in a small droplet of fluid.

In phagocytosis the cell takes in a solid then uses a lysosome (small enzyme packet) to break down the solid.

In receptor-mediated endocytosis substances bind to receptors on the cell membrane. The membrane responds by forming a vesicle and taking the substance into the cell. Substances can exit cells by the same method (exocytosis).

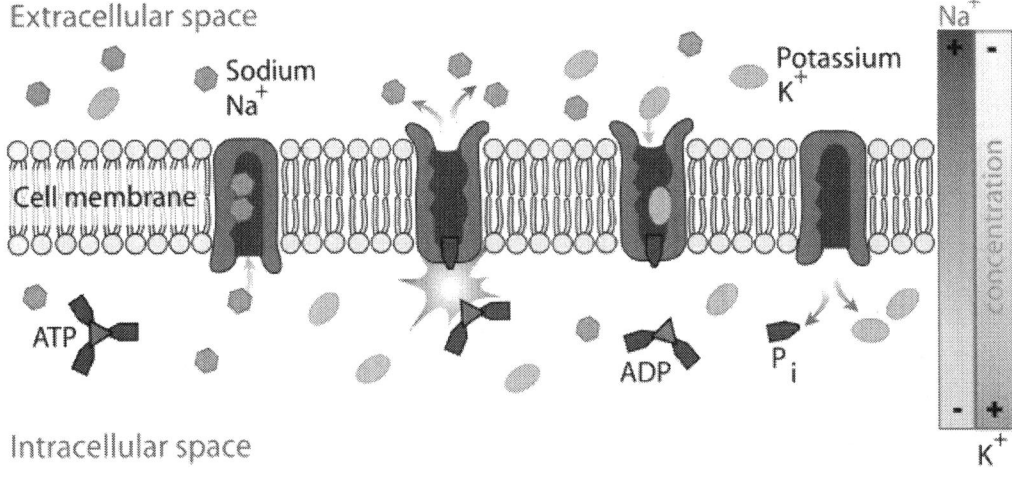

Figure 2.10. The sodium-potassium pump. Sodium moves out of the cell against its gradient while potassium moves into the cell (also against its gradient). The carrier protein must use energy in the form of ATP. One ATP moves three sodium ions out of the cell and two potassium ions in.

Endocytosis

Phagocytosis — solid particle, Pseudopodium, Phagosome (food vacuole)

Pinocytosis — Vesicle

Receptor-mediated endocytosis — Coated pit, Receptor, Coat protein, Coated vesicle

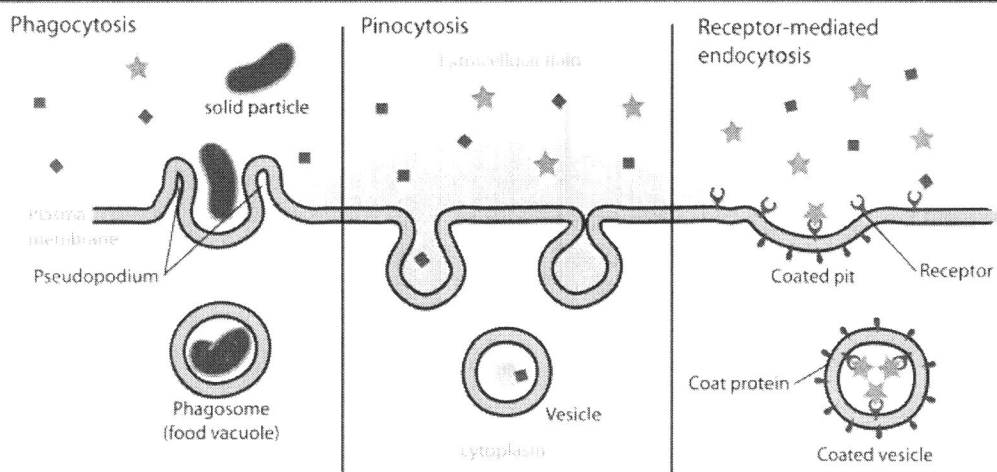

Figure 2.11. In phagocytosis the cell reaches out and engulfs a particle bringing it into the cell for destruction. The cell brings in fluid via pinocytosis. Substances attach to receptors on the cell membrane in receptor-mediated endocytosis.

Review Questions

1. Which of the following organelles is involved in protein synthesis:

   a. Mitochondrion
   b. Endoplasmic reticulum
   c. Vesicle
   d. Centrosome

2. Which of the following best describes the structure of a phospholipid:

   a. Hydrophilic phosphate head and a hydrophobic lipid tail
   b. Hydrophobic lipid head and hydrophilic phosphate tail
   c. Hydrophilic phosphate head and hydrophobic amino acid tail
   d. Hydrophobic amino acid head and hydrophilic lipd tail

3. Which type of substances can move across a phospholipid bilayer membrane:

   a. Water soluble
   b. Amino acids
   c. Lipid soluble
   d. Glucose

4. Which of the following cell organelles forms the mitotic spindle:

   a. Mitochondrion
   b. Golgi apparatus
   c. Centrosome
   d. Vesicle

5. Which of the following organelles produces ATP:

   a. Mitochondrion
   b. Endoplasmic reticulum
   c. Centrosome
   d. Vesicle

6. Which of the following structures is a protein that can move substances across the surface of cells:

   a. Microtubule
   b. Flagellum
   c. Cilium
   d. Vesicle

7. In diffusion substances move from areas of higher to lower concentration until what happens:

   a. Reaching equilibrium
   b. ATP runs out
   c. The protein channel collapses
   d. Nothing, they continue to move

8. Which of the following transport mechanisms uses ATP:

   a. Diffusion
   b. Facilitated diffusion
   c. Osmosis
   d. Active transport

9. Which of the following transport mechanisms relies on fluid pressure changes:

   a. Diffusion
   b. Facilitated diffusion
   c. Filtration
   d. Active transport

10. Which of the following transport mechanisms involves water moving across a semipermeable membrane:
    a. Diffusion
    b. Facilitated diffusion
    c. Osmosis
    d. Active transport

11. In osmosis water always moves toward:

    a. An area of lower concentration of solute
    b. An area of more water
    c. An area of higher concentration of solute
    d. An exit

12. In which stage of mitosis do the chromosomes line up in the center of the cell:
    a. Prophase
    b. Anaphase
    c. Metaphase
    d. Telophase

13. In this stage of mitosis the nuclear membrane forms and the chromosomes uncoil:

    a. Anaphase
    b. Telophase
    c. Metaphase
    d. Prophase

14. A DNA nucleotide consists of:

    a. Ribose sugar, phosphate, nitrogen containing base
    b. Phosphate, amino acid, nitrogen containing base
    c. Triglyceride, phosphate, nitrogen containing base
    d. Ribose sugar, amino acid, phosphate

15. Which of the following structures carries the information from DNA outside of the nucleus:

    a. Amino acids
    b. Messenger RNA
    c. Transfer RNA
    d. Ribosomal RNA

16. Where is the information from DNA translated:

    a. Nucleus
    b. Mitochondrion
    c. Golgi apparatus
    d. Ribosome

17. The 3-base sequence on tRNA that codes for an amino acid is known as the:

    a. Anticodon
    b. Code
    c. Codon
    d. Transcipter

18. Ribosomes consist of:

    a. 1 ribosomal subunit
    b. 2 ribosomal subunits
    c. 3 ribosomal subunits
    d. 4 ribosomal subunits

Chapter 3

Monstrous Metabolism

Chapter 3

Monstrous Metabolism

The section on cellular metabolism is the scourge of many a student. It is easy to get lost in the nitty gritty details on this one.

What the Heck is metabolism?

Metabolism is the sum total of all biochemical reactions in the body. There are two basic reactions; anabolic and catabolic. In anabolic reactions larger molecules are made from smaller molecules. Think of those illegal anabolic steroids used by athletes to build muscles. Catabolic reactions are characterized by the breaking down of larger molecules into smaller molecules. Think about how catabolic sounds like cannibal. Every time you consume a food your body uses catabolic reactions to break the food down into smaller molecules. For example proteins are broken down into amino acids and complex carbohydrates are broken down into simple carbohydrates.

A general model for an anabolic reaction would be:

A + B -> AB

For a catabolic reaction:

AB -> A + B

Many of the metabolic reactions in the body have to do with the production of ATP. Energy is extracted from the foods we eat and used to store energy in ATP.

The Big Picture: Energy Systems

Energy for the body is produced by 3 categories of energy systems.

One way to look at how energy is produced in the body is to examine where the majority of energy comes from in different activities. Let's use an example to illustrate this concept. Let's say our friend Hal is going on a long bicycle ride. Hal wants to make good time so he pedals vigorously.

During the first 30 seconds of Hal's ride the energy comes from a process known as the immediate energy system (ATP-phosphocreatine system)(fig. 3.1). There are molecules of phosphocreatine (PCr) stored near Hal's muscles. The PCr contains a phosphate that easily lends itself to phosphorylate ADP (stick a phosphate on it) to make ATP to power Hal's muscles. There is only a small supply of PCr so the energy only lasts for about 30 seconds. The amount of PCr is what limits the system.

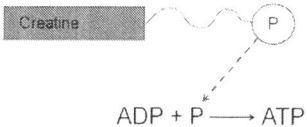

Figure 3.1 The immediate energy system. Molecules of phosphocreatine (creatine + phosphate) are stored around muscles. The phosphate lends itself to phosphoryllating ADP to make ATP.

Hal continues to ride beyond 30 sec. During the next 150 seconds of intense activity Hal's energy comes from the short term energy system which is known as glycolysis. Glycolysis means "sweet dissolution" which refers to a series of reactions that break down glucose and extract the energy for making ATP. Glycolysis is a bit more complicated than the ATP-PCr system.

Glycolysis occurs in the cytoplasm of Hal's cells. Glycolysis uses two ATPs to get things going and produces four ATPs and two molecules of another energy storing molecule called NADH2. This means that glycolysis produces a net gain of two ATPs to help power Hal's muscles. Other products of glycolysis include two molecules of pyruvic acid (fig. 3.2).

As Hal continues to cycle beyond 180 seconds his body switches to the aerobic energy systems. Instead of producing pyruvic acid the system produces pyruvate which is now converted to another molecule that enters a mysterious series of reactions known as the KREBS cycle. The KREBS cycle occurs in a cell organelle called the mitochondrion.

The KREBS cycle produces some ATPs as well as other energy storing NADH2 and FADH2 molecules. The NADH2 and FADH2 molecules are passed to another energy system known as the electron transport chain.

The electron transport chain is a series of enzymes that pass electrons from one to another. The enzymes pass the electrons along to lower energy levels. The energy is extracted to phosphorylate ADP (to make more ATPs).

The Krebs cycle and glycolysis do not require oxygen directly but are still part of the aerobic metabolism of glucose. This is due to the use of oxygen by the electron transport chain. The last enzyme in the chain gives up a pair of electrons that combine with hydrogen ions and oxygen to form water. Oxygen is the last electron acceptor in the chain.

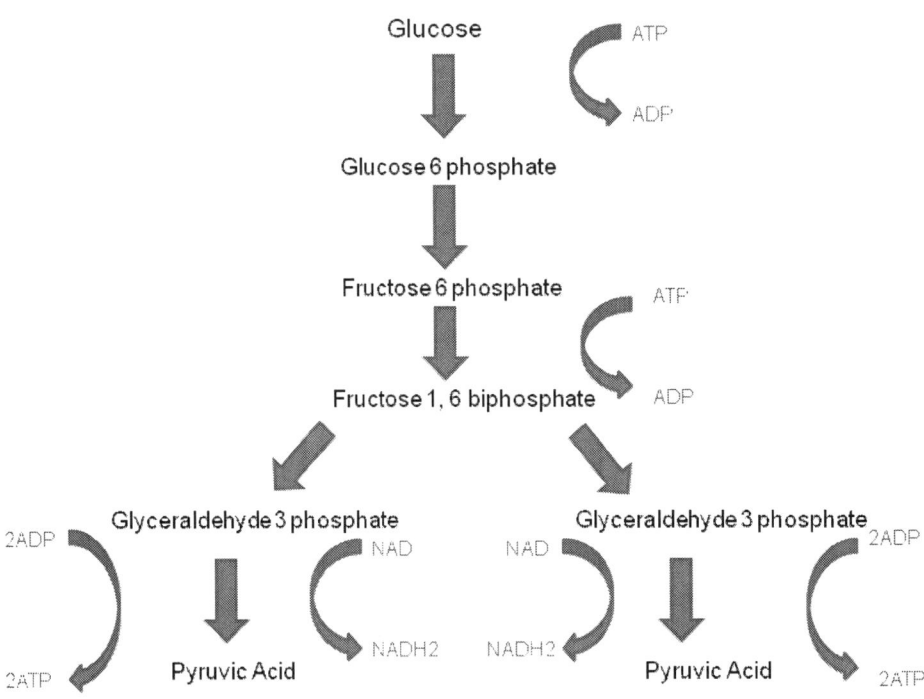

Figure 3.2. Glycolysis (summary). Notice that glycolysis uses 2ATP and produces 4ATP and 2 NADH2 molecules from 1 molecule of glucose.

So when Hal cycles longer than three minutes, most of the energy comes from the aerobic metabolism systems. These same systems provide energy when Hal is at rest. The products from these reactions include ATP, NADH2 and FADH2. We know that ATP carries energy in its phosphate bond. The other two molecules also carry energy that can be extracted to make more ATPs.

The Big Picture: Glycolysis

Glucose and ATP go in, pyruvic acid, NADH2 and more ATPs come out.

As we said, when Hal exercises vigorously for more than about 30 seconds his immediate energy system runs out of fuel. He then needs to activate the second energy system. This system is called glycolysis. It is anaerobic and uses glucose as a fuel. Glycolysis also occurs in the cytosol of the cell.

The big picture for glycolysis is that glucose goes in and if there is no oxygen present pyruvic acid, two ATPs, and two NADH2's come out. Glucose (a 6-carbon molecule) is split into two 3-carbon molecules of pyruvate that converts to pyruvic acid. Pyruvate can either enter the next energy system (KREBS) or can be converted to lactic acid by way of pyruvic acid. This will depend on whether the body is exercising anerobically (making lactic acid) or aerobically (pyruvate enters the KREBS cycle).

Getting into the nitty gritty means we have to do a bit of chemistry. Here's how it works.

First of all, glucose needs a bit of help to get things going so a couple of ATPs are needed to get it ready.

Glucose + 2ATPs -> Fructose, 1, 6 diphosphate

This molecule is then split into two molecules. You might remember that glucose is a 6 carbon molecule. When it splits it splits into two 3-carbon molecules.

Fructose 1, 6 diphosphate-> 2 glyceraldehyde 3 phosphate

The two molecules of glyceraldehyde 3 phosphate lose hydrogen atoms (oxidized) and gain phosphates to form two molecules of:

1, 3 diphosphoglycerate

A byproduct of this reaction is the formation of two NADH2 molecules that will be used later. The two molecules of 1, 3 diphosphoglycerate each give up one phosphate to phosphorylate ADP making two ATPs and converting to two molecules of:

3-phosphoglycerate

The phosphates move to another carbon forming:

2-phosphoglycerate

Water is removed from these molecules (dehydration) forming 2 molecules of:

Phosphophenolpyruvate

The phosphates are removed and added to ADP to make two more ATPs forming two molecules of:

Pyruvate (which can convert to pyruvic acid)

If the pyruvic acid is converted to lactic acid the two molecules of NADH2 are used. If not, pyruvate is converted to acetyl-coenzyme A and enters the KREBS cycle (coming up next). The conversion of pyruvate to acetyl coenzyme A produces two more NADH2 molecules (fig.3.3).

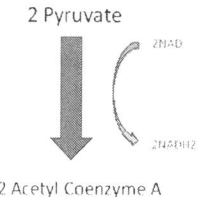

Figure 3.3 Conversion of pyruvate to acetyl coenzyme A produces two NADH2.

The Big Picture: The KREBS Cycle

Products from glycolysis go in, energy comes out.

Many students dread learning the KREBS (citric acid) cycle because it represents a good deal of chemistry and hey, we are anatomy students not chemistry majors. However, it is a good idea to at least get the big picture when it comes to this phenomenal series of reactions.

Let's summarize what we just said about Hal. Hal's energy source is primarily the food he eats. His food is broken down into fats, carbs and proteins. All of these can be used by the KREBS cycle to make good ole ATP. If ATP is not made directly in KREBS, its counterpart, the electron transport chain finishes the job and eeks out a whole lotta ATPs from carbs, fats, and proteins.

The Big Picture Metabolic Water Works

To begin our journey through the KREBS cycle we need to visit the Big Picture Metabolic Water Works (fig. 3.4). The water works operates much like an old fashioned water works complete with a big water wheel that turns and produces energy. We can extract the energy at various points around the wheel and either use it or send it off to the chute (electron transport chain) where we can extract even more energy. In the cell the KREBS cycle and electron transport chain are located in the mighty mitochondrion.

Hal's food is the water coming in at the top of the wheel. In order to better illustrate the process we will use glucose (a simple sugar) as an example. The goal is to eek out as many ATPs as we can from one molecule of glucose. In our water works world, the water represents glucose.

Notice that water comes in above the wheel. This reminds us of the concept of potential energy. Since the water is above the wheel we can use the potential energy of the water to drive the wheel. We can also extract energy at various locations around the wheel.

Notice that three things come off the wheel (fig. 3.5). We know that we can directly use ATP for energy but what about the other molecules (NADH2, FADH2)? These are also energy containing molecules that can be used to make more ATPs by the "chute" (electron transport chain) (fig. 3.6).

For every turn of the wheel we get:

- 1 Molecule of ATP
- 3 Molecules of NADH2
- 1 Molecule of FADH2

Since the glucose was split into two molecules the wheel goes around twice for every glucose molecule that enters. So for two turns of the wheel we get:

- 2 Molecules of ATP
- 6 Molecules of NADH2
- 2 Molecules of FADH2

Monstrous Metabolism 47

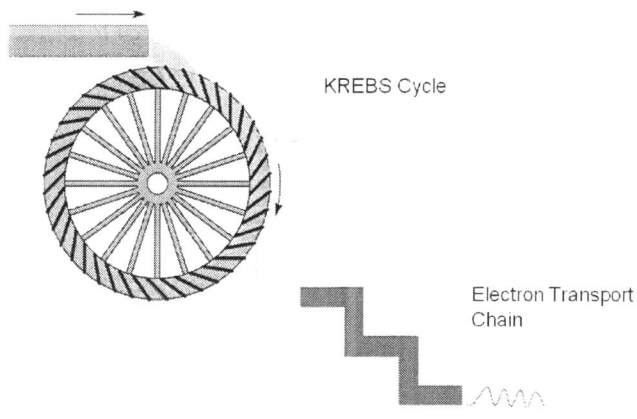

Figure 3.4. The Big Picture Metabolic Water Works. Notice the water from the water wheel flows into the electron transport chain staircase and out.

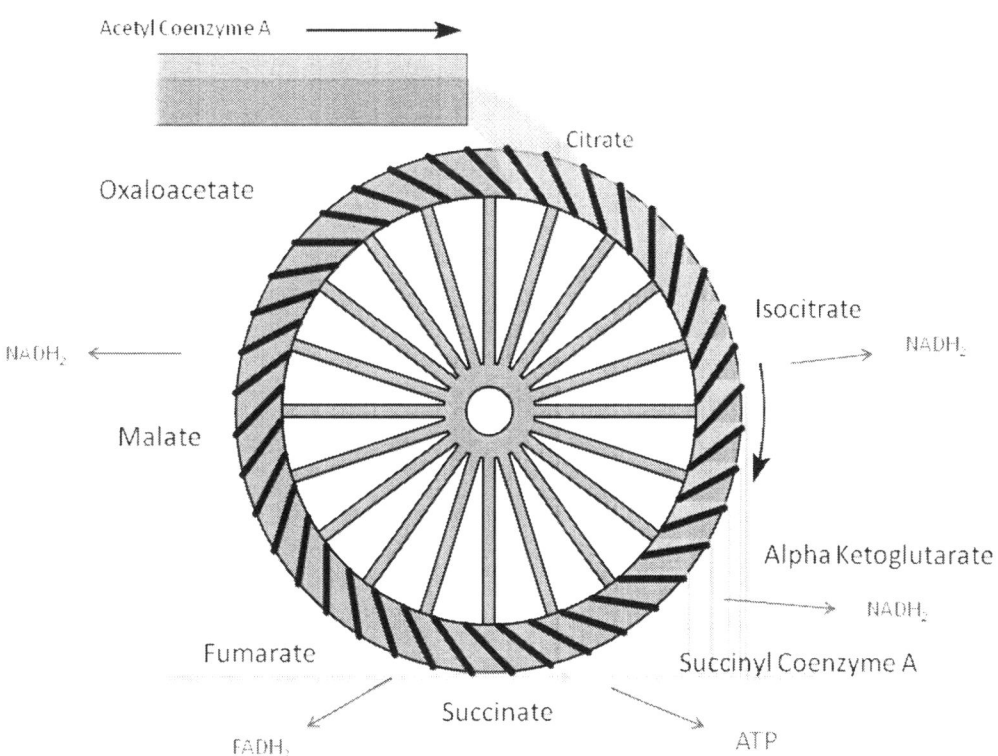

Figure 3.5. Detail of the KREBS cycle waterwheel. Notice the red energy molecules exiting the waterwheel. With the exception of ATP these will be used by the electron transport chain staircase to make more ATPs.

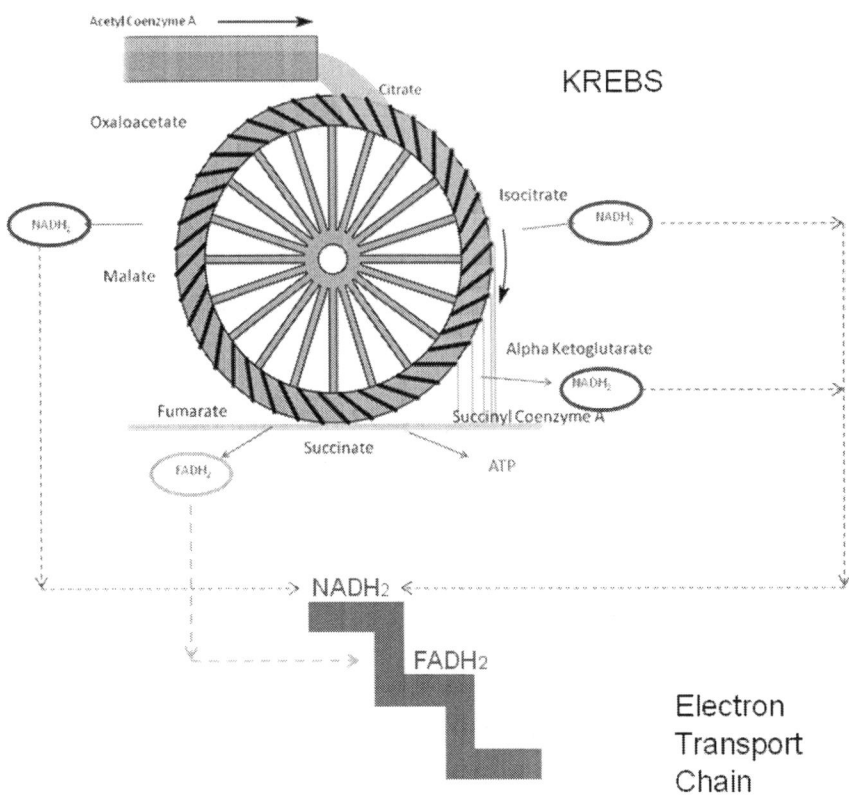

Figure 3.6. Energy molecules produced by the KREBS cycle enter the electron transport chain. Notice that NADH2 enters at the top step and FADH2 enters at the middle step.

So, how do we remember all of these steps? Here is a handy mnemonic that may help:

Crime Is Killing Some Super Females More Often

Citrate

Isocitrate

Ketoglutarate (alpha ketoglutarate)

Succinyl Coenzyme A

Succinate

Succinate

Fumarate

Malate

Oxaloacetate

The Big Picture Metabolic Water Works Part 2: The Electron Transport Chain

Notice that NADH2 and FADH2 enter the electron transport chain at different levels (fig. 3.7). The electron transport chain is going to extract some energy from these molecules and use it to make ATPs by adding a phosphate to

ADP. NADH2 has enough energy to make three ATPs while FADH2 only has enough energy for two ATPs. This is a bit like water flowing down a staircase. The key is to understand that water at a higher level in the staircase has a higher potential energy. This energy can be used to make the ATP energy molecules.

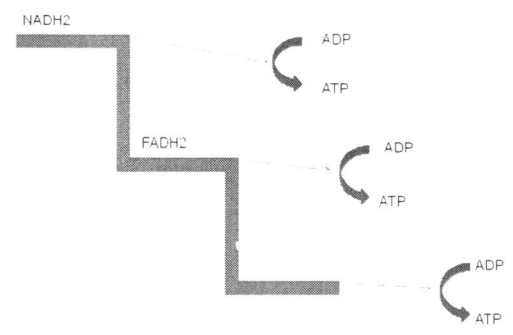

Figure 3.7. Electron Transport Chain. Notice how NADH2 can phosphorylate 3 ADPs to make 3ATPs while FADH2 can only make 2 ATPs.

How does it Work?

We know that energy is stored in the phosphate bond in ATP but how do the molecules of NADH2 and FADH2 store energy? The energy is stored in electrons. Remember that these are tiny objects that "orbit" around the nucleus of an atom. If an atom is *excited* (takes on energy) the electrons move to the outer orbital shells. If the atom *calms down* (by releasing energy) the electrons move to the inner shells.

It turns out that NADH2 *donates* electrons to the system. When a molecule loses electrons it is called oxidation. NADH2 loses two hydrogen atoms (which are also called protons). Take a look:

NADH2 -> NAD$^+$ + 2H$^+$

Now, all of this stuff happens in the good ole mitochondrion. Remember that the mitochondrion is the powerhouse of the cell. On the inside of the mighty mitochondrion is a folded membrane called the cristae. The inside of the membrane is called the matrix while the area outside of the membrane is called the intramembraneous space. Located in the membrane is a set of five special proteins.

The big picture is that the first four proteins extract the energy from the electrons from NADH2 and FADH2 and use this energy to pump protons across the membrane (from the matrix to the intramembraneous space). The protons build up on one side of the membrane forming a proton gradient. The proton gradient is used by the fifth membrane protein to add phosphates to ADP. This is called phosphoryllation of ADP which makes ATP.

That may be all you need to know about this complex process. However, if you need to known a bit more, read on.

Protein I

NADH2 encounters the first protein and loses two hydrogens and two electrons (oxidized). The two hydrogens are picked up by a molecule in the protein (FMN->FMNH2) which then passes the electrons from the hydrogens to iron (Fe). The hydrogens pick up their lost electrons and they are transferred to a third molecule (ubiquinone aka coenzyme Q10). The hydrogens again separate from their electrons and the electrons are again passed to iron. Iron again passes the electrons to another ubiquinone located outside of the protein and in the membrane. In order to do this the hydrogens must recombine with their lost electrons.

So, in a nutshell the first membrane protein acts as an active transport pump that pumps hydrogens from the matrix, across the membrane and into the intramembraneous space of the mitochondrion.

Protein II

This protein works with FADH2 generated from the KREBS cycle. The hydrogens are then stripped from FADH2 forming FAD+ and are combined with iron. The iron transfers the electrons to ubiquinone where they recombine with hydrogens.

Protein III

The membrane bound ubiquinone releases electrons that are picked up by the third protein. This protein passes the electrons via an electron carrier called cytochrome C. The cytochrome C transports electrons to the 4th protein.

Protein IV

At the fourth protein the electrons combine with hydrogen and oxygen (from breathing) to form water. We say that oxygen is the final electron acceptor (this is important and often a question on exams).

So what's the point of all of this moving electrons around? Well the energy allowing the electrons to move from carrier to carrier is used to pump protons (H+) across the membrane. NADH2 can pump more protons than FADH2. The protons build up and form a proton gradient. This gradient is used by the final protein to make ATP.

Protein V

The final protein is actually an enzyme called ATP synthase. This enzyme uses the proton gradient to add a phosphate to ADP (this is called phosphoryllation).

Since NADH2 releases its electrons on one side of the inner membrane the hydrogens (protons) build up. This creates a proton gradient whereby the protons move from one side of the membrane to the other.

Total ATPs From Glucose

So, if you've made it this far you're probably wondering just how many molecules of ATP can be eeked out of one molecule of glucose. Well, let's add em up:

Glycolysis = net gain of 2ATP

KREBS cycle (2 turns) = 2ATP

Okay so that's 4 ATPs but what about the electron transport chain using NADH2 and FADH2 to make more ATP? Remember that there is enough energy in one molecule of NADH2 to make 3ATPs. Likewise there is enough energy in one molecule of FADH2 to make 2 ATPs. So let's total them up.

NADH2s from conversion of pyruvate to acetyl coenzyme A = 2

 2 X 3 = 6 ATPs

NADH2s from glycolysis = 2

2 X 3 = 6 ATPs

NADH2s from KREBS (2 turns) = 6

 6 X 3 = 18 ATPs

FADH2s from KREBS (2 turns) = 2

 2 X 2 = 4 ATPs

So the grand total is—drum roll---

 2+2+6+6+18+4 = 38 ATPs from one molecule of glucose!

Review Questions

1. When Hal is at rest most of the ATP to power his muscles comes from:

   a. Anaerobic glycolysis
   b. ATP-Phosphocreatine system
   c. Cytosol
   d. KREBS and ETC

2. If Hal were to lift a heavy weight for a few seconds which system would generate the most ATP:

   a. ATP-Phosphocreatine
   b. Anaerobic glycolysis
   c. KREBS and ETC
   d. Mitochondria

3. If intense activity ceases at about 2 minutes which of the following occurs:

   a. Pyruvic acid is converted to lactic acid
   b. Pyruvic acid is converted to acetyl coenzyme A
   c. Acetyl coenzyme A enters the KREBS cycle
   d. Lipids and proteins are converted to pyruvic acid

4. Anaerobic glycolysis produces a net gain of how many ATPs:
   a. 1
   b. 2
   c. 3
   d. 4

5. For each turn of the KREBS cycle how many NADHs are produced:

   a. 1
   b. 2
   c. 3
   d. 4

6. The energy from each NADH can be converted to phosphorylate how many ATPs:

   a. 1
   b. 2
   c. 3
   d. 4

7. Extraction of energy from NADH and FADH occurs where:

   a. Anaerobic glycolysis
   b. ATP-Phosphocreatine
   c. KREBS
   d. Electron transport chain

8. Building a protein from amino acids is an example of which type of reaction:

   a. Catabolic
   b. Exothermic
   c. Anabolic
   d. Endothermic

9. During the first couple of steps of glycolysis 2 ATP are used for:

   a. Phosphoryllating ADP
   b. Adding phosphates to glucose
   c. Taking phosphates off of glucose
   d. Energy

10. KREBs and ETC occur where:
    a. Cytosol
    b. Endoplasmic reticulum
    c. Mitochondrion
    d. Nucleus

Chapter 4

Don't Be Terrified of Tissues

Chapter 4

Don't be Terrified of Tissues

I still remember peering into a microscope at tissue slides and exclaiming "they all look the same!" to my fellow students. They were all equally perplexed and completely blind to the nuances of histological exploration. Eventually, I did learn all those darn tissues by doing what most students do, i.e., the brute force method. This means I spent countless hours viewing slides until I just about had every cell on every slide memorized. This actually worked somewhat, that is, until the professor added some neon stained slides that were completely different colors than our "memorized" slides. The idea she was trying to drive home was that we weren't supposed to memorize the slides (hey isn't all A&P memorization?). We were supposed to learn something about tissue structure.

Lo and behold a little bit of knowledge about structure goes a long way in the world of tissues. I hope you will take some time to learn a bit about how these buggers are put together so that you can save a few brain cells for some of the more difficult concepts to come.

Learn the Main Categories

You can certainly use a bit of a formula for learning tissues. The idea is to summarize the structure so you can at least identify the right category of tissue. Here are some Big Picture pointers:

1. Epithelial tissue—consists of rows of cells. Sometimes there's just one row (we call this simple). Other times there are multiple rows (we call this stratified). Epithelium is also a covering so there is usually a space next to the tissue. So if you see a space, look for the tissue next to it. If it consists of rows of cells then yippee, it's epithelium.

2. Connective tissue—consists of cells and fibers embedded in a matrix. Kind of like the old jello mold with canned fruit suspended in a block of jello. The fruit is like the connective tissue cells and the jello is a glycoprotein matrix. Think of the fibers as long strands of carrot (my mom used to make this) suspended in the jello.

3. Muscle—long red cells connected together. Think of your long leg muscles consisting of rows of long cells hooked up in a straight line. Two out of the three types of muscle tissues are also striated or have a striped appearance.

4. Nervous—in nervous tissue you have some big ole cells to do your thinking called neurons. So look for big cells with a bunch of little cells around them. The little cells are called neuroglia.

Once you can identify the categories, then you can get a wee bit more specific. Too bad there is not just one kind of epithelium, there are a number of different types. There are some tricks to pinning them down though.

The Big Picture: Epithelium

Epithelium is a covering characterized by rows of cells of different shapes.

To review, we said epithelium consisted of rows of cells usually next to an open space (because epithelium covers things) (fig. 4.1). We also said you can have one row (simple) or more than one row (stratified). Well the next thing to do is to identify the shape of the cell. There are 3 basic shapes:

Squamous—I usually call these the squashed squamous cells. From the side they look like fried eggs.

Cuboidal—These are a bit thicker because they are cubes.

Don't be Terrified of Tissues 54

Columnar—Even thicker because they are rectangular.

So now we can combine what we know about rows (simple versus stratified) with what we know about cell shapes.

Simple squamous—one thin layer of squamous cells. Since it's so thin this is a great tissue for substance exchange. We see this in areas where we need to transport substances like in capillaries, lungs, and kidneys.

Stratified squamous—multiple layers of squamous cells. This one is a bit tricky in that the cells at the bottom of all of the layers are actually thicker (more cuboidal). As they migrate to the top layers they become thinner (more squashed). We see this tissue in the skin.

Simple cuboidal—one layer of cube-like cells. This is a thicker layer than simple squamous and usually is found lining tubes in the body (ducts).

Stratified cuboidal—a thicker layer that still lines ducts. Found in mammary, sweat, salivary glands and in the pancreas.

Simple columnar—one layer of thicker cells.

Stratified columnar—an even thicker layer.

Now we can learn about the oddball epithelium tissue. There are 2 types of oddball epithelial tissues. These include pseudostratified columnar and transitional epithelium.

Pseudostratified columnar—this one looks like multiple layers but it's not. It looks like it's stratified because the nuclei of the cells don't line up. In real stratified columnar the nuclei would form nice rows but in pseudostratified they are all over the place. This one also has hair (cilia) and a special type of mucous secreting cell called a goblet cell. Both work together to move stuff along the surface of the cells. You'll find this one in the respiratory and digestive tracts.

Transitional—kind of like stratified squamous but with a big difference. Remember in stratified squamous epithelium we had big cells at the bottom getting squashed on the top. Well in transitional we have big cells on the bottom and big cells on the top. No squashed squamous cells here. We find transitional epithelium in the urinary bladder.

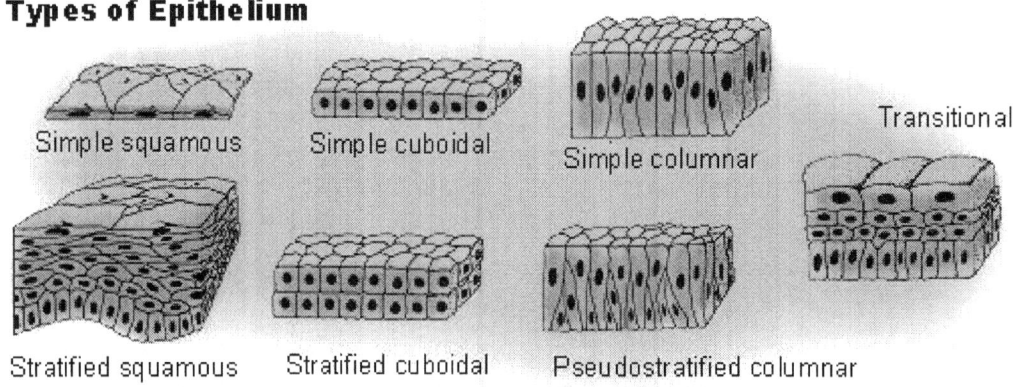

Figure 4.1. Examples of epithelium. Notice the tissue acts as a covering.

Don't be Terrified of Tissues 55

The Big Picture: Connective Tissue

Connective tissue is characterized by cells imbedded in a matrix.

There's a bunch of different kinds of connective tissue and each one has a specific characteristic. Let's see if we can narrow things down a bit.

Loose—looks like it sounds. Loosely organized tissue with some cells and fibers scattered about a matrix. Found in the deeper layers of the skin (dermis) and throughout the body. This tissue is sometimes called areolar (fig. 4.2).

Dense—now we are getting somewhere with regard to some structure. Anything called dense has gotta have some kind of structure to it. It does, with some thick collagen fibers running lengthwise through the tissue. Found in the tendons and ligaments. This tissue is sometimes called fibrous (fig. 4.2).

Adipose—adipose means fat. So wherever there's fat in the body there's adipose tissue. It consists of big fat cells containing fat (fig. 4.2)

Reticular—sounds complicated and it kinda is. It forms delicate branching networks in the lymphatic system.

Now that you have a handle on these let's look at some oddballs.

There's a category of connective tissue called "special." This includes bone, blood and cartilage. Here's the big picture:

Bone—nuthin else in the body looks like bone (figs. 4.3, 4.4). Bone (osseous tissue) looks something like the inside of a tree trunk that was cut down. Those circular units inside are called Haversian Systems or osteons and are oriented along the lines of force of the bone.

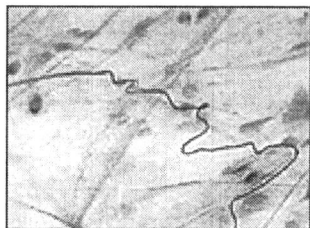
Areolar connective tissue

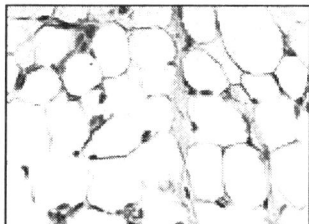
Adipose tissue

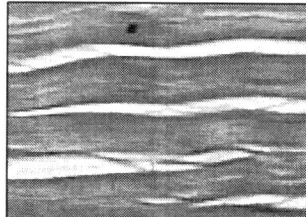
Fibrous connective tissue

Figure 4.2 Examples of connective tissue.

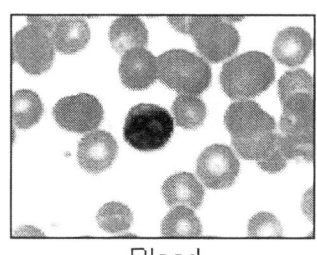
Blood

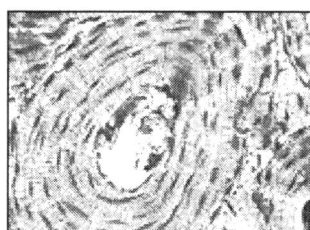
Osseous tissue

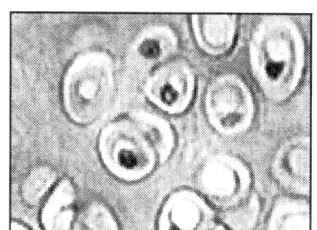
Hyaline cartilage

Figure 4.3. More examples of connective tissue.

Blood—another unique looking tissue and strangely categorized as a liquid connective tissue (fig. 4.3). The matrix in blood is the plasma and the cells include red and white blood cells and fragments of cells called platelets.

Cartilage—this one is a bit tricky. You need to look for the characteristic cell with the hole around it. The cell is called the chondrocyte and the hole is called the lacunae. You then look at the matrix. The matrix is either smooth looking (looks like ground glass) or contains fibers. If it's smooth with no fibers the cartilage is called hyaline cartilage (figs. 4.3, 4.5). If it contains fibers then you need to see if they are straight (collagen fibers in fibrocartilage) or branching (elastic fibers in elastic cartilage).

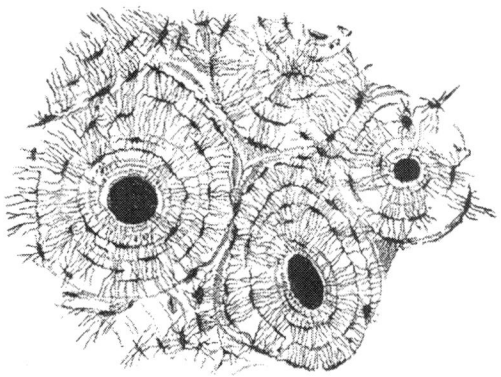

Figure 4.4. Bone is highly organized and contains structural units called Haversian systems. These are oriented along the long axis of bones to give it strength.

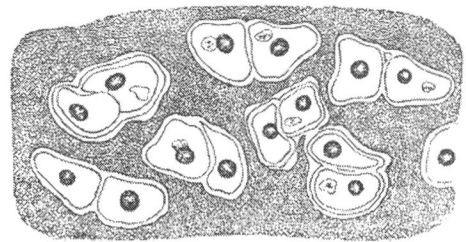

Figure 4.5. Hyaline cartilage has a smooth matrix. The chondrocytes are surrounded by a space called a lacuna.

Muscle tissue consists of long red cells that are either striped or not.

We already mentioned that muscle consists of long red cells. There are three types of muscle tissue as well. These include skeletal, cardiac and smooth. So in addition to long red cells we have some other distinguishing characteristics. For example skeletal (fig. 4.8) and cardiac muscle tissue are both called striated because they contain what looks like stripes (fig. 4.6). The striations come from the densely packed protein filaments.

So if both skeletal and cardiac muscles are striated then how can we tell these apart? The trick lies in an additional structure found in cardiac muscle called the intercalated disc. This structure appears as a dark line between cells. It is actually a cell junction that interconnects the cells. The disc helps to conduct electrical impulses across the heart muscle.

Smooth muscle is not striated and appears more loosely organized. Smooth muscle does not need to contract as forcefully as skeletal muscle so its fibers are a bit more random (fig. 4.6).

The Big Picture: Nervous Tissue

Nervous tissue consists of big cells called neurons and little cells called glia.

You need those big cells to do some thinkin and that's what you see in nervous tissue (fig. 4.7). Big cells called neurons surrounded by smaller cells called glial cells. Usually in the tissue section you only need to identify nervous tissue. Later in the nervous system you will learn more about the function of this tissue.

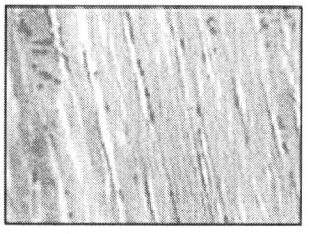 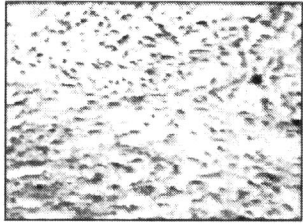 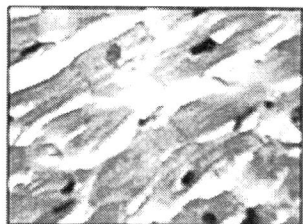

Skeletal muscle   Smooth muscle   Cardiac muscle

Figure 4.6. Three types of muscle tissue.

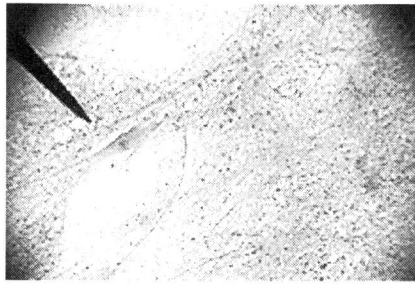

Figure 4.7. Nervous tissue contains large cells called neurons (stained dark) and supportive glial cells.

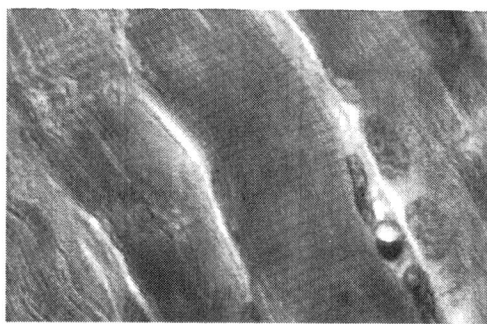

Fig. 4.8 Striated muscle tissue.

Review Questions

1. Which of the following is not an epithelial cell shape:

   a. Columnar
   b. Squamous
   c. Triangular
   d. Cuboidal

2. More than one layer of epithelial tissue is called:

   a. Layered
   b. Stratified
   c. Simple
   d. Complex

3. Which epithelial tissue is found in capillaries:

   a. Simple cuboidal
   b. Stratified squamous
   c. Simple columnar
   d. Simple squamous

4. Which tissue is found lining ducts:

   a. Loose connective
   b. Simple cuboidal
   c. Stratified squamous
   d. Fibrocartilage

5. Which tissue does bone develop from:

   a. Dense connective
   b. Hyaline cartilage
   c. Adipose
   d. Dense connective

6. Which tissue is found in the epidermis of the skin:

   a. Stratified squamous epithelium
   b. Transitional epithelium
   c. Loose connective
   d. Fibrocartilage

7. Which tissue has the chondrocyte in lacunae arrangement:

   a. Cartilage
   b. Bone
   c. Epithelium
   d. Connective

8. Which tissue is typically found in lymph nodes:

   a. Loose connective
   b. Hyaline cartilage
   c. Reticular connective
   d. Cuboidal epithelium

9. Blood is considered which type of tissue:

   a. Epithelium
   b. Cartilage
   c. Connective
   d. Blood is not a tissue

10. This tissue is found in the urinary bladder:

    a. Simple squamous epithelium
    b. Transitional epithelium
    c. Stratified squamous epithelium
    d. Loose connective

11. Which of the following types of muscle tissue contains intercalated discs:

    a. Skeletal
    b. Smooth
    c. Reticular
    d. Cardiac

12. Which of the following tissues contains Haversian systems:

    a. Cartilage
    b. Muscle
    c. Epithelium
    d. Bone

Chapter 5

Gimme Some Skin Man

Chapter 5

Gimme Some Skin Man

The skin (integument) is one big organ. Actually it is also one big membrane with a number of layers. Besides the obvious function of protecting our innards the skin also helps to regulate temperature, houses sensory receptors that send information to the nervous system, synthesize chemicals and excrete wastes. The skin also contains a good deal of immune system cells that help to protect our bodies against pathogens (nasty viruses and bacteria).

The Big Picture: Layers of the Skin

The skin has 3 layers. The top layer consists of rows of cells arranged in layers (strata), the middle layer contains connective tissue and accessory organs and the deepest layer contains connective tissue and fat.

The skin contains two layers and a subcutaneous layer. The superficial layer is called the epidermis. The epidermis consists of stratified squamous epithelium arranged in layers called strata. Deep to the epidermis is the dermis. The dermis consists of connective tissue and a number of other structures we will investigate later. The deepest layer is the subcutaneous layer that consists of loose connective tissue and adipose tissue (fat) along with blood vessels and nerves. In fact, when someone gets liposuction it is the subcutaneous fat that is sucked out (sluuurrrpp!).

The epidermis consists of stratified squamous epithelium arranged in layers or strata (fig. 5.1). The layers are:

- Stratum corneum
- Stratum lucidum
- Stratum granulosum
- Stratum spinosum
- Stratum basale

Here's a mnemonic for remembering these layers:

Come, Let's, Get, Sun, Burned

Corneum, Lucidum, Granulosum, Spinosum, Basale.

The epidermis is anchored to the dermis by means of a basement membrane. The epidermis does not contain any blood vessels. The cells of the stratum basale are nourished by the blood vessels in the dermis. These cells can divide and move toward the surface pushing the old cells off of the superficial layers.

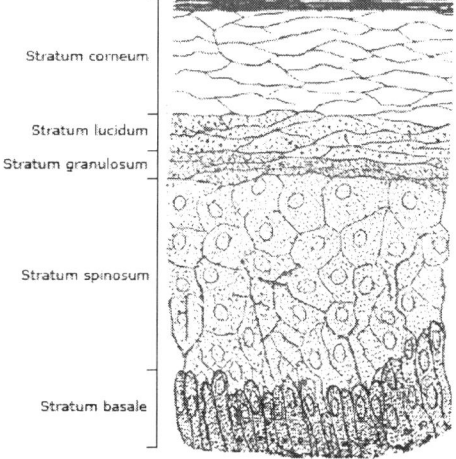

Figure 5.1. The epidermis is arranged in layers.

The stratum corneum is the most superficial layer of the epidermis. It consists of hardened cells that have been hardened with keratin. Keratin is secreted by cells located in the deep layers of the epidermis called keratinocytes.

The stratum lucidum is an additional layer that is found only in the palms of the hands and soles of the feet. It provides an added thickness to these layers.

The stratum granulosum contains cells that have lost their nuclei. These cells remain active and secrete keratin. The cells contain granules in their cytoplasm that harbor keratin.

The stratum spinosum contains cells called prickle cells. These cells have small radiating processes that connect with other cells. Keratin is synthesized in this layer.

The stratum basale or basal cell layer contains epidermal stem cells. This is the deepest layer of the epidermis. It consists of one layer of cells that divide and begin their migration to the superficial layers. This is the layer where basal cell cancer develops.

As we have seen, there are a good number of keratinocytes located in the epidermis.

An abnormality of keratinocytes is known as psoriasis—keratinocytes abnormally divide rapidly and migrate from stratum basale to stratum corneum. Many immature cells reach the stratum corneum producing flaky, silvery scales (mostly on knees, elbows and scalp).

The epidermis also responds to the environment. Friction causes the formation of corns and calluses.

Another kind of cell found in the epidermis is the melanocyte. This cell produces the pigment melanin that gives skin its color. Melanocytes are located in the deepest portion of the epidermis and superficial dermis.

The color of the skin results from the activity of the melanocytes, not the number. Melanocytes are located in the deepest layer of the epidermis. They respond to ultraviolet radiation by producing more melanin pigment which turns skin a darker color. Melanocytes respond to UVB radiation (approximately 320 nm wavelength). Melanocytes are also found in the hair and middle layer of the eye. A condition known as malignant melanoma can develop in melanocytes.

Vitamin D

The skin also helps to synthesize vitamin D. Vitamin D (aka cholecalciferol) is synthesized when a precursor molecule known as 7-dehydrocholesterol absorbs ultraviolet radiation. This molecule then travels to the liver and kidney where it is converted to the active form of vitamin D (1,25 hydroxycholecalciferol).

Vitamin D is an important substance in the body. It functions to help the body absorb calcium. It also helps with calcium transport in the intestines.

The Dermis

The middle layer of the integument is known as the dermis. The dermis consists of loose connective tissue and houses a number of accessory structures of the skin. The dermis connects to the epidermis by means of wavy structures called dermal papillae (fig. 5.2).

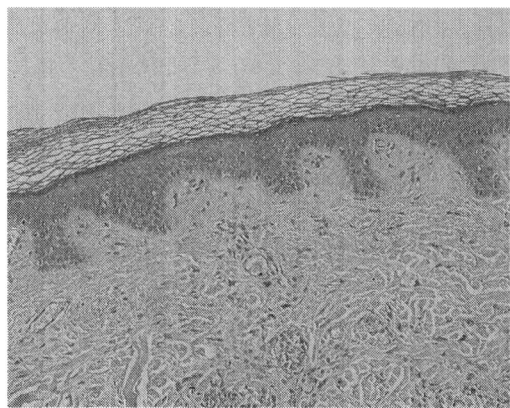

Figure 5.2. The integument. The epidermis (top layer) connects to the dermis via wavy structures called dermal papillae.

Structures of the Dermis

The dermis contains a variety of accessory structures of the integument (fig. 5.3). These include:

- Hair follicles
- Arrector pili muscles
- Sweat glands
- Sebaceous glands
- Sensory receptors
- Blood vessels

Hair Follicles/Sebaceous Glands

The human body has approximately 2.5 million hairs on its surface. Hair is not found on the palms of the hands and soles of the feet as well as on the lips, parts of the external genitalia and sides of the feet and fingers. Hair is produced by hair follicles. Hair is not alive and develops from old dead cells that are pushed outward by new cells. The cells contain keratin for hardness and melanin for color. Hairs can be very sensitive. This is due to a tiny plexus of nerves that surround each hair follicle. Hair is so sensitive that you can feel the movement of even a single hair.

A band of smooth muscle is connected to each hair follicle. This structure is called an arrector pili muscle and is capable of moving each follicle causing it to stand up in times of sympathetic nervous system activity such as emotional stress.

Hair begins to grow at the base of the hair follicle in a structure called the hair bulb. The hair bulb is surrounded by a hair papilla that contains blood vessels and nerves. The cells of the hair bulb divide and push the cells toward the surface along the hair root and shaft.

Hair grows at a rate of about .33 mm per day. Normal adults lose about 50 hairs per day. A loss of over 100 hairs per day will cause a net loss of hair. This can happen especially in males due to changing levels of sex hormones (male pattern baldness).

There are two types of hair. Vellus hairs are the fine hairs located on much of your body's surface. Terminal hairs are thicker, more pigmented and are found on your head as well as genitals and axillary region.

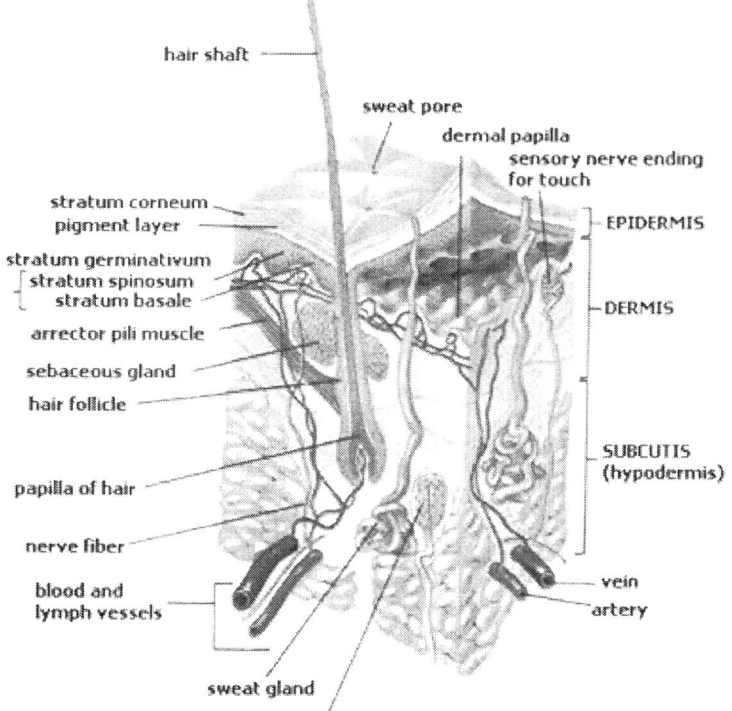

Figure 5.3. Integument. The dermis contains a number of accessory structures.

A small gland surrounds each hair follicle. This gland is called a sebaceous gland. The sebaceous glands secrete an oily substance known as sebum. The substance is secreted in response to contraction of the arrector pili muscle. Sebum contains triglyceride, protein, cholesterol and some electrolytes. Sebum makes the hair more flexible and hydrated.

Sweat Glands

Sweat glands (aka sudoriferous glands) are also located in the dermis. There are two types of sweat glands. Apocrine sweat glands secrete their substances into the hair follicles. The secretions of apocrine glands can develop an odor. The odor can increase because the secretion acts as a nutrient for bacteria that enhance the odor. Apocrine glands begin to secrete substances at puberty and are located in the axilla and genital regions.

Eccrine sweat glands secrete their substances directly onto the surface of the skin. They are coiled tubular glands that secrete a substance that mostly consists of water with a trace of some electrolytes and a peptide with antibiotic properties. The eccrine sweat glands primary function is to help to regulate body temperature. The sweat can evaporate and carry away heat. The sweat also excretes water and electrolytes.

Nails

The nails exist at the distal portions of the fingers and toes. The visible portion of the nail is called the nail body and it sits over the nail bed. The nail begins deep in the skin proximal to where it is seen. It extends out to beyond an area of thickened epidermis called the hyponychium. The nail begins to grow at the nail root which is close to the bone. A portion of the superficial epidermis (stratum corneum) extends over the proximal portion of the nail forming the eponychium or cuticle. Blood vessels deep to the nail give it a pink color. These vessels may be obscured leaving a white area known as the lunula.

Temperature Regulation

The skin is very important in regulating body temperature. The skin helps keep in heat produced by skeletal muscles and liver cells. When the body gets too hot the skin opens up the sweat pores so that the sweat can carry the heat away by evaporation.

Heat can be lost by the body in a number of ways. Heat always moves along a gradient from warmer to cooler temperatures. Heat can radiate from the body to the surrounding areas at lower temperatures. In conduction, heat moves via molecules from the warmer body to cooler objects. An example of conduction would be to lean against a cooler concrete wall. The heat flows from your body into the wall. In convection, heat moves via air molecules circulating around body. In evaporation, fluid on the surface of the body carries heat away.

Body temperature is primarily regulated by an area in the brain known as the hypothalamus. The hypothalamus sets the body's temperature and controls it by opening and closing sweat glands and contracting muscles.

Let's say that it is a hot summer day and you are working hard mowing the lawn. As your body's temperature rises the hypothalamus senses this and sends a message to your sweat glands to open. The sweat evaporates off of your skin and you begin to cool down.

Now let's say that you've finished mowing the lawn and you go inside of your air conditioned home. Your body's temperature will begin to drop. The hypothalamus senses this and sends a message to your sweat glands to close. If your body's temperature continues to drop the hypothalamus may send a message to your muscles to contract or shiver. The muscles will generate heat to help maintain your core temperature. In more severe cases of cold your blood vessels will constrict in your extremities in an attempt to conserve heat at the core of your body for survival.

If your core body temperature continues to drop you may develop a condition called hypothermia. You will progress from feeling cold to shivering, experiencing mental confusion, lethargy, loss of reflexes and eventually loss of consciousness and shutting down of organs.

Conversely if your core body temperature increases too much, you can develop hyperthermia. This can develop in humid conditions because of lack of evaporation. The signs of hyperthermia include light headedness, dizziness, headaches, muscle cramps, fatigue and nausea.

Skin Repair

The skin has remarkable healing properties. It can heal cuts, bruises and burns.

A cut is known as a laceration. If the cut extends only into the epidermis the epidermal cells will divide rapidly to repair the skin.

If the cut extends into the dermis broken blood vessels form a clot. The clot forms from a protein called fibrin which is a product of blood cells. Fibroblasts (cells) collect in the injured area and grow new collagen fibers.

Burns are also a type of skin injury. There are three categories of burns.

First degree burns are known as superficial partial thickness burns. Only the epidermis is affected. First degree burns usually heal quickly because growth occurs from the deeper layers of the dermis.

Second degree burns are known as deep partial thickness burns. In second degree burns the epidermis and some of the dermis is damaged. Fluid accumulates between the dermis and outer layer of epidermis forming blisters. The skin becomes discolored from dark red to waxy white. Healing depends on the accessory organs of skin because new cell growth emerges from these layers.

Third degree burns involve the epidermis, dermis and accessory organs. Third degree burns are called full thickness burns. In third degree burns there is no new cell growth from the damaged area. Growth can only occur from the margins of the burn. Skin substitutes can be used to cover the skin while healing. These include amniotic and artificial membranes and cultured epithelial cells. Skin grafts can also be used.

Review Questions Chapter 5

1. Which of the layers of the epidermis only exist on the palms of the hands and the soles of the feet:

    a. Stratum corneum
    b. Stratum lucidum
    c. Stratum basale
    d. Basement membrane

2. Which of the layers of the epidermis contains hardened keratinized cells:

    a. Stratum lucidum
    b. Stratum basale
    c. Stratum granulosum
    d. Stratum corneum

3. Which of the following chemicals is responsible for skin color:

    a. Keratin
    b. Melanin
    c. Vit D
    d. Collagen

4. Vit D is synthesized in the skin by the action of _____

    a. Melanin
    b. UV radiation
    c. Keratin
    d. Vit A

5. Sweat glands consist of 2 types including:

    a. Eccrine and appocrine
    b. Holocrine and eccrine
    c. Appocrine and sudoris
    d. Sebaceous and eccrine

6. Which of the following is not a constituent of sebum:

    a. Triglyceride
    b. Protein
    c. Electrolytes
    d. Sucrose

7. A structure located in the dermis that allows for hair to stand on end is known as:

    a. Arrector pili muscle
    b. Levator papillae muscle
    c. Tertiary protein
    d. Erector muscle

8. The cuticle of a nail is also known as the:

    a. hyponychium
    b. lunula
    c. nail body
    d. eponychium

9. A white area at the base of a nail is called the:

    a. Hyponychium
    b. Body
    c. Lunula
    d. Keratin

10. Body temperature is regulated by:

    a. Hypothalamus
    b. Sensory receptors in the skin
    c. Brain stem
    d. Thalamus

11. Which of the following should not happen in response to a lower than normal body temperature:

    a. Shivering
    b. Vasoconstriction in extremities
    c. Opening of sweat glands
    d. Closing of sweat glands

12. The type of burn where healing must occur from the outer margins is called:

    a. First degree
    b. Second degree
    c. Third degree
    d. Fourth degree

Chapter 6

Bones, Bumps and Holes, Oh My!

Chapter 6

Bones, Bumps and Holes, Oh My!

Learning the human skeleton can be a daunting task for any anatomy student. I remember looking at Halloween skeletons during my student days and thinking that learning those bones can't be all that bad. What I didn't know was that I had to learn all of the bumps, grooves and holes on the bones *in addition to* the names.

Why do we need to know all of these structures? Well one reason is that ligaments and tendons connect to these bony bumps and they can be a location for injuries. Some bumps are popular anatomical landmarks used for doing medical procedures like giving injections. Also, important structures go through those darned holes in the bones including arteries, veins and nerves.

Bone Structure

No, we are not talking about those beautiful people on the front cover of magazines. We are referring to how actual bones are put together. We need to learn some of these basic concepts before moving on to learning the actual bones.

Basically there are four types of bones categorized according to shape:

- Long bones are long. They have a long longitudinal axis (fig.6.1).

- Short bones are short. They have a short longitudinal axis and are more cube-like.

- Flat bones are flat. They are thin and curved such as some of the bones of the skull.

- Irregular bones are often found in groups and have a variety of shapes and sizes.

There are also two types of bone tissue in different amounts in bones. Compact bone (sometimes called cortical bone) is very dense. Cancellous bone (sometimes called spongy bone) looks more like a trabeculated matrix (fig. 6.1). It is found in the central regions of some of the skull bones or at ends (epiphyses) of long bones. The bone forming cells (osteocytes) get their nutrients by diffusion.

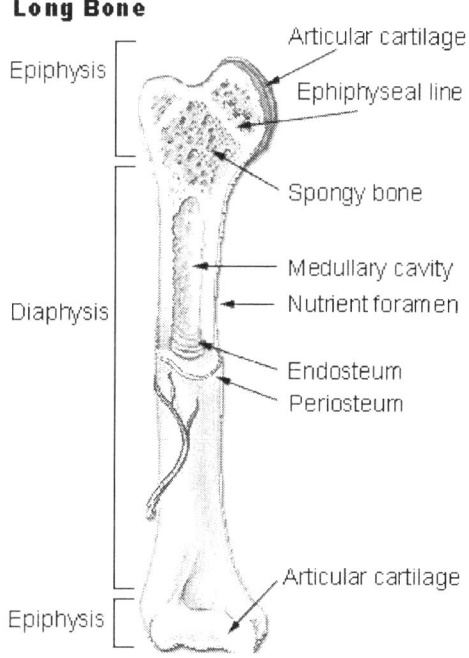

Figure 6.1. Parts of a long bone. Notice the long shaft or diaphysis in the middle of the bone. The diaphysis contains compact bone surrounding a medullary cavity containing bone marrow. On either end is an epiphysis containing cancellous or spongy bone. The epiphyseal line is a remnant of the growth plate. The epiphyses also contain hyaline cartilage for forming joints with other bones. Surrounding the bone is a membrane called the periosteum. The periosteum contains blood vessels and cells that help to repair and restore bone.

Compact bone is organized according to structural units called Haversian systems or osteons (fig. 6.2). These are located along the lines of force and line up along the long axis of the bone. The Haversian systems are connected together and form an interconnected structure that provides support and strength to bones.

Haversian systems contain a central canal (Haversian canal) that serves as a pathway for blood vessels and nerves. The bone is deposited along concentric rings called lamellae. Along the lamellae are small openings called lacunae. The lacunae contain fluid and bone cells called osteocytes. Radiating out in all directions from lacunae are small canals called canaliculi. Haversian systems are interconnected by a series of larger canals called Volksmann's canals (perforating canals).

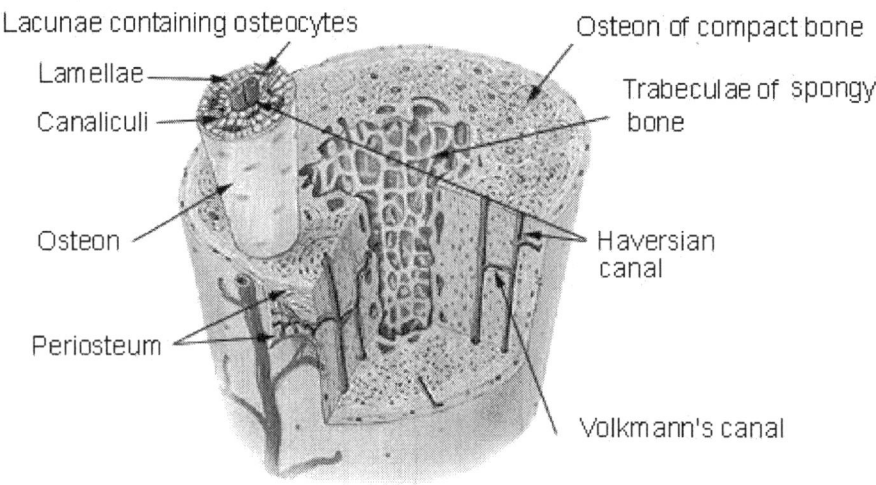

Figure 6.2. Compact and Spongy Bone.

Bone Cells

There are three basic types of cells in bone. Osteoblasts undergo mitosis and secrete a substance that acts as the framework for bone. Once this substance (called osteoid) is secreted, minerals can then deposit and form hardened bone. Osteoblasts respond to certain bone forming hormones as well as from physical stress. Osteocytes are mature osteoblasts that cannot divide by mitosis (fig. 6.3). Osteocytes reside in lacunae. Osteoclasts (another type of bone cell) are capable of demineralizing bone. They free up calcium from bone to make it available to the body depending on the body's needs.

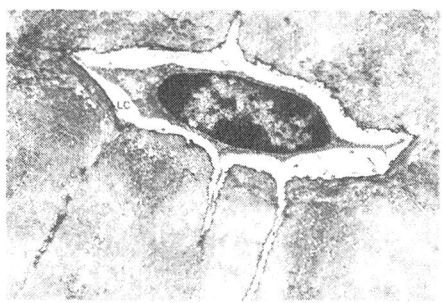

Figure 6.3. Osteocytes are mature osteoblasts that reside in a lacuna. They are surrounded by bony matrix.

Bone Marrow

Bone marrow is located in the medullary (marrow) cavity of long bones and in some spongy bones. There are two kinds of marrow. Red marrow exists in the bones of infants and children. It is called red because it contains a large number of red blood cells. In adults the red marrow is replaced by yellow marrow. It is called yellow because it contains a large proportion of fat cells. Yellow marrow decreases in its ability to form new red blood cells. However, not all adult bones contain yellow marrow.

The following bones continue to contain red marrow and produce red blood cells:

- Proximal end of humerus
- Ribs
- Bodies of vertebrae
- Pelvis
- Femur

Bone Growth

Bones begin to grow during fetal development and complete the growth process during young adulthood. There are two bone forming processes. Flat bones called intramembraneous bones develop in sheet like layers. Tubular bones called endochondral bones develop from cartilage templates.

Intramembranous Ossification

Big Picture: Intramembranous Ossification

Flat bones form from sheets of connective tissue by cells that secrete bone in all directions.

Flat bones such as some of the bones of the skull develop from a process called intramembranous ossification. During this process bones form from sheet-like layers of connective tissue. These layers have a vascular supply and contain bone forming cells called osteoblasts. The osteoblasts secrete bony matrix in all directions around the cell. The matrix unites with that secreted by other osteoblasts as the bone forms. Eventually the osteoblasts may be walled off by the bony matrix. At this point the osteoblast is called an osteocyte.

Endochondral Ossification

Big Picture: Endochondral Ossification

Tubular bones form from cartilage templates containing primary and secondary ossification centers.

Tubular bones develop from a process known as endochondral ossification. During this process bones develop from hyaline cartilage templates. The template is surrounded by an area called the perichondrium. The perichondrium will become the periosteum as the bone develops. Chondrocytes in the cartilage begin to secrete bony matrix and eventually wall themselves off in lacunae. Next blood vessels extend into the bone and transport osteoblasts and osteoclasts from the perichondrium forming a primary ossification center. The bone continues to grow in a cylindrical fashion.

Eventually blood vessels enter the calcified matrix of the epiphyses and form secondary ossification centers. Osteoclasts remove matrix from the center of the diaphysis to form a medullary cavity. The secondary ossification centers form about one month before birth. Bone continues to form from the cartilage until all of the cartilage is replaced except for the epiphyseal plates. These will complete their calcification in young adulthood (fig. 6.4).

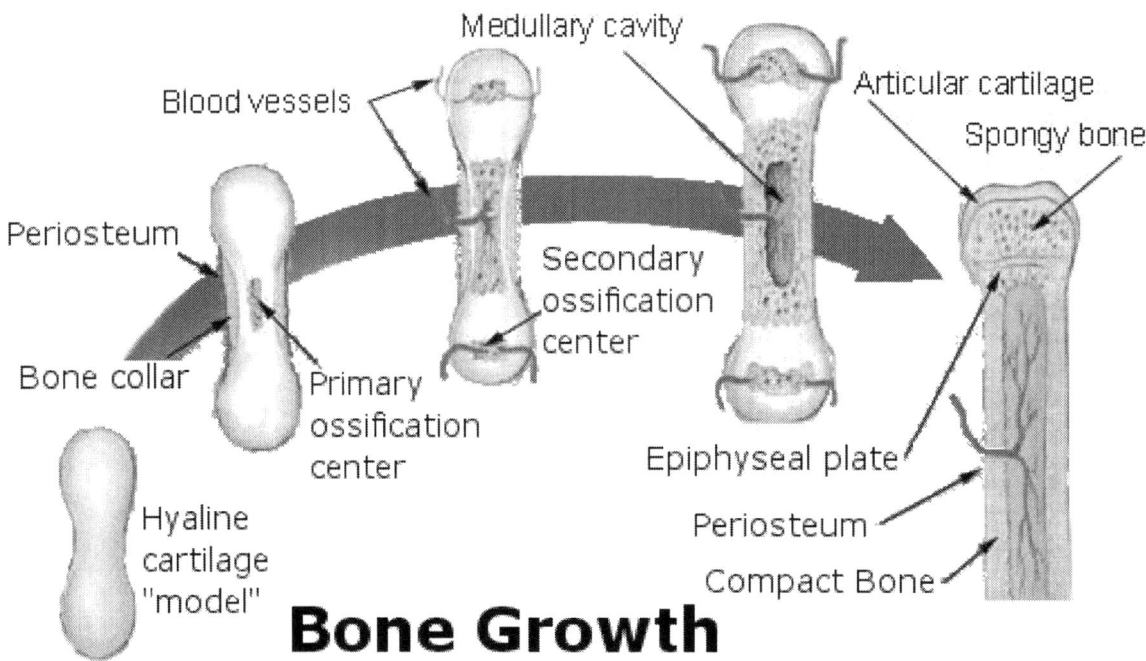

Figure 6.4. Endochondral ossification.

Epiphyseal Plates

Bone grows longitudinally as the epiphyseal plates secrete bony matrix. There are four zones in epiphyseal plates:

1. Zone of resting cartilage. This zone contains chondrocytes that do not divide rapidly.

2. Zone of proliferation. This zone contains active chondrocytes that produce new cartilage.

3. Zone of hypertrophy. In this zone chondrocytes from the zone of proliferation mature and enlarge.

4. Zone of calcification. In this zone the enlarged chondrocytes are replaced by osteoblasts from the endosteum. The osteoblasts secrete bone that calcifies the area.

As the chondrocytes produce cartilage and hypertrophy the bone grows on the diaphyseal side of the plate. The plate remains the same thickness because ossification on both sides of the plate occurs at the same rate. The epiphyseal plates complete their growth and ossify between the ages of 12-25 years depending on the bone.

Bone Growth Factors

The length of bones and subsequent height of an individual is determined genetically. However there are other factors that affect the expression of genes that in turn can affect bone growth. These include certain hormones, nutrition, and exercise.

Growth hormone is a hormone secreted by the anterior portion of the pituitary gland. Growth hormone stimulates protein synthesis and growth of cells in the entire body including bones. Thyroxine is secreted by the thyroid gland and increases osteoblastic activity in bones. Calcitrol is secreted by the kidneys and helps the digestive tract absorb calcium. The synthesis of calcitrol depends on vitamin D (see the integumentary system section). Sex hormones from the ovaries and testes also stimulate osteoblastic activity.

Vitamins such as vitamin D, C, A, K and B12 are also important in bone growth. Vitamin C is required for collagen synthesis and stimulates osteoblastic activity. A lack of vitamin D can lead to a condition called Rickets in children or osteomalacia in adults. Rickets is characterized by malformed bones. Calcium and phosphorus must be adequately supplied by the diet for use in boney matrix. Vitamin A stimulates osteoblastic activity and vitamins K and B12 are needed for protein synthesis in bone cells.

Bone grows according to the imposed demands of the body. This is known as Wolf's law. In other words the body produces bone along lines of force. For example, weight bearing exercises will increase the strength of bones. Likewise bones that are casted during the healing process for fractures will be weaker. This is one reason that weight-bearing exercise is recommended for those predisposed to osteoporosis.

Back to Bumps and Holes

Now that we know a bit about how bones are structured and formed. Let's get into learning actual bones and their bumps and holes.

Just what these bumps and holes called? Here is a list:

- Tubercle---a small, knoblike process.

- Tuberosity—a knoblike process larger than a tubercle.

- Trochanter—a large process.

- Styloid process—a pointed process.

- Suture—an interlocking union between two bones.
- Foramen—an opening in a bone (usually a passageway for vessels).
- Sinus—a cavity within bone.
- Condyle—a rounded process (usually articulates with another bone).

The Big Picture: Skeleton

The axial skeleton consists of the skull, spine, ribcage and sacrum. The appendicular skeleton consists of the arms and legs.

Learn General Structures First, Specific Structures Second

Begin with the whole skeleton (fig. 6.5). The skeleton is divided into two sections. These are the axial and appendicular skeletons. The axial skeleton consists of the skull, spine, ribcage and sacrum. The appendicular skeleton consists of the arms and legs (upper and lower extremities).

Now let's learn about the skull. Start with the bones of the cranium (figs. 6.6, 6.7).

The eight bones of the cranium include:

1. Frontal
2. Occipital
3. Right Parietal
4. Left Parietal
5. Right Temporal
6. Left Temporal
7. Sphenoid
8. Ethmoid

Drill yourself on these by using the diagrams.

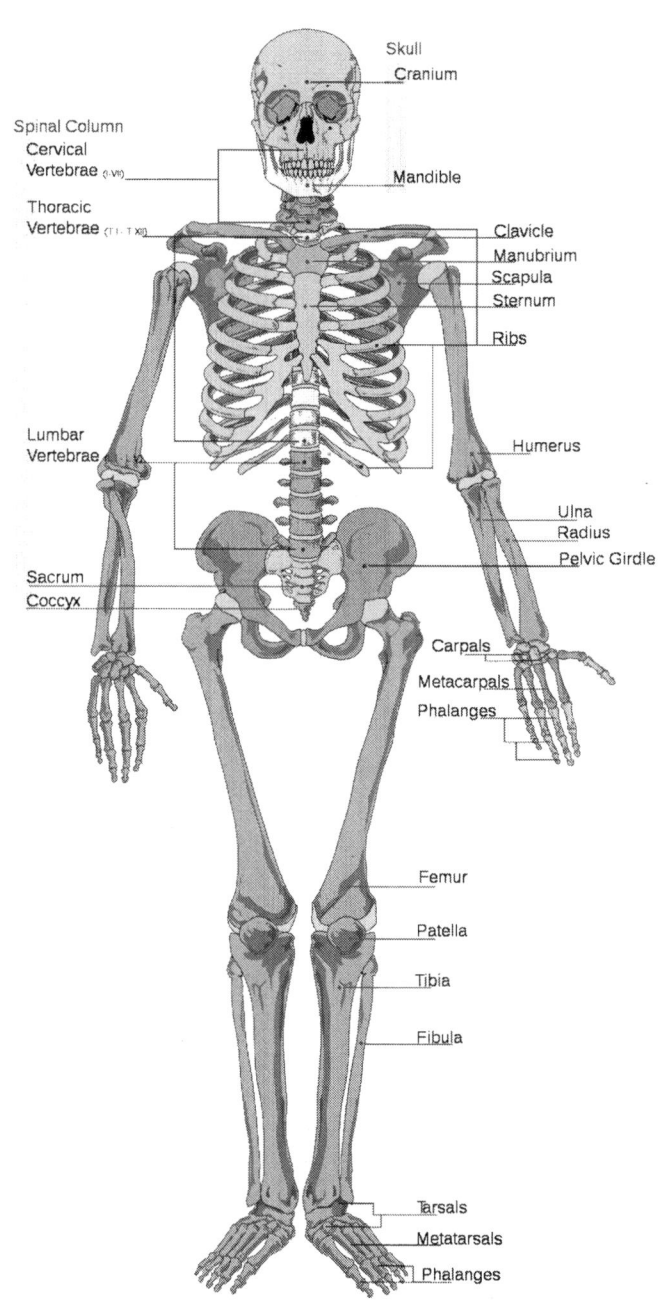

Figure 6.5. The skeleton.

Bones, Bumps and Holes, Oh My! 75

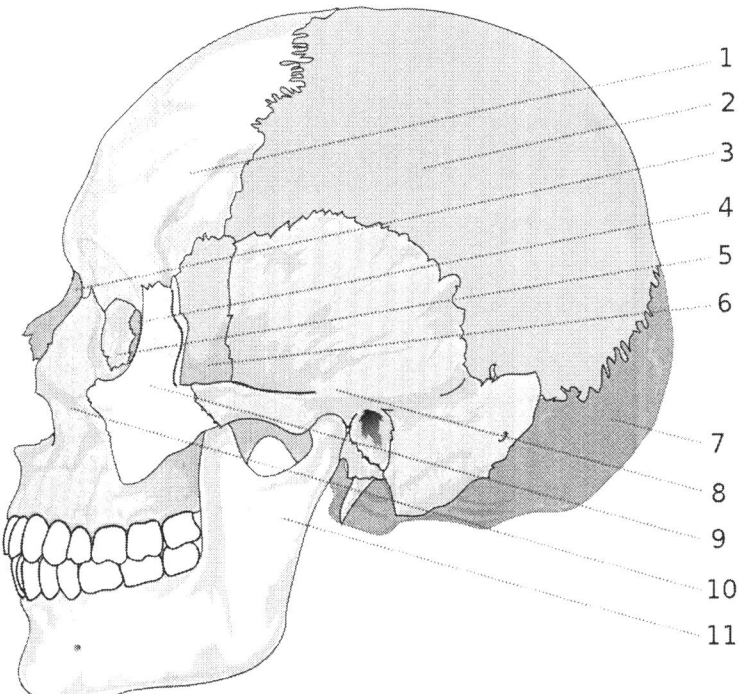

Figure 6.6. The skull.

1. Frontal
2. Parietal
3. Nasal
4. Lacrimal
5. Ethmoid
6. Sphenoid
7. Occipital
8. Temporal
9. Zygomatic
10. Maxilla
11. Mandible

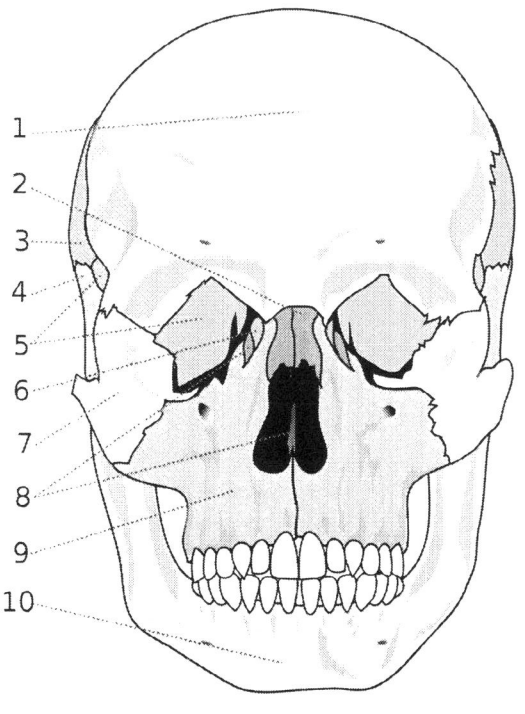

Figure 6.7.

1. Frontal
2. Nasal
3. Parietal
4. Temporal
5. Sphenoid
6. Ethmoid
7. Zygomatic
8. Ethmoid
9. Maxilla
10. Mandible

Now move on to the remaining skull bones.

1. Maxilla
2. Mandible
3. Zygomatic
4. Palatine
5. Vomer

Once you know the locations of these bones on diagrams you can learn more of the nuances about each bone. Also, remember that the skull is three dimensional so you may want to get your hands on a skull or a series of pictures that show different views of the skull to get the full picture of how these bones go together.

Don't Memorize--Learn

Avoid This Common Mistake!

It is tempting to just memorize the pictures in your text. Often your textbook contains a number of labeled pictures of the skeleton. Some students feel they can just memorize those pictures. The problem is that an arrow can be in a different location and still point to the same structure which tends to happen in exams. You need to learn the three dimensional aspects of the bones. Look at where one bone begins and ends on the skull. Use different pictures with varying views of the skull and try to identify the bones. ***The idea is to learn, not to memorize.***

Study the Tricky Bones

There are two classically tricky bones in the skull. They are tricky because their structure is complex and they reside inside the skull. You can see them from a number of different views so you need to work to learn the 3D aspects of these bones.

These bones are the sphenoid and ethmoid bones. I remember puzzling over these for a long time when I first saw them. Take some time to really get a handle on these as they are favorites of many professors.

Sphenoid Bone

This one sits at the base of the cranium and extends out to the lateral walls. A favorite view is that of looking down at the base of the cranium from above (figs 6.8, 6.9). When you look at it this way the bone looks something like a bat. Not a baseball bat but one of the flying, scary types. In fact it actually has wings called the greater and lesser wings. The middle part of the bat contains one of those all important structures that ends up on just about every skeletal anatomy test. In fact it makes most anatomy instructors giddy when they put this one on lab exams. This structure is called the sella turcica, aka Turkish saddle. What sits in the saddle? None other than the good ole pituitary gland.

So what else do we need to know about the sphenoid? Well, there are some holes that the optic nerve goes through called the optic foramen. Sometimes these will also show up on exams.

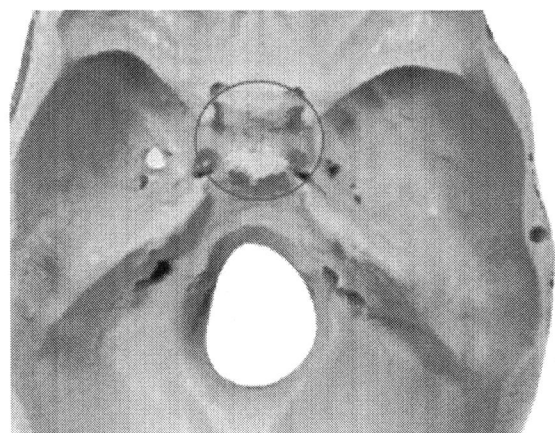

Figure 6.8. The sella turcica (Turkish saddle) is located in the central region of the sphenoid bone.

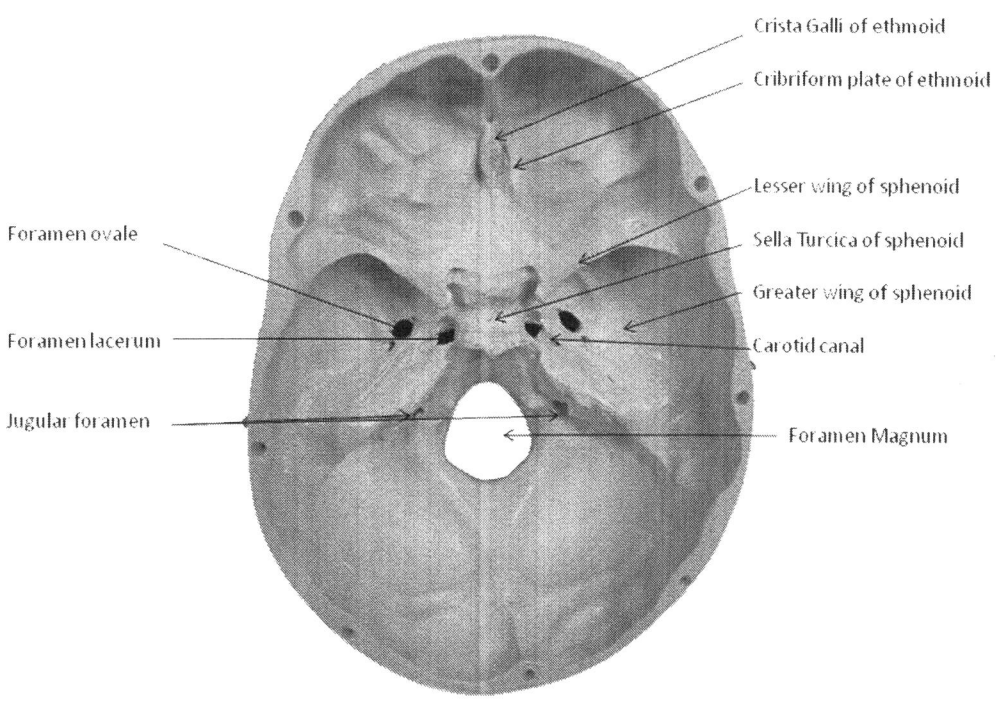

Figure 6.9. Internal view of the skull.

Ethmoid Bone

The other tricky skull bone is the ethmoid bone (fig. 6.10). Like the sphenoid this bad boy sits in the skull and you can see it from a number of views. You really need to learn the 3D aspects of this one too.

One of the favorite views of the ethmoid is the same as the sphenoid. Again, standing above the cranium and looking down (fig. 6.9). The ethmoid is anterior to the sphenoid in the front part of the floor of the cranium. You can also see it in the orbit (more on that later).

This one also contains some famous landmarks. These include the crista galli and cribriform plate. Again looking down from above the skull will help you to see these.

The crista galli is a ridge that sticks up. It is a connection point for the coverings of the brain called the meninges. The cribriform plate is just like it says, a plate. But this plate has holes in it. Why, you ask? Well, they are for the fibers of the olfactory nerve that makes its way to the brain. The nerve fibers for the sense of smell have to get to the brain somehow and that way is right through the ethmoid bone. Then you can smell grandma's apple pie and think about how wonderful it tastes.

Speaking about smell brings up another important part of the ethmoid bone. If you look right in the nasal area of the skull you will see a ridge of bone right in the middle. This is called the perpendicular plate and it forms part of the nasal septum.

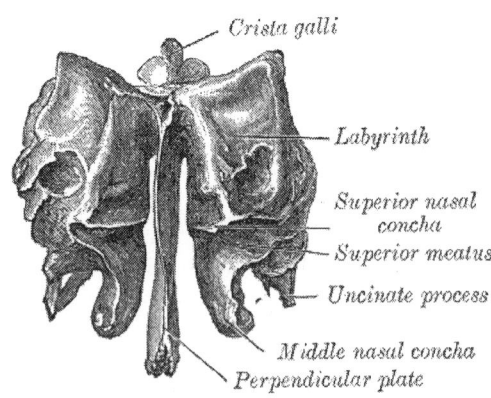

Figure 6.10. Ethmoid bone.

Getting Specific with the Skull: Some Nitty Gritty Structures

So now you know a lot about the skull already. It's time to get into the nitty gritty. Here we will look at each bone and go through a few landmarks for each. You might want to check out appendix one at the back of the book for more detailed diagrams.

Here is a list of bones with their important landmarks. Your professor should guide you as to what you should know unless you are in one of those super difficult anatomy classes where you have to know everything. If so I feel sorry for you! Also, you may need to find pictures of some of these structures since I've tried to keep the pages to a minimum.

Frontal Bone

The frontal bone contains sinuses (frontal sinuses) that secrete mucous to help flush the nasal cavity.

Landmarks

The supraorbital margin is a thickened process above the orbits that helps to protect the eye.

The lacrimal fossa located on the superior and lateral aspect of the orbit is a small landmark for the lacrimal (tear) gland.

The suprorbital foramen is a passageway for blood vessels supplying the frontal sinus, eyebrow, and eyelid.

Occipital Bone

Landmarks

The occipital condyles are rounded processes that articulate with the first cervical vertebra (atlas) of the neck.

The foramen magnum is a passageway for the spinal cord.

The jugular foramen lies between the occipital and temporal bones and provides a passageway for the internal jugular vein.

Temporal Bones

The temporal bones form the inferior-lateral margins of the cranium. They house the inner ear structures and articulate with the mandible.

Landmarks

The zygomatic process forms the posterior portion of the zygomatic arch. It articulates with the temporal process of the zygomatic bone.

The mastoid process is a site of muscle attachments for some of the neck muscles. It also contains small air cavities called air cells that connect with the middle ear. These can be a site of infection called mastoiditis.

The styloid process is a pointed process that attaches to ligaments that support the hyoid bone.

The external auditory meatus (canal) is a tube-like structure that houses structures for the external and middle ear.

The carotid canal is a passageway for the internal carotid artery that supplies the brain.

The foramen lacerum is a narrow slit-like structure located between the temporal and sphenoid bones. It carries small blood vessels that supply the inner portion of the cranium.

Maxilla

The maxilla is actually two bones that have fused.

Landmarks

The alveolar process holds the upper teeth.

The infraorbital foramen provides passage for the infraorbital artery and nerve.

The palatine process forms the anterior portion of the hard pallete.

The maxillary sinus is a hollow area lined with a mucous membrane. This cavity opens to the nasal passages.

Mandible

The mandible forms the lower jaw. It is actually two bones that have fused.

Landmarks

The alveolar process holds the lower teeth.

The mandibular foramen provides passage for the inferior alveolar nerve (a division of the trigeminal nerve). It is located on the medial aspect (inside) of the mandible.

The mental foramen contains fibers of the inferior alveolar nerve.

The condyles form the lateral part of the temporomandibular joint (TMJ). They articulate with the temporal bone.

The mental protuberance is a ridge of bone that extends anteriorly and is located in the central region of the mandible. It forms the chin.

Zygomatics

The zygomatic bones are located in the anterior portion of the skull. They connect with the maxilla, frontal and temporal bones and form the cheeks.

Landmarks

The temporal process is an extension of bone that connects with the zygomatic process of the temporal bone to form the zygomatic arch.

Palatine

The palatine bone is one of the bones that form the hard palette. It connects with the palatine process of the maxilla to form the posterior portion of the hard palette. It is located between the maxilla and sphenoid bones.

Bones of the Orbit

Some professors make you learn the bones of the orbit.

The orbit is formed by the following bones:

- Frontal
- Lacrimal
- Ethmoid
- Zygomatic
- Maxilla
- Palatine
- Sphenoid

You can use a mnemonic to learn these. Think of a flea with mumps. The mnemonic is FLEZMPS.

Sutures and Fontanels

The bones of the skull connect together by way of special joints called sutures (figs. 6.11, 6.12). There are four major sutures in the skull. If you look at the top of the skull you will see two sutures. One runs from front to back and connects both parietal bones while the other runs from side to side connecting the parietal bones with the frontal bones.

If you remember the anatomical planes these two will be a breeze. The back to front suture lies in the sagittal plane and is aptly named the sagittal suture. The side to side suture lies in the coronal plane and is named the coronal suture.

The other two primary sutures are the squamosal that surrounds the temporal bone and the lambdoidal that connects the occipital bone with the parietal bones. The sutures are not fully developed in the infant skull so there are some gaps. These are called fontanels.

The fontanels serve a useful purpose in allowing for compression of the fetal skull during birth (appendix one).

The anterior fontanel is located at the junction of the developing frontal and parietal bones. The anterior fontanel can be palpated for up to age two years. The posterior or occipital fontanel is located at the junction of the parietal and occipital bones. There are also sphenoidal and mastoid fontanels on the lateral sides of the skull. The sphenoidal fontanel is located at the junction of the frontal, parietal, temporal and sphenoid bones. The mastoid fontanel is located at the junction of the parietal, temporal and occipital bones. The remaining fontanels usually ossify by the end of the first year (appendix one).

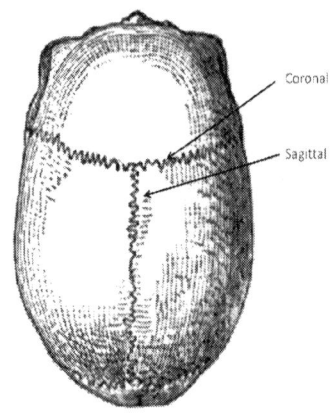

Figure 6.12. The coronal suture unites the frontal and parietal bones. The sagittal suture unites both parietal bones. Both sutures run in their respective planes.

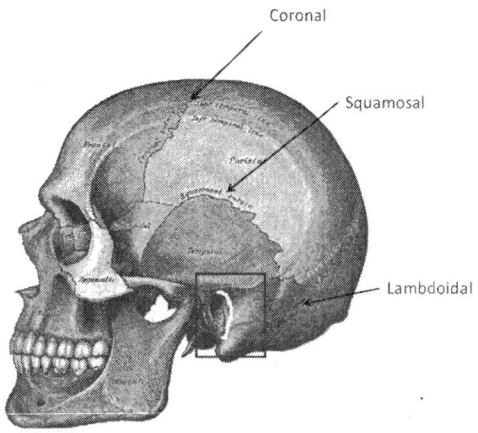

Figure 6.11. Sutures of skull.

The Spine

The spine consists of 24 vertebra "stacked" one on the other forming a column. The spine provides support for the head and trunk and houses the spinal cord. There are three basic sections of the spine. The cervical spine consists of seven vertebrae and has two very unique vertebrae called the atlas and axis. The thoracic spine consists of twelve vertebrae that articulate with ribs. The lumbar spine consists of five large vertebrae. The vertebrae are numbered according to their location from top to bottom. For example C2 is the second cervical vertebra, T5 is the fifth thoracic vertebra and L5 is the fifth lumbar vertebra.

How to Tell the Differences between Vertebrae

Most vertebrae have a similar construction with some slight differences. Vertebrae generally consist of a body with two strut-like structures called pedicles extending laterally connecting to transverse processes. Structures called lamina complete the ring and fuse at the spinous processes (see appendix one).

Bones of the cervical spine have small bodies and large appearing spinal canals. They have foramen in their transverse processes that contain the vertebral artery and vein. They also have a forked or bifid spinous process. The atlas appears as a ring of bone. The axis has a large process extending superiorly called the dens or odontoid process.

The thoracic vertebrae are larger than the cervical vertebrae. Their bodies are larger and contain flat spots known as articulating facets which serve as connection points for ribs.

The lumbar vertebrae are the largest because they bear the most weight. Their spinal canals appear smaller.

Know the Curves of the spine

There are actually four spinal curves (figs. 6.13, 6.14). These include cervical, thoracic, lumbar and pelvic curves. The cervical and lumbar curves are both known as lordotic curves (example = cervical lordosis). A lordotic curve is characterized by having its convexity anterior. Lordotic curves are considered secondary curves because they develop after birth when humans begin to hold their heads up, sit up and walk. The cervical and lumbar areas of the spine are considerably more mobile than the thoracic or a pelvic area because of the latter's connection to the bony pelvis and ribs. The thoracic and pelvic curves are ca.led kyphotic curves (example = thoracic kyphosis). Kyphotic curves are characterized as being concave anteriorly. The kyphotic curves are considered primary curves because they are present at birth.

An increased curvature of the cervical or lumbar spine is called a hyperlorosis. A decreases curvature is called a hypolordosis.

An increased curvature of the thoracic spine is called a hyperkyphosis and a decreased curvature is called a hypokyphosis.

A lateral curvature is called a scoliosis. Sometimes, if the curve is not severe the curves are described as increased lateral convexities or concavities.

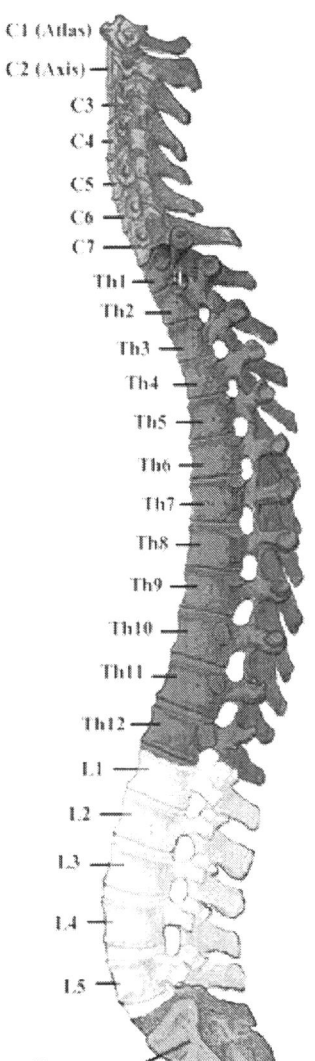

Figure 6.13. Spine.

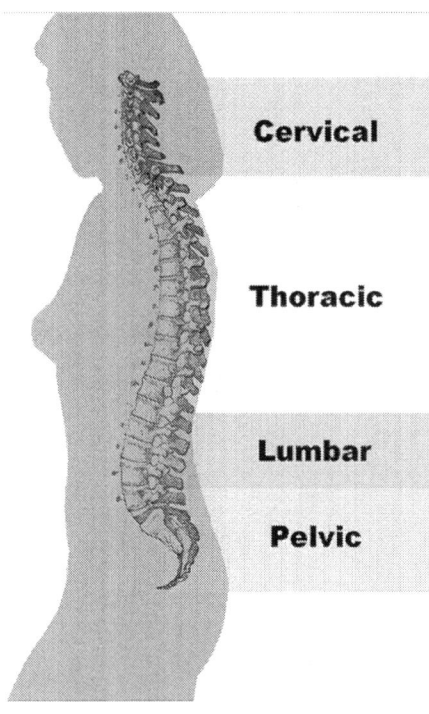

Figure 6.14. The spine is divided into cervical, thoracic, lumbar and pelvic sections.

The Ribcage (Finishing up the Axial Skeleton)

The ribcage consists of twelve pairs of ribs (fig. 6.15). There are true, false and floating ribs. Usually ribs 1-7 are true ribs with ribs 8-10 being false ribs. Ribs 11-12 are floating ribs. The ribs attach to the vertebra in the back and the sternum in the front by way of cartilage connections (costochondral cartilage). True ribs connect directly to the sternum by way of their cartilage connections. False ribs connect to the cartilage of true ribs and floating ribs only connect to the vertebrae in the back. There is no anterior connection to floating ribs.

Sternum

The sternum has three parts. The most superior portion is called the manubrium. Just inferior to this is the body and the most inferior portion is called the xiphoid process which consists of cartilage. The sternum also articulates with the clavicle.

Hyoid Bone

The hyoid bone is located in the anterior region of the throat (fig. 6.16). It supports the larynx. A number of muscles that extend to the larynx, pharynx and tongue attach to the hyoid bone.

The Appendicular Skeleton

Now that you know a lot about the axial skeleton you can move on to the appendicular skeleton (arms and legs). Again, learn the main bones first then the picky landmarks.

The bones of the appendicular skeleton include:

Shoulder and Arm

- Clavicle (collar bone)
- Scapula (wing bone)
- Humerus (arm bone)
- Radius (forearm bone on thumb side)
- Ulna (forearm bone on pinky side)
- Carpals (wrist bones)
- Metacarpals (hand bones)
- Phalanges (finger bones)

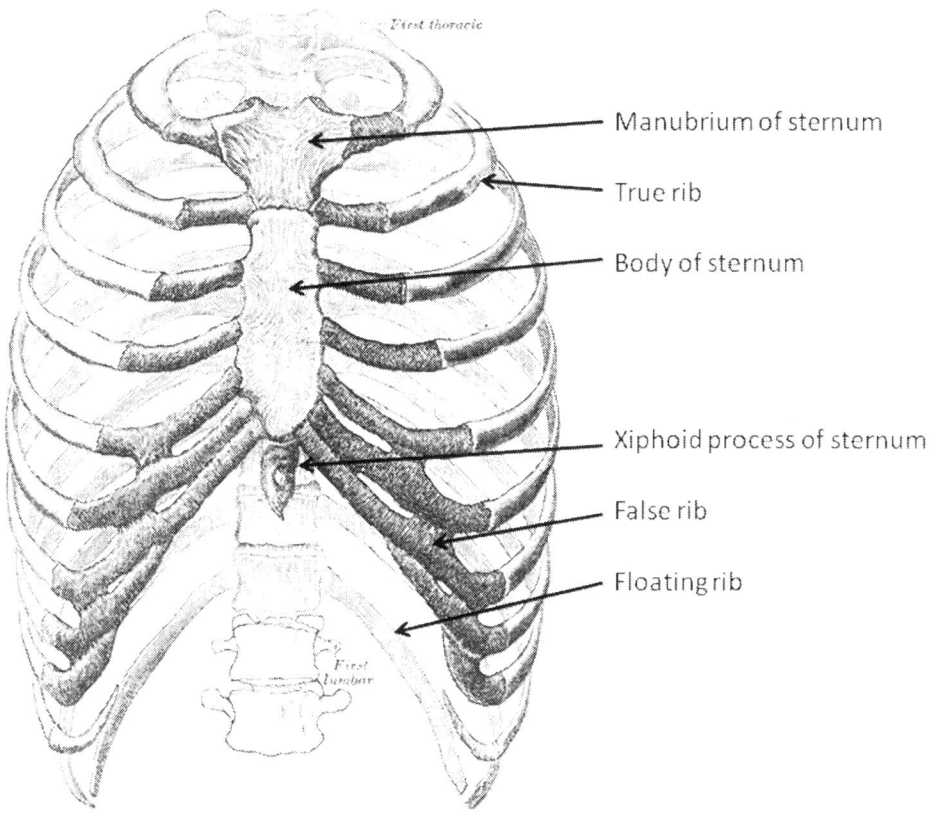

Figure 6.15. Ribcage

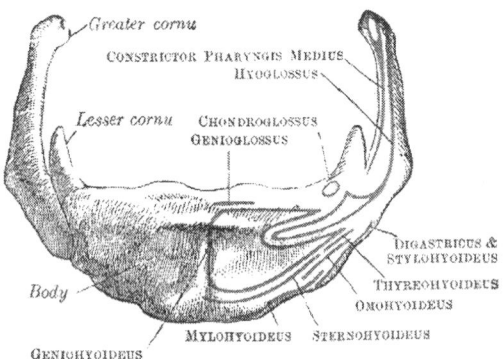

Figure 6.16. Hyoid bone.

A number of muscles attach to the hyoid bone.

Hip and Leg
- Coxal (pelvic bone)
- Femur (thigh bone)
- Patella (knee bone)
- Tibia (shin bone)
- Fibula (outer shin bone)
- Calcaneus (heel bone)
- Talus (ankle bone)
- Cuboid (foot bone that looks like a cube)
- Navicular (foot bone)
- Cuneaforms (three more foot bones)
- Metatarsals (foot bones analogous to hand bones)
- Phalanges (toe bones)

The upper extremity begins with what is called the pectoral girdle (aka shoulder girdle). This consists of the clavicle and scapula. The pectoral girdle acts as a support for the arms (hey they have to attach somewhere). The pectoral girdle attaches to the axial skeleton where the clavicle attaches to the sternum (sternoclavicular joint). This is the only direct attachment of the arm to the body. However there are a number of muscles that also help to stabilize the connection.

Clavicle (Collar Bone)

The clavicle is the only S-shaped bone in the body and connects to the sternum and scapula (fig. 6.17).

Scapula (Wing Bone)

The scapula is a triangular bone located in the posterior portion of the thoracic area (figs. 6.18, 6.19). It articulates with the clavicle and the humerus. It is important to study the scapula from both posterior and anterior sides.

Landmarks
The glenoid cavity (fossa) is a concave surface on the lateral aspect of the scapula. It forms the "socket" of the ball and socket joint of the shoulder.

The spine of the scapula is located on the back surface. It is a ridge of bone extending superiorly from medial to lateral.

The acromion process is the terminal end of the spine of the scapula. It is a large process and articulates with the clavicle. The acromion process marks the highest point of the shoulder. This is an important one.

The coracoid process is located on the anterior aspect of the scapula. This process is smaller than the acromion process and is located anterior and inferior to it. Another important one.

The supraspinous fossa is an indentation on the posterior portion of the scapula. It lies just above the spine.

The infraspinous fossa is located just inferior to the spine of the scapula.

Bones, Bumps and Holes, Oh My! 85

Figure 6.17. Clavicle

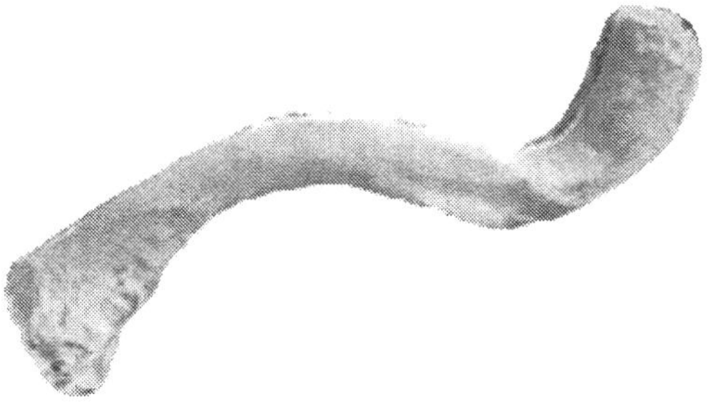

Fig. 6.18. Anterior view of scapula.

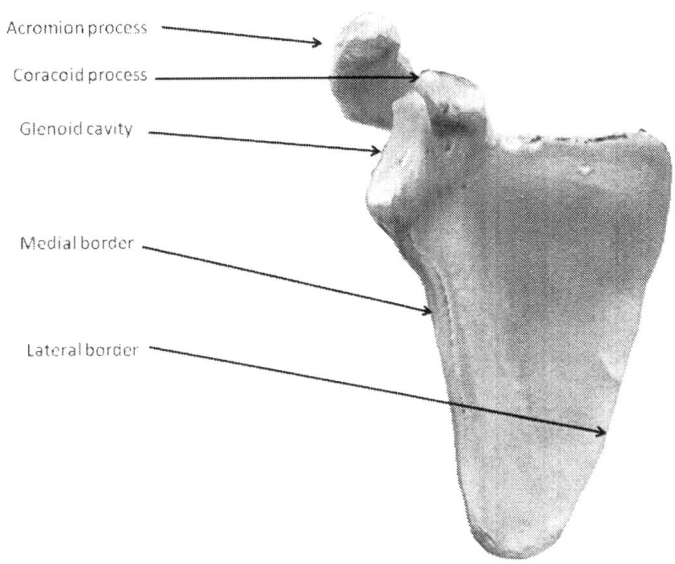

Figure 6.19. Posterior view of scapula.

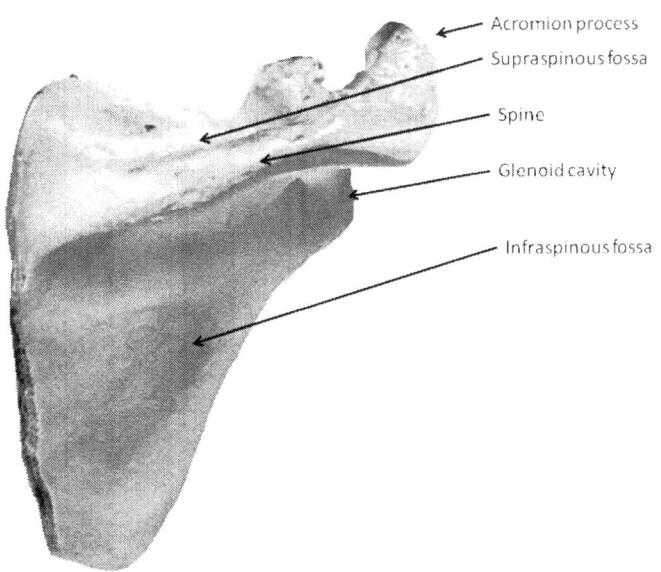

Humerus (Arm Bone)

The humerus is the proximal bone of the arm. It is a long tubular bone that articulates proximally with the scapula and distally with the radius and ulna (figs. 6.20).

Landmarks

There are a lot of potential landmarks on the humerus. Hopefully your professor will just pick a few of these.

The head of the humerus is the proximal rounded end of the bone.

The anatomical neck of the humerus is a small region that marks the end of the joint capsule between the humerus and the scapula.

The greater tubercle is a rounded process on the lateral aspect of the proximal humerus.

The lesser tubercle is a smaller rounded process on the medial aspect of the proximal humerus.

The intertubercular groove (sulcus) is a groove between the greater and lesser tubercles.

The deltoid tuberosity is a bump with a gradual slope on the lateral aspect of the humerus and is the site of attachment of the deltoid muscle.

The lateral epicondyle is a widened area on the lateral aspect of the distal humerus. It is an important site of muscle attachments for the wrist extensor muscles.

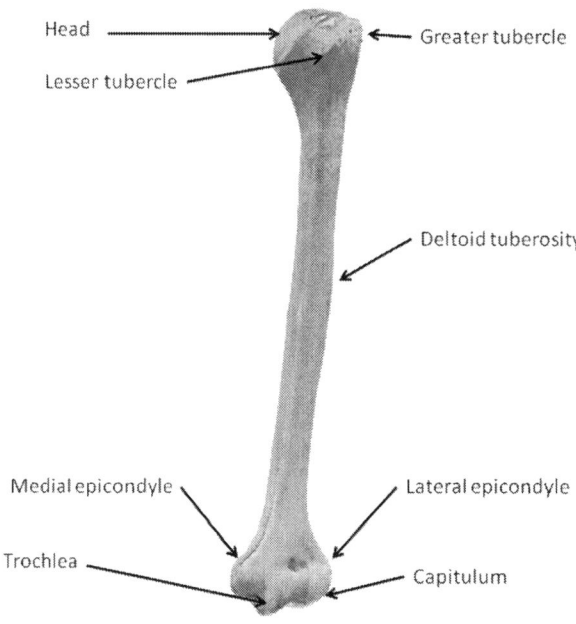

Figure 6.20. Humerus

The medial epicondyle is a widened area on the medial aspect of the distal humerus. This is where the wrist flexor muscles attach.

The capitulum is a small rounded process at the distal end of the humerus on the lateral side. It articulates with the radius.

The trochlea is a small spool shaped process at the distal medial end of the humerus. It articulates with the ulna.

Ulna (Forearm bone on pinky side)

The ulna and radius both support the forearm (antebrachium). The ulna is on the medial side of the forearm (fig. 6.21). The bump on your elbow is actually the olecranon process of the ulna. The ulna articulates with the trochlea of the humerus and forms a hinge joint.

Landmarks

The olectranon process is a rounded process on the proximal end of the ulna.

The trochlear notch of the ulna articulates with the trochlea of the humerus.

The radial notch is a flat spot that articulates with the radius.

The styloid process of the ulna is a needle-like process at the distal end.

Radius (Forearm bone on thumb side)

The radius is also located in the forearm. It articulates with the ulna and carpal bones (figs. 6.22). The radius allows for rotation of the forearm .

Landmarks

The head of the radius articulates with the capitulum of the humerus. This joint can rotate.

The radial tuberosity is a bump on the proximal aspect of the radius. The biceps muscle attaches there.

The styloid process is a needle-like process on the distal aspect of the radius.

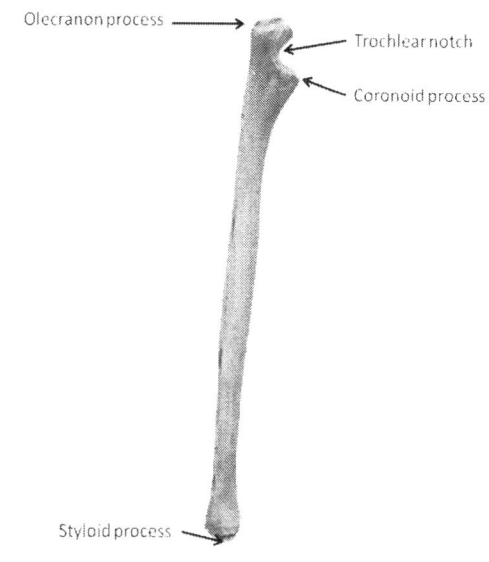

Figure 6.21. Ulna.

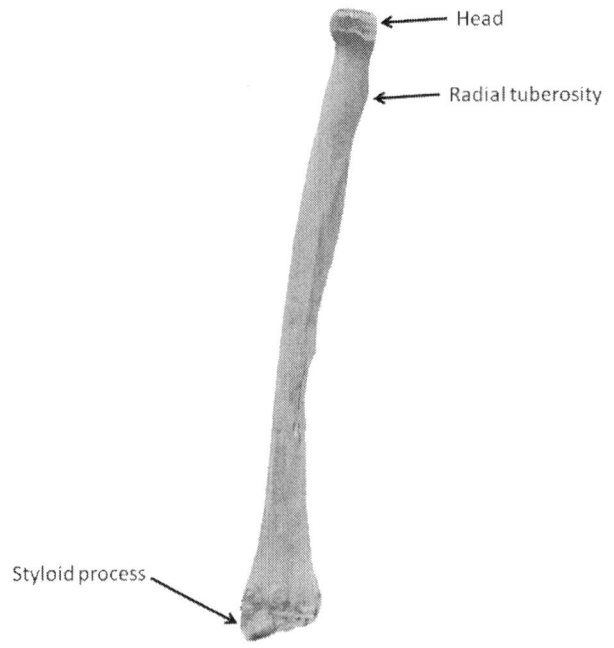

Figure 6.22. Radius

Carpals (Wrist bones)

The carpal bones are located in the wrist (fig. 6.23). They consist of eight bones.

The eight carpal bones:

- Scaphoid
- Lunate
- Triquetrum
- Pisiform
- Trapezium
- Trapezoid
- Capitate
- Hamate

Here is a mnemonic to help you remember the carpal bones:

Some Lovers Try Positions That They Can't Handle

Proximal Row (lateral to medial): Scaphoid-Lunate-Triquetrum-Pisiform
Distal Row: Trapezium—Trapezoid—Capitate—Hamate

Bones, Bumps and Holes, Oh My! 89

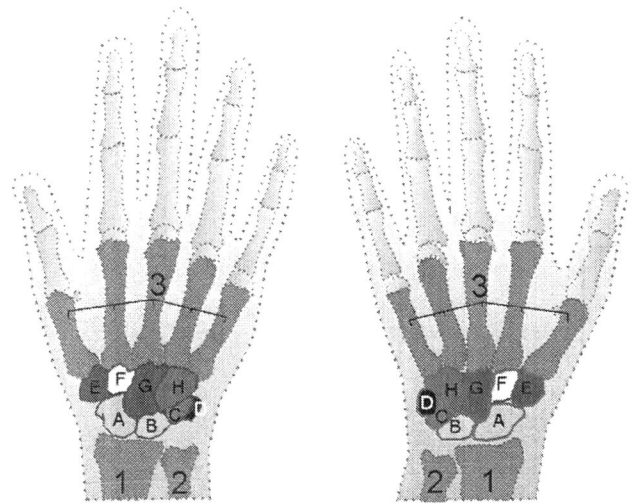

Figure 6.23. Carpals.
A. Scaphoid
B. Lunate
C. Triquetrum
D. Pisiform
E. Trapezium
F. Trapezoid
G. Capitate
H. Hamate

Figure 6.24. Metacarpals

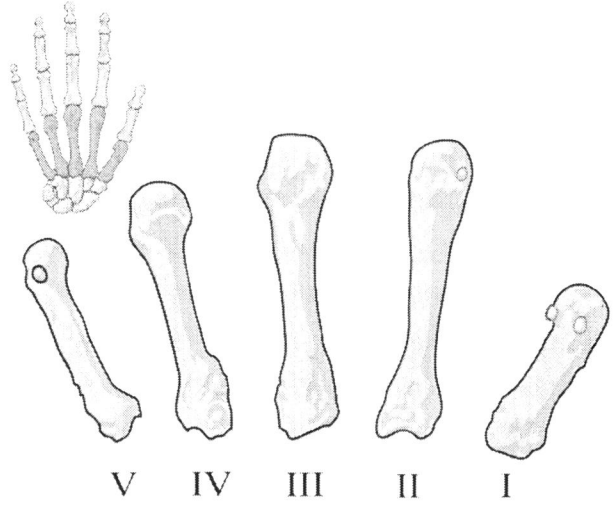

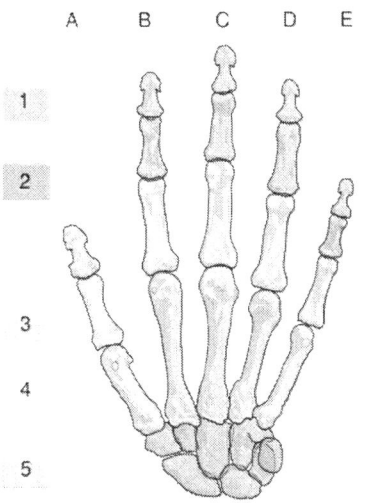

Figure 6.25. Bones of the hand.

1. Distal phalanx
2. Middle phalanx
3. Proximal phalanx
4. Metacarpals
5. Carpals
A. First
B. Second
C. Third
D. Fourth
E. Fifth

Notice the thumb only has proximal and distal phalanges. The remaining fingers have proximal, middle and distal phalanges.

Metacarpals (Hand Bones)

The metacarpals are tubular shaped bones that lie distal to the carpals (6.24). There are five metacarpals numbered accordingly from the thumb (1) to the little finger (5).

Phalanges (Finger Bones)

The phalanges comprise the fingers. They are numbered the same as the metacarpals and named for their location (fig. 6.25). The thumb has only a proximal and distal phalanx. The remaining fingers have proximal, middle and distal phalanges.

Coxal Bone (Pelvic Bone)

The pelvis consists of the sacrum and two coxal bones (fig. 6.26, 6.27). The coxal bones are actually three bones fused together (fig. 6.27). The three bones are the ilium, ischium and pubis. The coxal bones articulate with the sacrum at the sacroiliac joints and the femurs at the hip joints (fig. 6.28).

Landmarks

The acetabulum is a socket-like concave structure that articulates with the head of the femur to form the hip joint.

The iliac crest is the most superior structure of the coxal bone. It is a ridge of bone that extends along the superior margin of the ilium.

The anterior superior iliac spine is a bump on the anterior portion of the ilium. The iliac crest terminates here. This is an important site of muscle attachments.

The posterior superior iliac spine is a bump on the posterior aspect of the ilium. The iliac crest terminates here posteriorly.

The obturator foramen is a space that is formed by the pubis and ischium.

The symphysis pubis is a fibrocartilaginous disc that forms a fibrous joint between the 2 pubic bones.

The pubic tubercle resides on the anterior superior aspect of the pubis.

The ischial tuberosity is a thickened area of bone located on the posterior aspect of the ischium. The hamstring muscles attach here.

The pubic arch is the angle between the pubic bones.

Male and Female Differences

Generally the differences between male and female pelves are due to functions of childbirth. The pubic arch is greater in females and the ilia may be more flared. The sacrum tends to be more curved in males. The female pelvis is wider in all directions.

Femur (Thigh bone)

The femur is the longest bone in the body (figs. 6.29, 6.30). It connects with the acetabulum of the coxal bone proximally and with the patella and tibia distally.

Landmarks

The head of the femur is a rounded process on the proximal end.

The neck of the femur is the area that connects the head with the shaft.

The greater trochanter is a large process located on the proximal lateral aspect of the femur.

The lesser trochanter is a smaller process located on the proximal medial aspect.

There are two large rounded processes on the distal aspect of the femur called the medial and lateral condyles.

The linea aspera is a roughened area on the posterior aspect. It is a site of muscle attachments.

Bones, Bumps and Holes, Oh My! 91

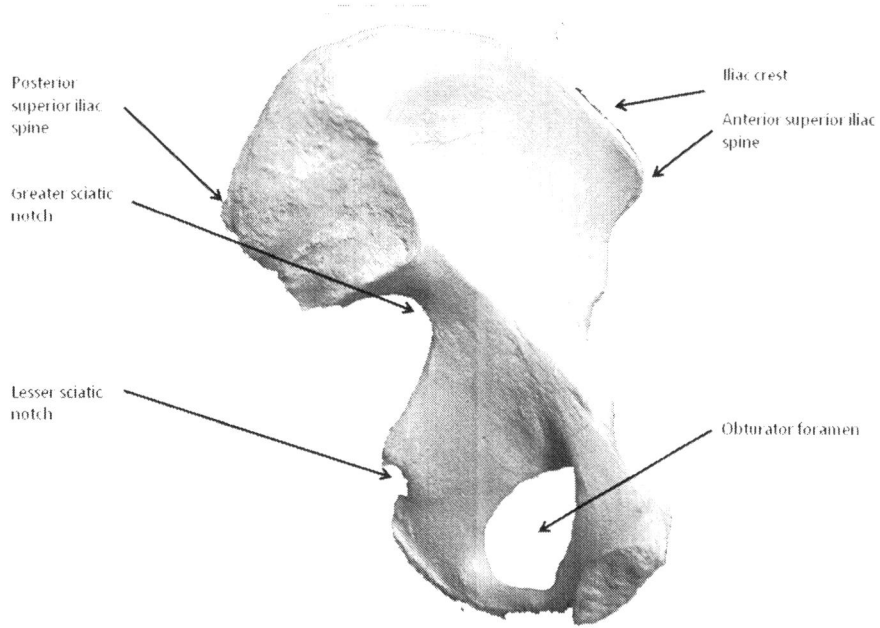

Figure 6.26. Medial view of the coxal bone.

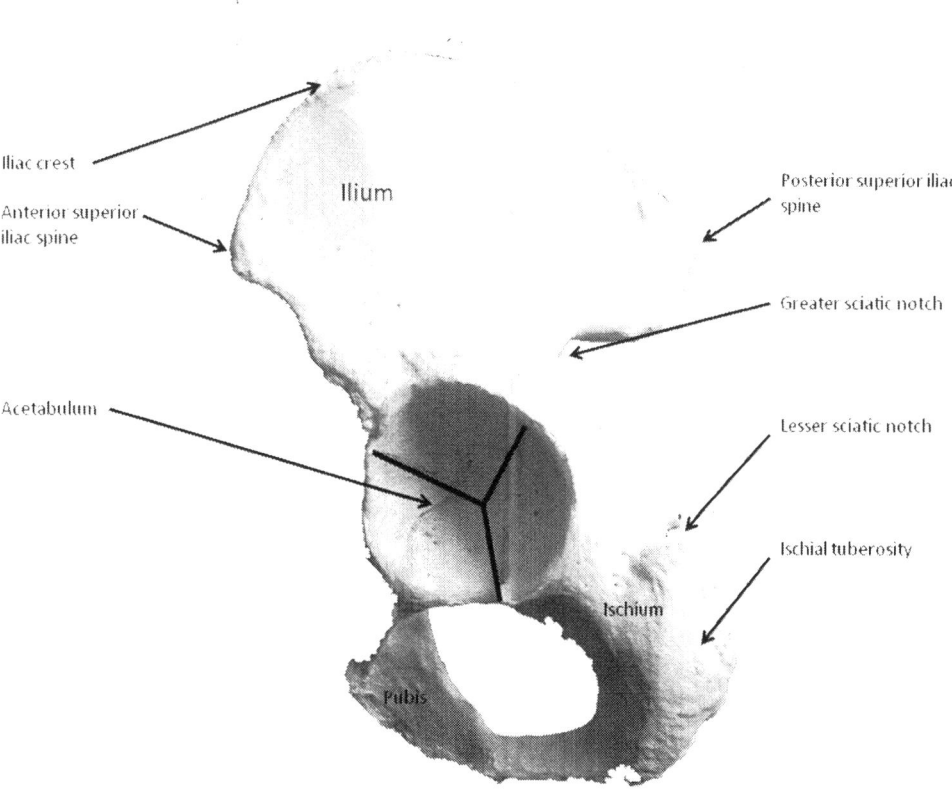

Figure 6.27. Lateral view of the coxal bone.

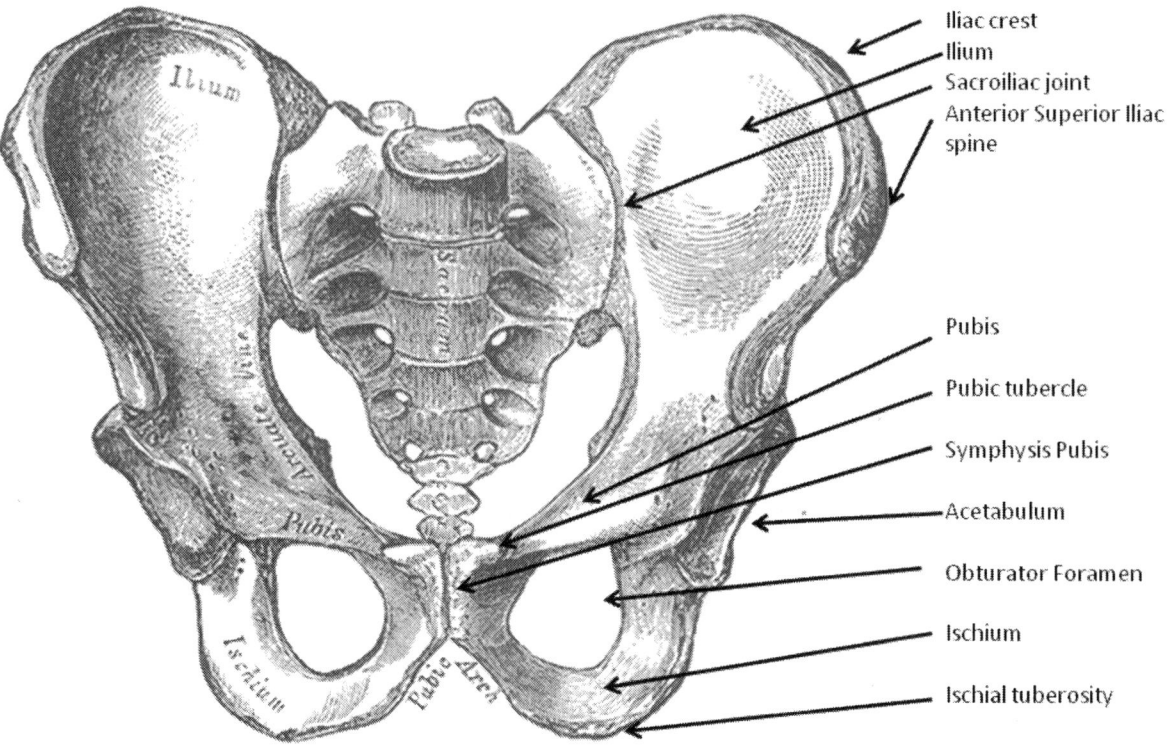

Figure 6.28. Pelvis. The pelvis consists of the two coxal bones and the sacrum.

Tibia (Shin bone)

The tibia is the larger of two bones of the lower leg (figs. 6.31, 6.32).

Landmarks

The tibial condyles are large rounded processes on the proximal aspect of the tibia. The two condyles are named medial and lateral.

The tibial tuberosity is a broad bump on the anterior aspect of the tibia.

The medial malleolus is a rounded process on the distal medial aspect of the tibia. It is the bump on the inside of the ankle.

Fibula (Outer shin bone)

The fibula is the lateral bone in the lower leg.

Landmarks

The fibular head is a rounded process on the proximal end of the bone.

The lateral malleolus is a rounded process on the distal end of the bone. It forms the lateral ankle.

Calcaneus and Talus (Heel and ankle bones)

The calcaneus or heel bone is the largest of the tarsals (fig. 6.33). The talus forms the ankle joint with the tibia and fibula. These bones articulate with the navicular and cuboid bones. There are 3 cuneiform bones named for their position which articulate with the metatarsals.

Metatarsals and Phalanges (Foot and Toe Bones)

There are five tubular metatarsals that are named for their position (1-5). The phalanges are similar to those in the fingers. The big toe only has proximal and distal phalanges while the remaining toes have proximal, middle and distal phalanges.

Ain't no more bones!

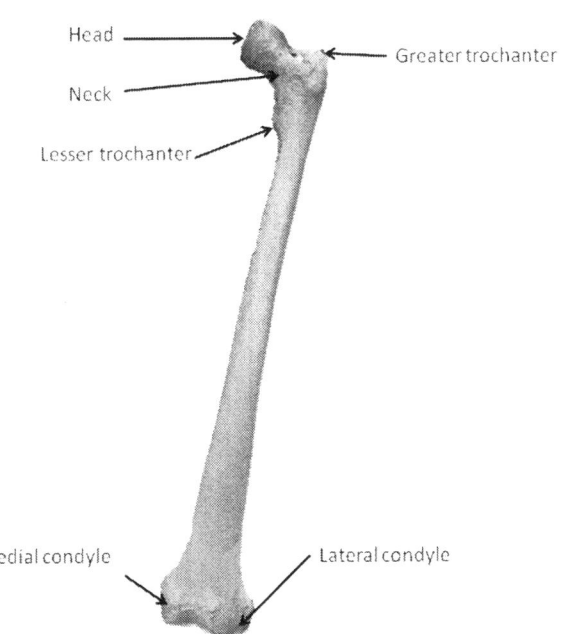

Figure 6.29. Anterior view of femur

Bones, Bumps and Holes, Oh My! 94

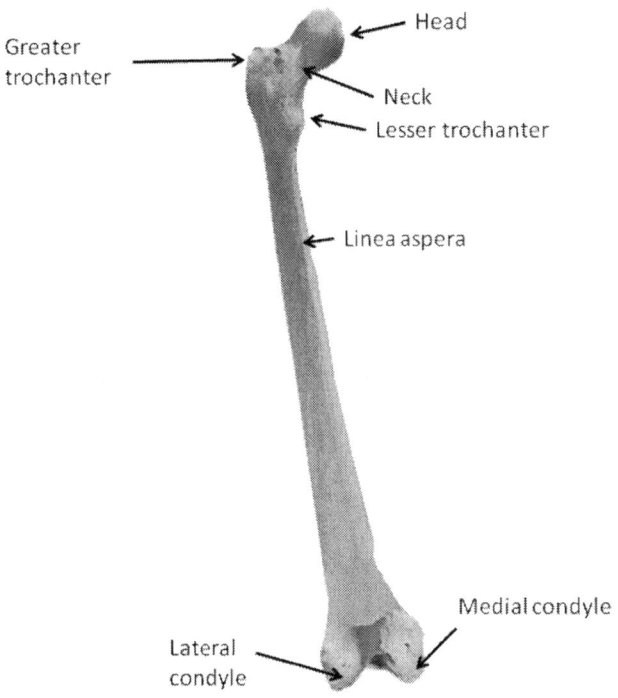

Figure 6.30. Posterior view of femur.

Figure 6.31. Anterior view of the tibia.

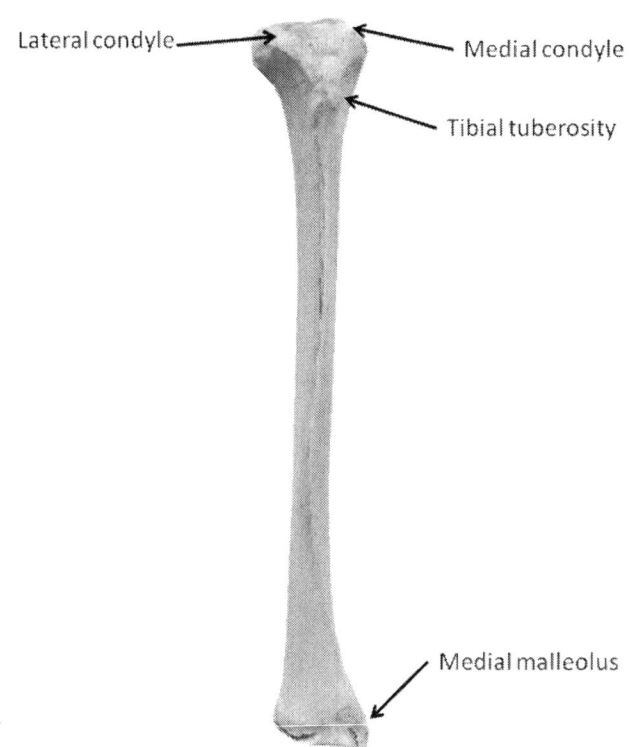

Bones, Bumps and Holes, Oh My! 95

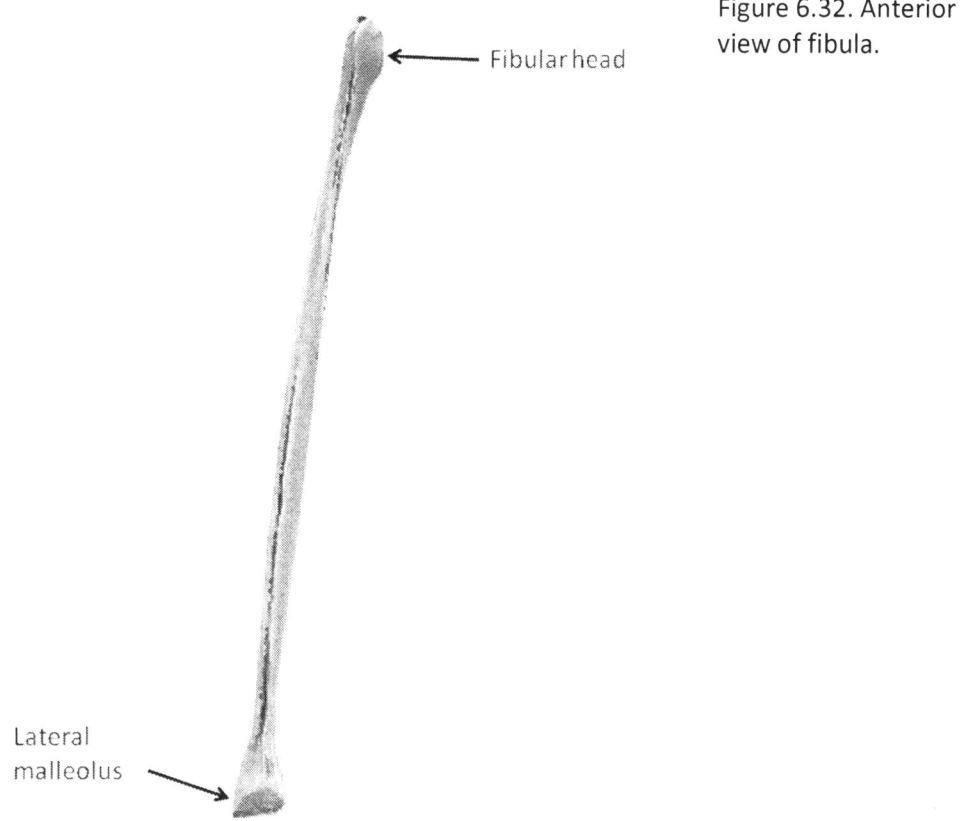

Figure 6.32. Anterior view of fibula.

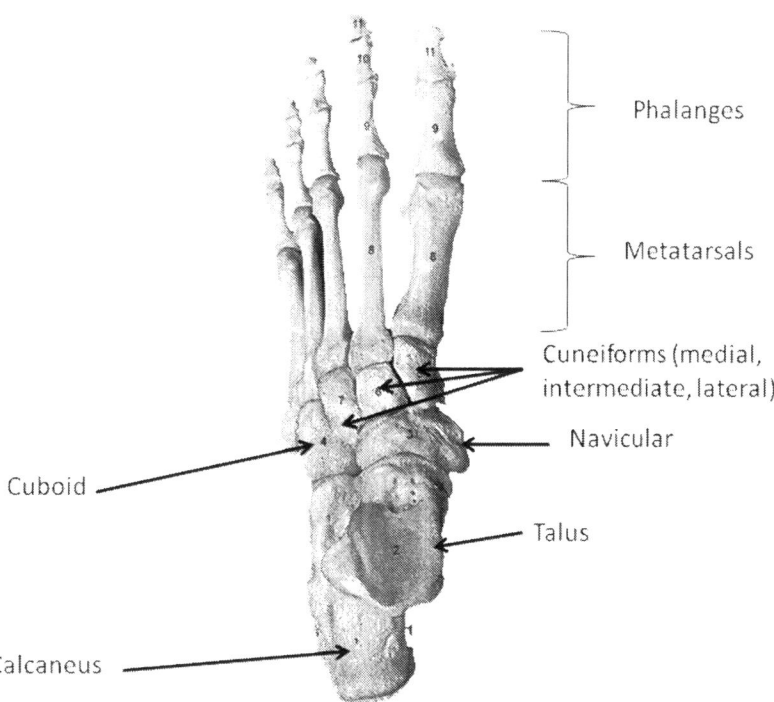

Figure 6.33. Foot and Ankle

Review Questions

1. Which of the following is not a part of an endochondral bone:

   a. Epiphysis
   b. Diaphysis
   c. Condyle
   d. Suture

2. Which of the following is not a stage in endochondral ossification:

   a. Ossification of a cartilage template
   b. Secondary ossification center forms
   c. Medullary cavity forms
   d. Bone grows from osteocytes in all directions

3. Which of the following bones is an intramembraneous bone:

   a. Femur
   b. Radius
   c. Metacarpal
   d. Parietal

4. Where is trabeculated bone found:

   a. Epiphysis
   b. Diaphysis
   c. Medullary cavity
   d. Growth plate

5. Which of the following bones is not part of the axial skeleton:

   a. Rib
   b. Humerus
   c. Frontal
   d. Cervical vertebra

6. Which bone contains the external auditory meatus:

   a. Frontal
   b. Temporal
   c. Parietal
   d. Sphenoid

7. The foramen magnum is located in which bone:

   a. Occipital
   b. Parietal
   c. Temporal
   d. Frontal

8. The sella turcica is located in which bone:

   a. Ethmoid
   b. Frontal
   c. Sphenoid
   d. Occipital

9. Fibers from the olfactory nerves pass through this skeletal structure:

   a. Sella turcica
   b. Cribriform plate
   c. Foramen magnum
   d. Foramen ovale

10. The coronal suture unites which bones:

    a. Frontal, parietal
    b. Parietal, temporal
    c. Temporal, occipital
    d. Parietal, occipital

11. Which of the following is not usually present in a typical cervical vertebra:

    a. Transverse foramen
    b. Bifid spinous process
    c. Large vertebral canal
    d. Large body

12. The dens is found on which vertebra:

    a. C1
    b. C2
    c. T1
    d. L5

13. Which 3 bones unite to form the coxal bone:

    a. Ilium, pubis, sacrum
    b. Ilium, ischium, pubis
    c. Pubis, coccyx, sacrum
    d. Ischium, ilium, sacrum

14. Where is the greater trochanter located:

    a. Tibia
    b. Humerus
    c. Femur
    d. Sacrum

15. Which bony process is the sharpest:
    a. Tubercle
    b. Tuberosity
    c. Styloid
    d. Trochanter

16. The head of the radius articulates with the:

a. Trochlea
b. Coronoid process
c. Coracoids process
d. Capitulum

17. The acromion process is located on which bone:

a. Humerus
b. Scapula
c. Sternum
d. Femur

18. The medial malleolus is part of which bone:

a. Tibia
b. Fibula
c. Humerus
d. Ulna

19. The most superior portion of the sternum is known as:

a. Body
b. Xiphoid
c. Head
d. Manubrium

20. The zygomatic arch consists of which 2 bones:
a. Zygomatic and parietal
b. Zygomatic and temporal
c. Zygomatic and maxilla
d. Zygomatic and mandible

Chapter 7

Managing Muscles

Chapter 7

Managing Muscles

We found out that there are a lot of bones in the body. Well it should be no surprise that there are also a lot of muscles. If you've ever watched modern dancers you might have appreciated the number of muscles needed to move the human body in such complex ways. Hopefully we can provide some help in learning about the muscular system. But before we dive into the many muscles of the body let's go over some background information.

Big Picture: Muscle Movements

Muscles move joints just like teeter-totters, wheelbarrows and fishing poles.

In many courses you will need to learn how muscles and joints act as levers. There are 3 types of levers (fig. 7. 1). These are:

Class 1—the teeter-totter

Class 2—the wheelbarrow

Class 3—the fishing pole

Levers contain three basic parts: the fulcrum, pull and weight. The object is to lift something (this is the weight). The fulcrum is the pivot point and the pull is where the force connects to the system. Let's take our three examples and explain them in a bit more detail.

The Teeter-totter

A class 1 lever is like a teeter-totter. Remember those cartoons where one character jumps off of a cliff and lands on one end of a teeter-totter causing another character to launch into the sky? Well the jumper is the pull and the poor unfortunate flyer is the weight. The fulcrum is in the middle. So we describe first class levers in terms of the fulcrum being between the pull and the weight.

The Wheelbarrow

A class 2 lever is like a wheelbarrow. Think of carting dirt to your garden. You lift the wheelbarrow by the handles (pull) and carry the dirt (weight) as the wheelbarrow glides on the wheel (fulcrum).

The Fishing Pole

A class 3 lever is like a fishing pole. Think of going to your favorite fishing spot and casting in the calm, glittering lake. The fulcrum is where you hold the pole. The pull is provided by your arm and the weight is the combination of hook, line and sinker at the end of the pole. So the fulcrum and weight are at opposite ends while the pull is in the middle.

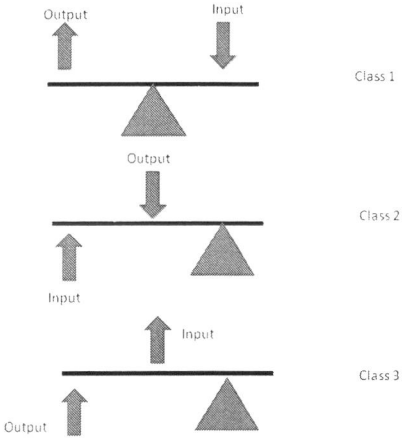

Figure 7.1. Three classes of levers.

Big Picture: Muscle Contractions

Isotonic: lifting weights at the gym.

Isometric: pushing against a wall.

Isokinetic: walking on a treadmill.

There are three types of muscle contractions. All three are used in treating injuries in rehabilitation and physical therapy settings.

It is important to learn what these terms actually mean. Iso means equal and tonic means tone or force. So in isotonic contractions (iso = equal, tonic = tone) the force remains the same but the length of the muscle changes. An example of an isotonic contraction is the classic biceps curl with a barbell. The force exhibited by the barbell does not change. However the length of the bicep muscle can change by shortening during the lifting phase and lengthening during the lowering phase. Isotonic exercises are used in many gyms in which participants use barbells and selectorized weight equipment.

In isometric contractions (iso = equal, metric = length) the force can change but the length of the muscle remains the same. In isometric contractions there is no movement of the joint since the muscle length does not change. An example of an isometric contraction would be pushing against an object that cannot be moved such as a wall. You can push with a little amount of force or a lot of force (force can change) but there is no movement of the joint

In isokinetic contractions (iso = equal, kinetic = motion) both the force and length of the muscle can vary but the contraction happens at a fixed speed. Isokinetic exercises are primarily used in rehabilitation settings. Sophisticated machines are used to control the speed of the exercise while allowing varying resistance. However a simple treadmill is a good example of isokinetic exercise. You can exercise at a fixed speed with varying degrees of force provided by the different incline angles of the treadmill.

Agonist—Antagonist

Some muscle terms (like agonist and antagonist) can be a bit confusing at first. The trick to learning these terms is to think of them as follows:

1. Name the joint and the movement.
2. Figure out which muscle is moving the joint.
3. Apply the right term.

Let's use an example. Let's say I finally got my behind in the gym and picked up a 10 lb. barbell to do a biceps curl. So how do I determine this agonist-antagonist stuff? Let's use the above steps:

1. Name the joint and the movement. Okay my elbow is moving into flexion during the curl.
2. Figure out which muscle is moving the joint. Okay I feel my biceps muscle contracting.
3. Apply the right term. Hmmm, that must mean my biceps is the agonist.

Great, so if the biceps is the agonist, which muscle is the antagonist? It's the one that opposes the movement (usually on the other side of the joint). This would be the triceps.

So, that's great but what if the movement changes? Do the agonist and antagonist change as well? Yes they can change. Let's go back to the gym to find out. Let's say that now I lift the weight from behind my head by straightening my elbow.

1. Name the joint and the movement. Okay my elbow is now going into extension.
2. Figure out which muscle is moving the joint. Okay I feel my triceps is contracting.
3. Apply the right term. Hmmm, that must mean my triceps is now the agonist while my biceps is now the antagonist.

See, use these steps and you will always get the right muscle.

Concentric—Eccentric Contractions

More muscle terms. These have to do with the shortening or lengthening of muscles. In concentric contractions the muscle shortens against a load. Eccentric contractions are just the opposite. Muscles lengthen against a load.

Back to the gym (I told you the gym was a great place to learn about muscles!). During my biceps curl exercise when I lift the weight (elbow flexion) my biceps is shortening so it is performing a concentric contraction. Likewise when I lower the weight (elbow extension) my biceps is lengthening so it is now performing an eccentric contraction.

Muscles are more prone to injuries during eccentric contractions. A few years ago I attended a training seminar where all of us attendees had to participate in an exercise sequence consisting of lunges.

Lunges are where you take a big step forward with say your right foot and then bend your right knee until your left knee touches the floor. Lunges incorporate an eccentric contraction of the hamstrings. Needless to say the next day I could hardly get out of bed because of my sore hamstrings!

Another classic example of an injury occurring from an eccentric contraction is the good ole rotator cuff tear. One of the jobs of the rotator cuff muscles is to slow down the internal rotation of the shoulder. Baseball pitchers (or dads who play ball with their kids) incorporate eccentric contractions of the rotator cuff muscles when they throw the ball. In some cases this can either aggravate or cause an injury to the rotator cuff.

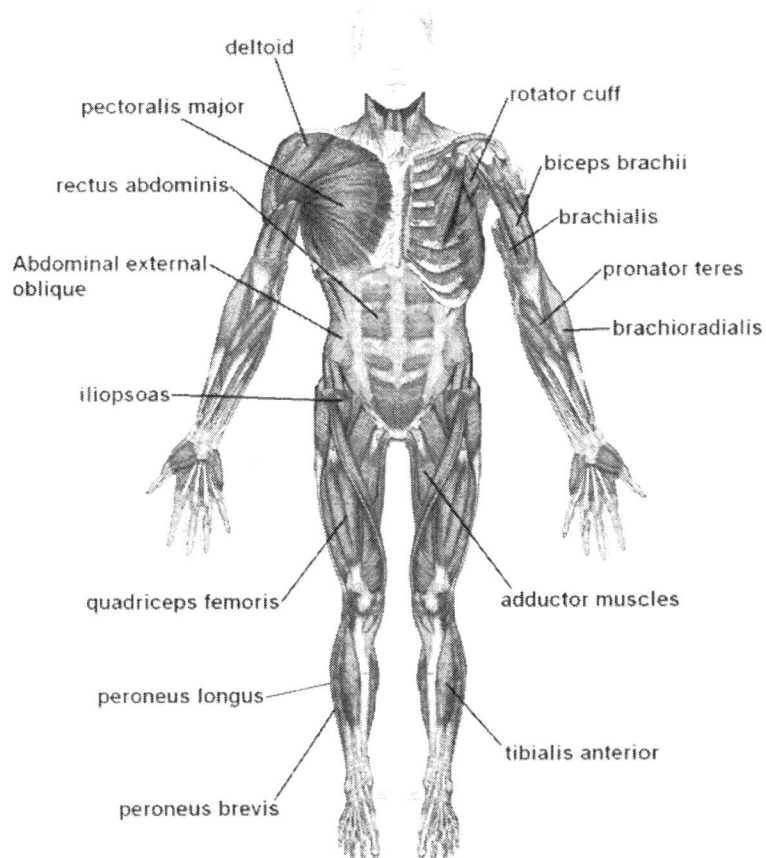

Figure 7.2. Overview of anterior muscles.

Managing Muscles 102

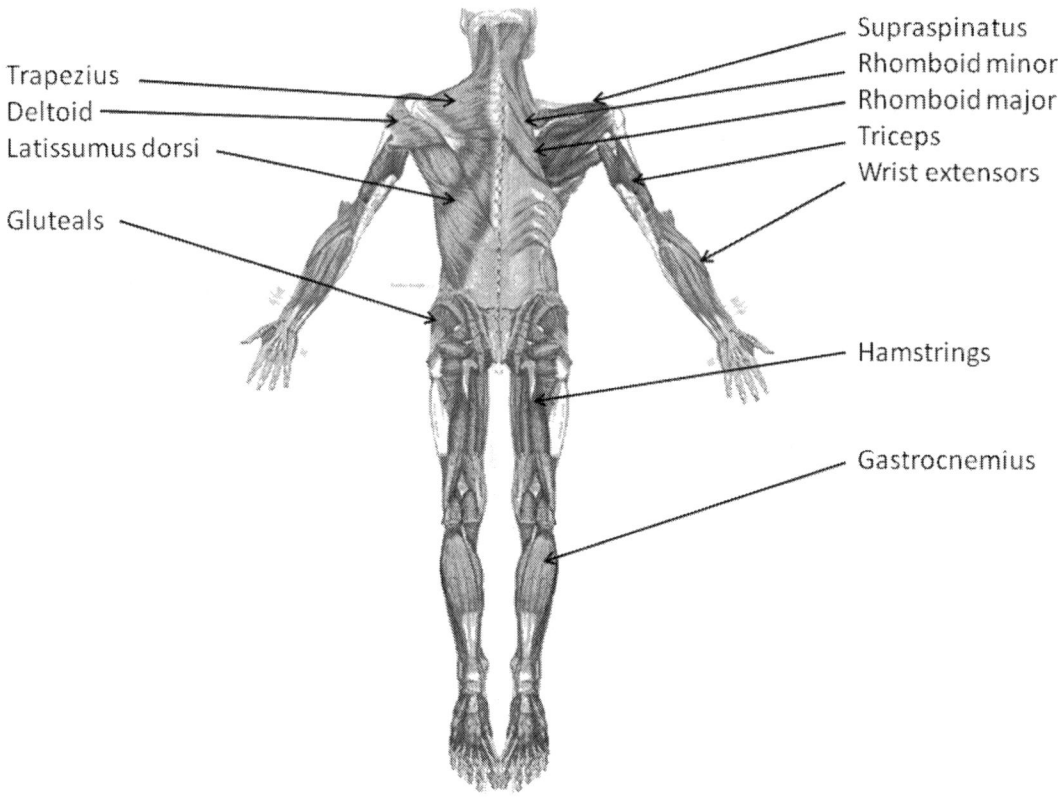

Figure 7.3. Overview of posterior muscles.

Overview of the Muscular System

How to Learn Muscles

Just as there are lots of bones in the body there are also lots of muscles that move the bones (figs. 7.2, 7.3). One trick to learning these is to group the muscles according to regions. For example learn the shoulder muscles, then arm muscles, forearm muscles and so on. It also helps to learn the superficial muscles first before learning the deep muscles. You can learn just a few muscles in a region. Once you learn these then learn a few more and so on. The more you repeat this process the more you will learn.

Another method to help you to learn muscles is to (heaven forbid) work out. Think about which muscles are active during certain movements like walking, reaching, bending at the waist and so on. Begin with the big ones and then move on to the more specific muscles.

Try to associate the muscle with the movement.

I use my biceps to bend my arm.

I use my triceps to straighten my arm.

I use my deltoid to raise my arm.

I use my wrist flexors to bend the wrist.

I use my wrist extensors to straighten the wrist.

I use my quadriceps to kick a ball.

I use my hamstrings to bend my knee.

I use my gastrocs to walk on my toes.

I use my tibialis anterior to walk on my heels.

I can lift my leg with my psoas.

When I walk my glutes will talk.

I bend forward with my abdominals.

I straighten up nice and erect with the erector spinae.

I shrug my shoulders with my traps.

Put your hands together and squeeze your pecs.

Here's a little rhyme to help you to remember the muscles and their movements.

I use my gastrocs to walk on my toes.

My biceps and deltoid help me touch my nose.

I bend my waist with my abs and wait...

Until my erector spinae make my spine straight...

My quads let me kick a ball...

My triceps let me push on a wall...

I lift my leg to take a step...

My psoas muscle has a lot of pep...

As I walk my leg falls behind...

The behind that moves my leg this way is of the gluteal kind...

I turn my head to the right and left and then...

Tell my SCM to stop contracting when...

I wave goodbye and as my wrist flops...

My wrist flexors and extensors stop.

The following is a table of lots of muscles. You should highlight the muscles you need to know then look them up in the following section or your course textbook.

Managing Muscles 104

Muscle	Origin	Insertion	Action
trapezius	Occipital bone, spines of C7 and all T vertebrae	spine and acromion of scapulaula	Extends head; elevates, depresses, rotates and retracts scapulaula
latissimus dorsi	lower vertebrae, iliac crest	intertubercular groove of humerus	Extends, adducts and medially rotates arm
serratus anterior	upper 8 ribs	anterior aspect of medial border of scapulaula	Protracts and rotates scapulaula
rhomboideus	spinous process of C1-T5	medial border of scapulaula	Retracts and rotates scapulaula
pectoralis major	clavicle, sternum, costal cartilages	greater tubercle of humerus	Flexes, adducts and medially rotates arm
pectoralis minor	ribs 3,4,5	coracoid process of scapulaula	Draws scapulaula anteriorly and inferiorly
supraspinatus	supraspinous fossa of scapulaula	greater tubercle of the humerus	abducts and stabilizes humerus
infraspinatus	infraspinous fossa	greater tubercle of the humerus	lateral rotation of humerus
triceps brachii	axillary border of scapulaula, posterior humerus	olecranon process of the ulna	extends forearm, stabilizes shoulder
biceps brachii	coracoid process, intertubercular groove of the humerus	radial tuberosity	flexes arm and forearm, supinates hand
brachialis	distal anterior humerus	coronoid process of ulna	flexes forearm
flexor carpi ulnaris	medial epicondyle of humerus, olecranon process	base of 5th metacarpal, pisiform and hamate	flexes wrist, adducts hand
palmaris longus	medial epicondyle of humerus	palmar aponeurosis	flexes wrist
flexor carpi radialis	medial epicondyle of humerus	base of 2nd and 3d metacarpals	flexes wrist, abducts hand
pronator teres	medial epicondyle of humerus	lateral radius	pronates and flexes forearm
brachioradialis	distal humerus	styloid process of radius	flexes forearm
extensor carpi radialis	lateral epicondyle of humerus	metacarpals II and III	extends and abducts wrist
extensor digitorum communis	lateral epicondyle of the humerus	posterior surfaces of distal phalanges of digits 2-5	Extends fingers and wrist, abbucts fingers

Muscle	Origin	Insertion	Action
extensor digitorum minimi	--	--	extends 5th digit
extensor carpi ulnaris	lateral epicondyle of humerus	metacarpal V	extends and adducts wrist
masseter	zygomatic arch	angle and ramus of mandible	elevates mandible
mylohyoideus	mandible	Hyoid	elevates hyoid
digastricus	mandible and mastoid process	hyoid bone	elevates hyoid and depress mandible (open mouth)
external oblique	lower 8 ribs	iliac crest and linea alba	flexion and rotation at waist
internal oblique	lumbodorsal fascia	lower 4 ribs	flexion and rotation at waist
transverse abdominis	iliac crest, cartilages of lowest ribs	linea alba and pubic crest	compresses abdominal wall
rectus abdominis	pubic crest and pubic symphysis	ribs 5-7 and xiphoid process	flexion at waist
tensor fascia latae	iliac crest and anterior superior iliac spine	iliotibial tract	flexes, abducts and medially rotates thigh
gluteus medius	ilium	greater trochanter of the femur	abduction and medial rotation of thigh
gluteus maximus	ilium, sacrum, coccyx	iliotibial tract, gluteal tuberosity of femur	extension and lateral rotation of thigh
sartorius	anterior superior iliac spine	tibia	flexes, abducts and laterally rotates thigh; flexes lower leg
gracilis	pubis	medial tibia	adducts thigh, flexes and medially rotates leg
adductor femoris	ischium and pubis	linea aspera of femur	adducts, flexes and laterally rotates thigh
biceps femoris	ischial tuberosity and femur	tibia and fibula	extends thigh and flexes lower leg
semitendinosus	ischial tuberosity	medial aspect of proximal tibia	extends thigh, flexes lower leg
semimembranosus	ischial tuberosity	medial condyle of tibia	extends thigh, flexes lower leg
vastus lateralis	linea aspera	patella and tibial tuberosity	extends lower leg, stabilizes knee

Managing Muscles 106

vastus medialis	linea aspera	patella and tibial tuberosity	extends lower leg
vastus intermedius	proximal femur	patella and tibial tuberosity	extends lower leg
rectus femoris	anterior inferior iliac spine	patella and tibial tuberosity	extends knee, flexes thigh
gastrocnemius	medial and lateral condyles of femur	via achilles tendon onto calcaneal tendon	flexes lower leg, plantarflexes foot
tibialis anterior	lateral condyle and tibial shaft	first cuneiform and first metatarsals	dorsiflexes and inverts foot
soleus	head of fibula and tibia	calcaneal tendon onto calcaneus	plantarflexes foot
fibularis longus	head of fibula	first metatarsal	plantar flexion
extensor digitorum longus	posterior tibia	distal phalanges of toes 2-5	Extend toes 2 - 5 and dorsiflexes ankle

Muscles, Muscles, and more Muscles!

A Big Picture Approach to Learning Muscles

So you need to memorize 50 muscles for the upcoming exam? Well, personally I am against the brute force memorization technique. In my opinion I think relating the muscles to something you already know is a better way to learn muscles. So what do you already know? Well, if you have been studying joint movements you are actually well on your way to learning muscles.

Hey, something's gotta move those joints like the shoulder, elbow, hip and knee. That something is a muscle (or groups of muscles). So let's learn some muscles by relating them to joint movements.

Arm, Forearm and Hand (figs 7.4-7.7)

Let's start with something a bit on the easy side. You probably know that you can bend your elbow (flexion) and straighten your elbow (extension). Well, there must be some muscles that pull on your bones to do this.

The forearm (antebrachium) can move into flexion, extension and rotation.

The flexors include the biceps brachii, brachioradialis and brachialis.

The biceps brachii attaches to the scapula and extends to the radial tuberosity. This muscle has two heads at its proximal region. It works to flex the elbow.

The brachialis lies deep to the biceps brachii and extends to the ulna.

The brachioradialis attaches to the humerus and extends to the radius.

There is only one muscle that functions in elbow extension. This muscle is the triceps brachii. This muscle is a large three headed muscle that attaches to the scapula and humerus and extends to the ulna. It is the only muscle on the back side of the arm.

The rotators of the forearm include the supinator, pronator teres and pronator quadratus.

The supinator attaches to the ulna and extends to the lateral aspect of the humerus. It works to move the wrist into supination (remember palms up—cup of soup—supination).

There are two pronators and their names include the word "pronation." The pronator teres attaches to the humerus and ulna and extends to the radius. It works to move the wrist into pronation.

The pronator quadratus attaches to the distal ulna and radius.

Now let's move to the wrist and hand. Which do you think can move in a more complex manner, the elbow or wrist? If you said wrist you were right. The elbow can only flex, extend and rotate but the wrist and hand can do some really complex things like write an essay or conduct an orchestra. So there are a few more muscles that are needed to do this.

First of all you can group the wrist and hand flexors together. The name of the muscle also tells you where it is located. For example the flexor carpi radialis longus can be translated into the "long flexor running to the wrist (carpi) on the side of the radius bone (radialis)". Likewise there is a sister flexor on the other (ulnar) side as well called the flexor carpi ulnaris. There is a flexor muscle running down the middle of the forearm called the palmaris longus that goes to, you guessed it, the palm of the hand.

Okay so now we have some flexor muscles that move the wrist and hand but what about the fingers? We have some deeper forearm muscles to take care of that as well. These are the flexor digitorum superficialis (superficial flexor to the digits) and the flexor digitorum profundus (profundus means deep).

The wrist area contains a large number of tendons from muscles that move the wrist and hand. There is a large flat tendon on the palmar side of the wrist known as the flexor retinaculum. This structure helps to form the infamous carpal tunnel. A few muscles travel through the carpal tunnel on their way to the fingers.

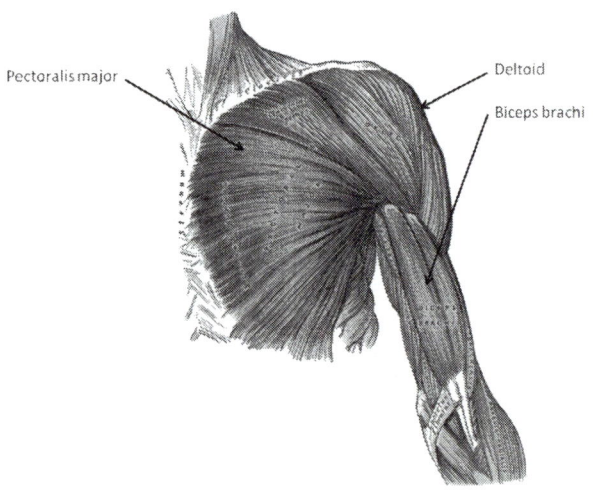

Figure 7.4. Muscles of anterior thorax and arm.

These muscles include:

- Flexor carpi radialis longus and brevis
- Flexor digitorum profundus
- Flexor digitorum superficialis

So much for the flexors, but what about the extensors? Well these are located on the back of the arm and forearm. The wrist extensors have a common origin on the lateral epicondyle. Wrist extensor tendonitis (lateral epicondylitis) known as tennis elbow can develop here.

The wrist and hand extensors include the extensor carpi radialis longus (long extensor muscle on the side of the radius bone), extensor carpi radialis brevis (short extensor muscle on the side of the radius bone), extensor carpi ulnaris, extensor digitorum and extensor digiti minimi.

Shoulder (figs 7.8, 7.9)

The shoulder is also a pretty complex joint so there are a good number of muscles needed to move it.

The arm and scapula work together to allow the arm to move.

There are a couple of really important muscles in the shoulder. One of these sits right on top of the shoulder. This is the deltoid muscle. If you see an athlete with broad shoulders chances are he has some well developed deltoids (or as they say in the gym "delts"). The deltoid is located on top of the humerus bone. It attaches to the spine of the scapula and acromion process and extends to the deltoid tuberosity of the humerus. The deltoid works to flex, abduct and extend the arm.

Another important muscle connects to the shoulder and spine. This one is the trapezius. It is a large diamond shape (trapezoidal) muscle that begins at the occipital bone and reaches out to the scapula and down to the end of the thoracic spine.

The trapezius has upper, middle and lower divisions. The trapezius attaches to the thoracic and cervical vertebra and extends upward to the occipital bone and laterally to the scapula. The upper portion raises the shoulder and scapula. The middle portion pulls the scapula toward the vertebral column and the lower portion pulls the scapula downward. The divisions of the trapezius are evidenced by the direction of the fibers.

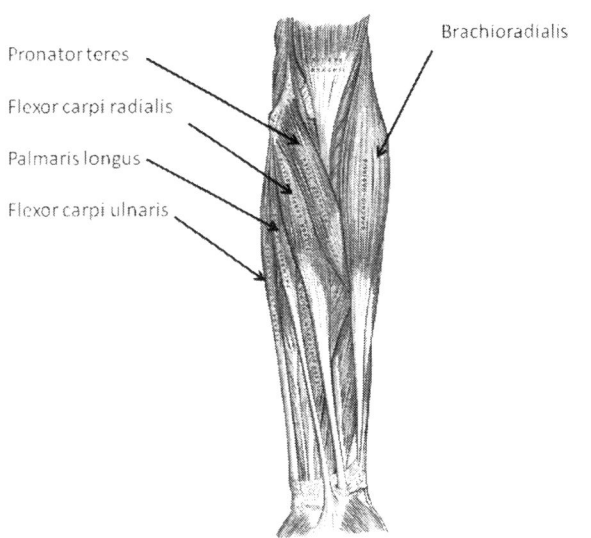

Figure 7.5. Anterior forearm muscles.

Fig. 7.6. Deep anterior forearm muscles.

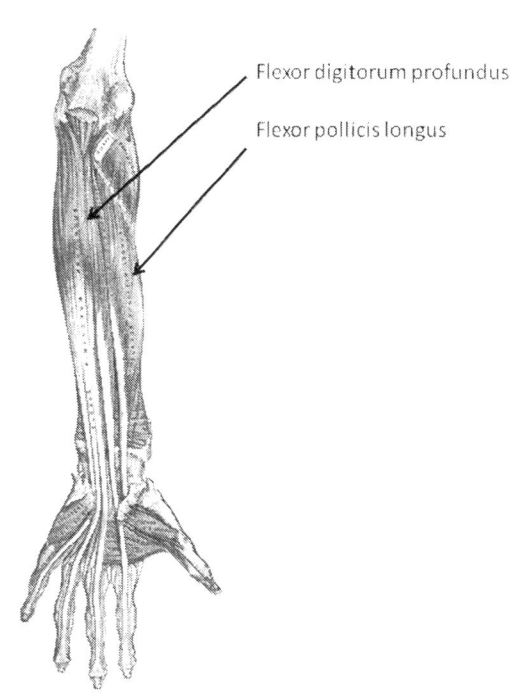

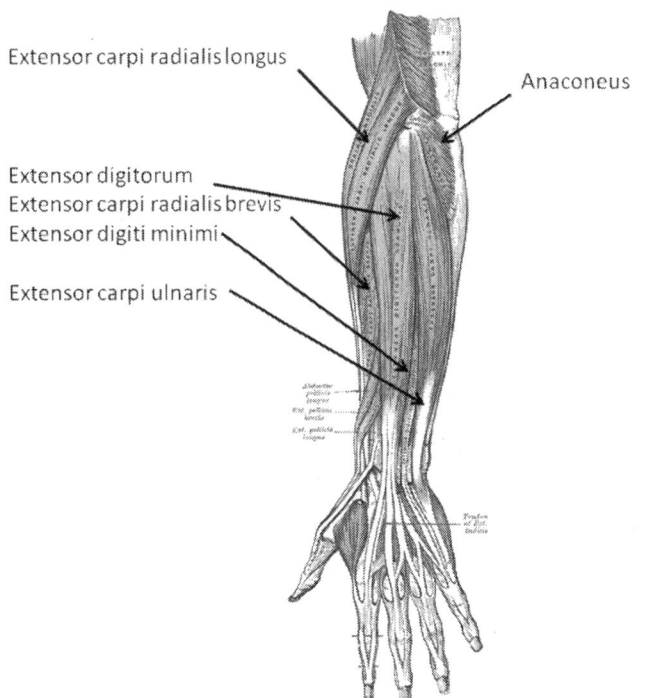

Figure 7.7. Posterior forearm muscles.

The arm can move into flexion, extension, adduction, abduction, internal and external rotation as well as combinations of these movements.

The flexors include the coracobrachialis, pectoralis major, and deltoid.

The coracobrachialis also tells you where it is located by virtue of its name. It attaches to the coracoid process of scapula (coraco) and extends to the shaft of the humerus (brachialis). It runs deep to the deltoid and biceps muscles. The pectoralis major (another great gym muscle—pecs) attaches to the clavicle, sternum and costal cartilages of ribs and extends to the intertubercular groove of the humerus.

The arm extensors include the teres major and latissumus dorsi. The teres major attaches to the lateral border of the scapula and extends to the intertubercular groove of the humerus. The latissumus dorsi attaches to the lower thoracic area to the iliac crest and extends to the intertubercular groove of humerus.

The arm abductors include the supraspinatus, and deltoid. The supraspinatus attaches to the posterior surface of scapula above spine of scapula and extends to the greater tubercle of the humerus. The deltoid was mentioned earlier.

The Rotator Cuff

The rotator cuff consists of four muscles, three of which are external rotators and one being an internal rotator. The first letter of each muscle can be taken to spell the acronym SITS which stands for supraspinatus, infraspinatus, teres minor and subscapularis.

The supraspinatus attaches to the superior portion of the scapula at the suprascapular fossa and extends to the greater tubercle of the humerus. It abducts as well as externally rotates the arm.

The infraspinatus attaches to the posterior portion of the scapula at the subscapular fossa and extends to the greater tubercle of the humerus. It externally rotates the arm.

The subscapularis attaches on the anterior surface of the scapula and extends to the lesser tubercle of the humerus. It is the only rotator cuff muscle that provides internal rotation.

The teres minor attaches to the lateral border of the scapula and extends to the greater tubercle of the humerus. It externally rotates the arm.

Here is a way to help you learn the rotator cuff muscles.

We can think of these as the **SITS** muscles.

Supraspinatus
Infraspinatus
Teres minor
Subscapularis

Since the rotator cuff muscles are often injured by pitching we can think of a baseball player who SITS out for the rest of the game and then gets sent to the minor league.

Another way to learn the rotator cuff is to stand next to someone facing the same direction as you and place your hand on their left shoulder so it ends up on the backside with the index finger on top. Your index finger is now over the supraspinatus. Your second finger is over the infraspinatus and your third finger is over the teres minor and subscapularis.

There are two rhomboid muscles that pull the scapula upward and medially. The larger rhomboid major is the inferior muscle of the two. The smaller rhomboid minor is superior to the major.

The levator scapula is a long thin muscle that attaches to the superior border of the scapula and extends upward to the occipital bone. As its name implies, the levator scapula works to elevate the scapula.

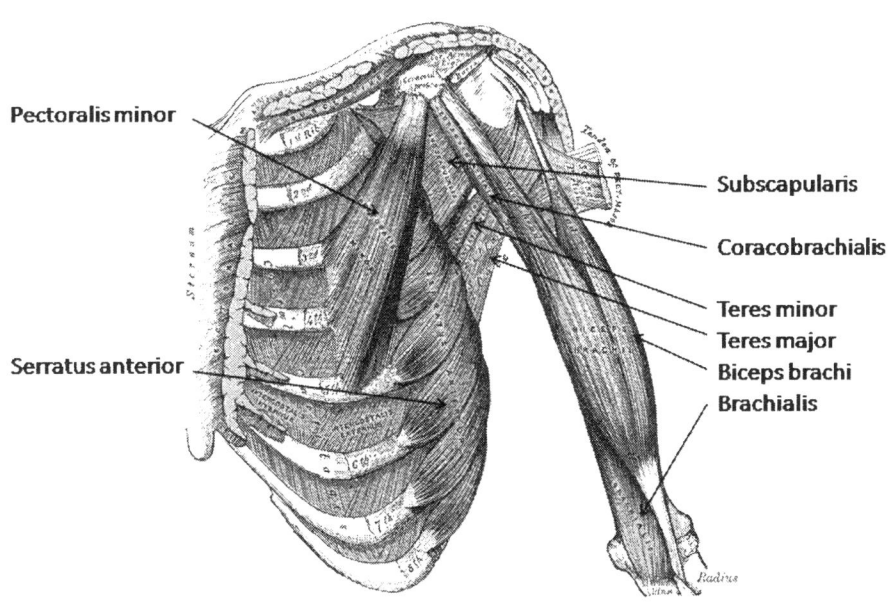

Figure 7.8. Deep muscles of the thorax and shoulder.

The serratus anterior attaches to the anterior surface of the scapula and extends to the ribs. The serratus anterior works to hold or stabilize the scapula against the ribcage.

The pectoralis minor muscle is located deep to the major. It attaches to the upper ribs and extends to the coracoid process of the scapula. It works to pull the scapula anterior and inferior. The pectoralis minor is also an accessory muscle of inspiration.

Muscles of the Head and Neck

The muscles of the head and neck move the face, larynx and tongue. A sample of muscles of facial expression follows.

Muscles of facial expression (figs. 7.10, 7.11)

Located on top of the head is a broad flat tendon called the epicranial aponeurosis.

There is one muscle with two parts attached to the anterior and posterior sections of this tendon. The muscle is called the occipitofrontalis. The anterior portion lifts the eyebrows. The posterior is a weak head extensor and can cause headaches.

There are two circular muscles called sphincters. The orbicularis oculi encircles the eye. It compresses the lacrimal gland and closes the eye. The orbicularis oris encircles the mouth. It causes the lips to pucker.

The buccinator is located in the cheek. It compresses the cheek against the teeth. The zygomaticus muscle has major and minor divisions and attaches to the orbicularis oris and zygomatic bone. It raises the lateral ends of the mouth when smiling. The platysma is a very thin and superficial muscle located under the chin. It causes the action of frowning when contracted.

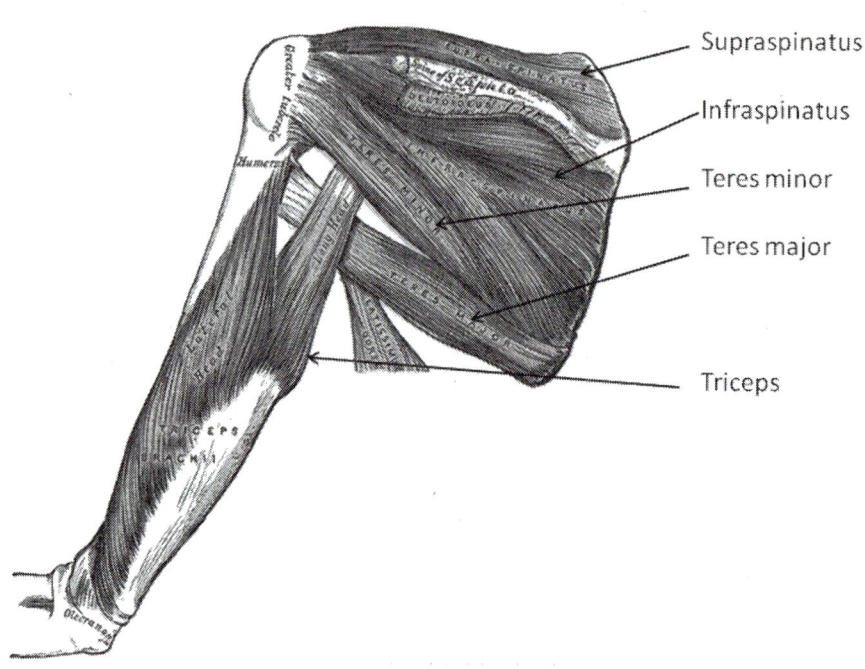

Figure 7.9. Posterior muscles of the arm.

Muscles of Mastication

The muscles of mastication (chewing) include the masseter, temporalis, medial and lateral pterygoids. The masseter muscles attach to the mandible and allow for closing the jaw. The temporalis is located in the lateral skull and attaches to the temporal bone. The temporalis aids in closing the jaw. In fact you can feel your temporalis muscles contract when touching the sides of your head when clenching your jaw. The medial and lateral pterygoids are deep muscles in the jaw. These can elevate, depress, protract and cause lateral movement of the mandible. These muscles are often involved in temporomandiblular joint (TMJ) disorder.

Head and Spine (figs 7.12, 7.13)

There are a number of muscles that attach to the spine and move the head. The sternocleidomastoid (SCM)(sterno—sternum, cleido—clavicle, mastoid—mastoid process on temporal bone) attaches to the mastoid process of the temporal bone as well as the clavical and sternum. It produces contralateral rotation when one muscle contracts and neck flexion when both muscles contract.

The splenius capitus is located in the posterior portion of the neck. It helps bring head into an upright position (head extension). It also causes ipsilateral rotation and lateral flexion when one muscle contracts.

The semispinalus capitus also produces head extension as well as lateral flexion and rotation. It connects to the occipital bone and vertebra of the cervical and thoracic spines.

The erector spinae group of muscles consists of several muscles running up and down the spine. These consist of the spinalis, longissumus, iliocostalis and semispinalis muscles. They are located in the cervical, thoracic and lumbar spines. The names of these muscles give you a good clue as to their locations. For example the spinalis muscles are located medially attaching directly to the spinal segments. The iliocostalis muscles attach to the ribs (iliocostalis thoracis) (costal = ribs). The longissumus muscles have long fibers and the semispinalis muscles run just lateral to the spinal segments.

Here is a trick to help you remember the erector spinae muscles:

To memorize these think "I Love Sex:"

Iliocostalis
Longissimus
Spinalis

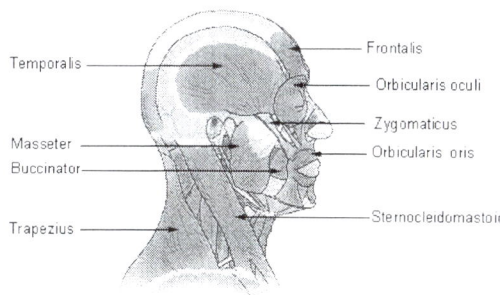

Fig. 7.10 Head and neck muscles.

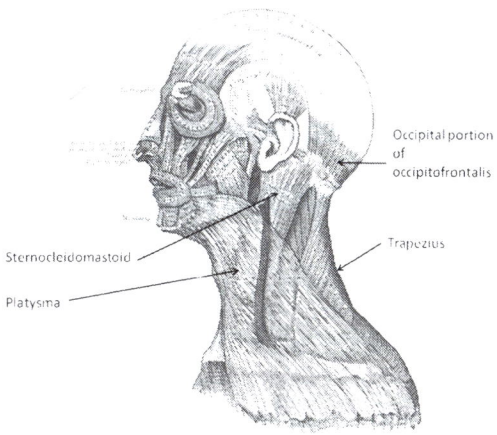

Figure 7.11. Lateral view of neck muscles.

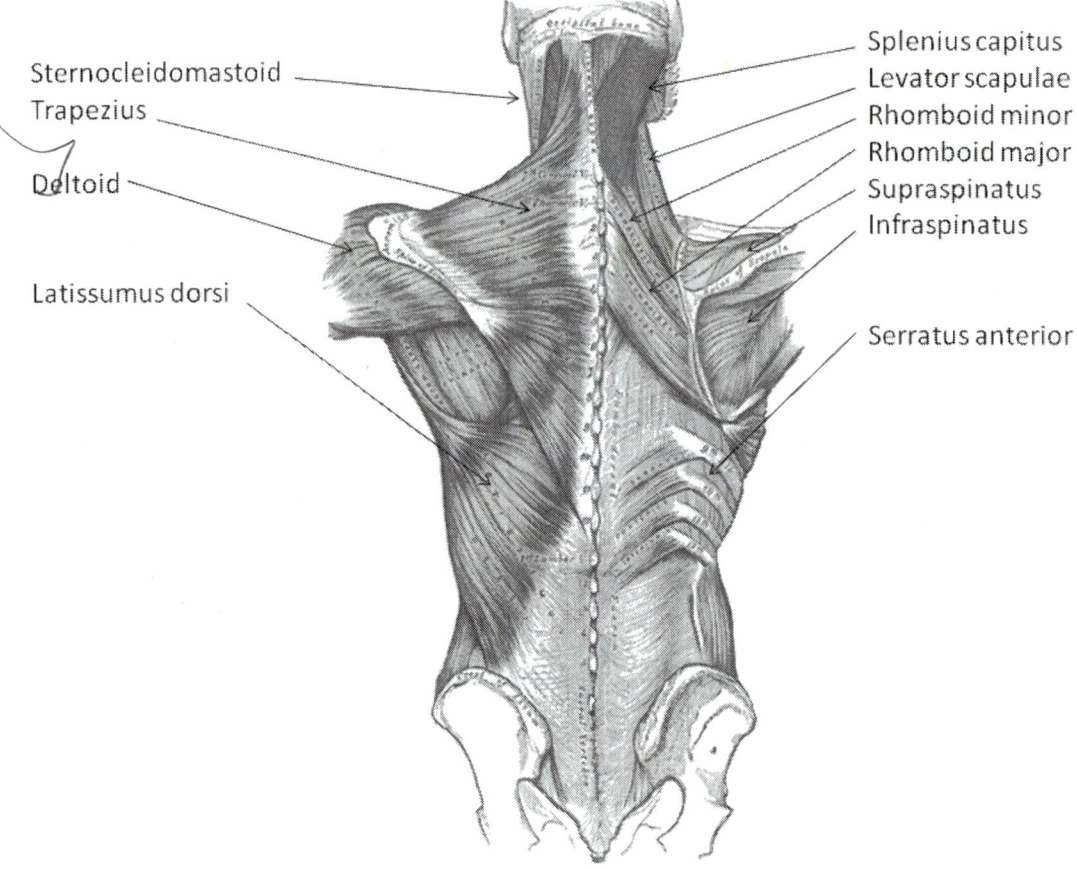

Figure 7.12. Posterior muscles of the thorax.

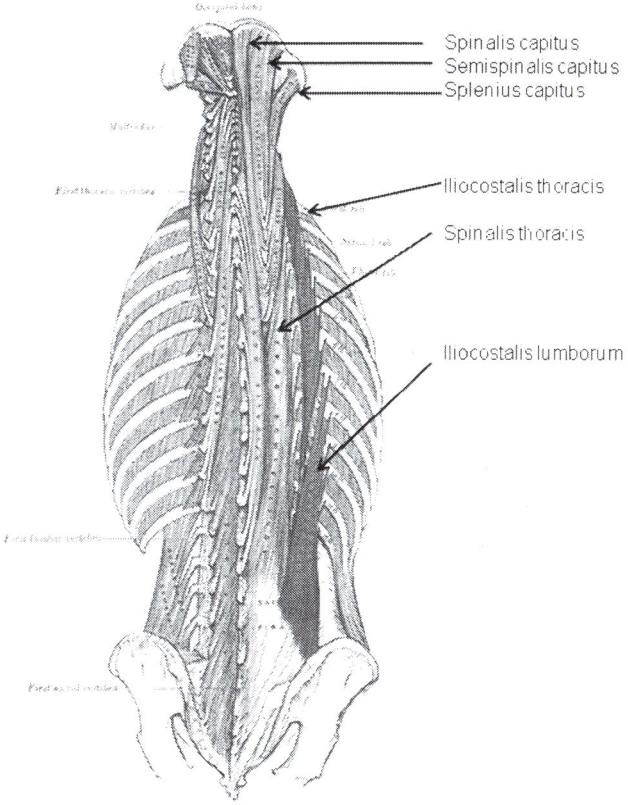

Figure 7.13. Erector spinae muscles

Muscles of the Tongue (fig. 7.14)

"Glossus" means tongue so the word glossus is found in a number of tongue muscles. The muscles of the tongue include the genioglossus that pulls the tongue to one side when one side contracts and protrudes tongue when both sides contract. The hyoglossus depresses the tongue while the styloglossus pulls the tongue superior and posterior.

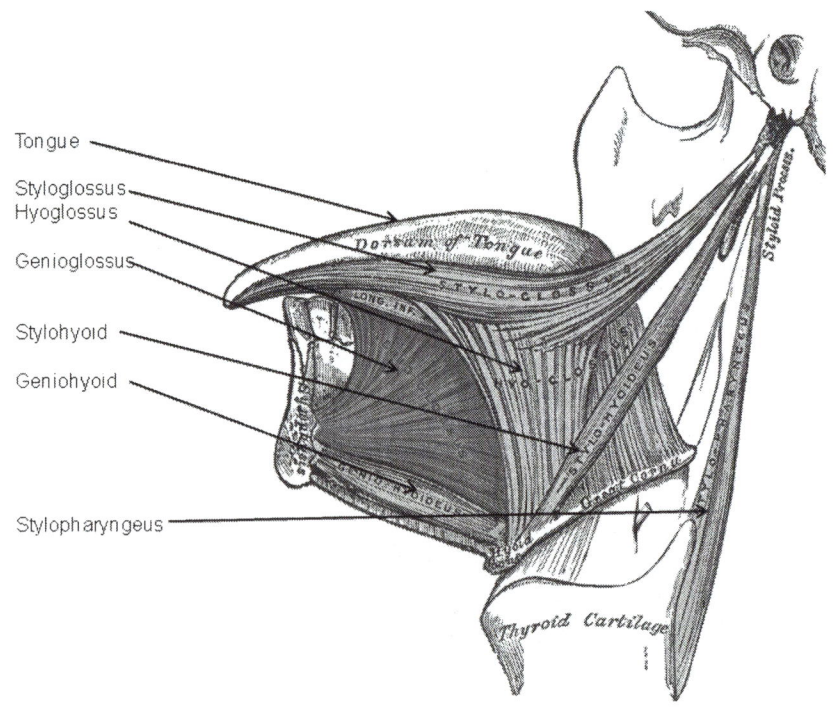

Figure 7.14. Muscles of the tongue.

Abdominal Wall Muscles (figs 7.15, 7.16)

The contents of the abdomen are protected by a band of muscles. There are three layers of abdominal muscles that include four muscles. The first layer consist of the rectus abdominus (six pack) which lies in front and center in the abdomen and the external obliques which are located on the sides of the abdomen. The second layer consists of the internal obliques which lie deep to the external obliques and the third layer consists of the transverse abdominus.

Some of the abdominal muscles attach to a broad dense band of connective tissue known as the linea alba. The linea alba extends from the xiphoid process to the symphysis pubis.

The abdominal muscles aid in trunk flexion. They also compress the contents of the abdominal cavity, increase intra-abdominal pressure and help to transmit force through trunk to protect the spine and contents of the abdominal cavity.

The transverse abdominus muscle is becoming a very important muscle in rehabilitation of low back injuries. This muscle acts as a natural back brace since its fibers run in a transverse plane.

Here is a trick to help you remember the abdominal muscles:

We can think of someone with a spare tire around their abdomen.

TIRE

Transversus abdominis
Internal obliques
Rectus Abdominis
External obliques

As far as the way the fibers run, think of your hands in your pockets. Your fingers point in the direction of the external obliques. The internal obliques go the opposite way.

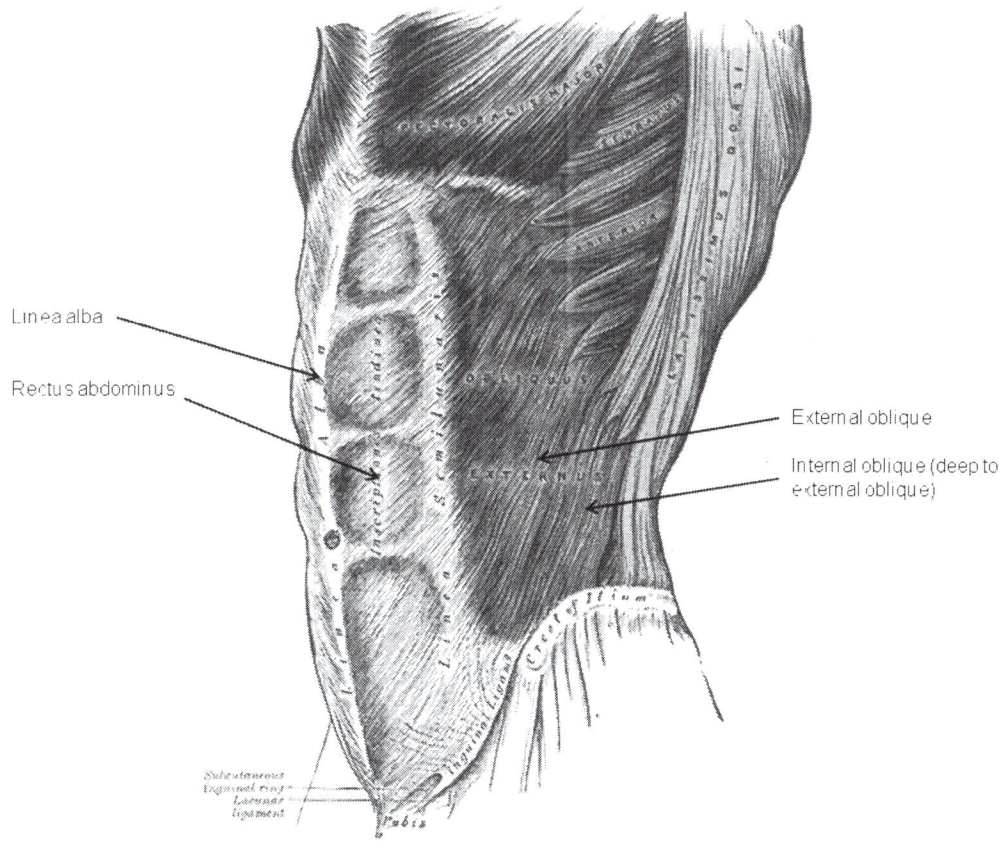

Figure 7.15. Superficial Abdominal Muscles

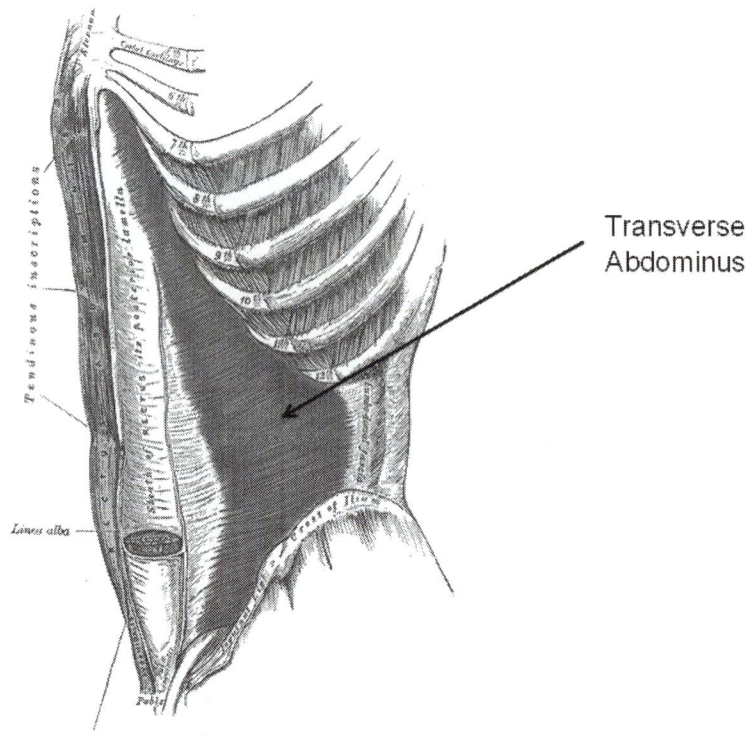

Figure 7.16. The transverse abdominus is the deepest abdominal muscle.

Muscles of the Pelvic Outlet (figs 7.17, 7.18)

The floor of the pelvic cavity is formed by the pelvic diaphragm which consists of a layer of muscles. The urogenital diaphragm lies superficial to this layer and forms a second layer. The pelvic diaphragm consists of the levator ani and coccygeus muscles. The uogenital diaphragm consists of the superficial transverses perinea, bulbospongiosus (males only), ischiocavernosus and the sphincter urethrae.

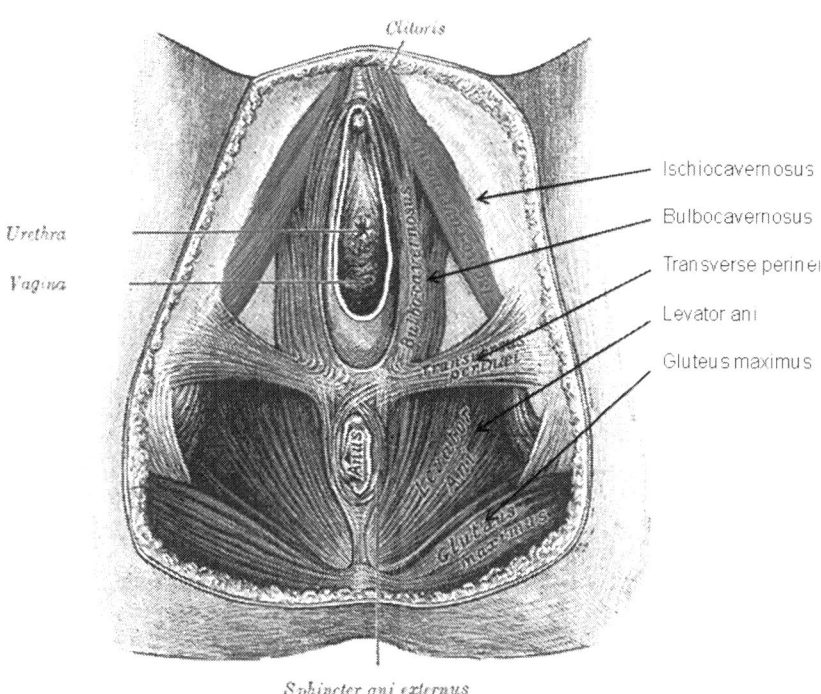

Figure 7.17. Muscles of the pelvic outlet (female).

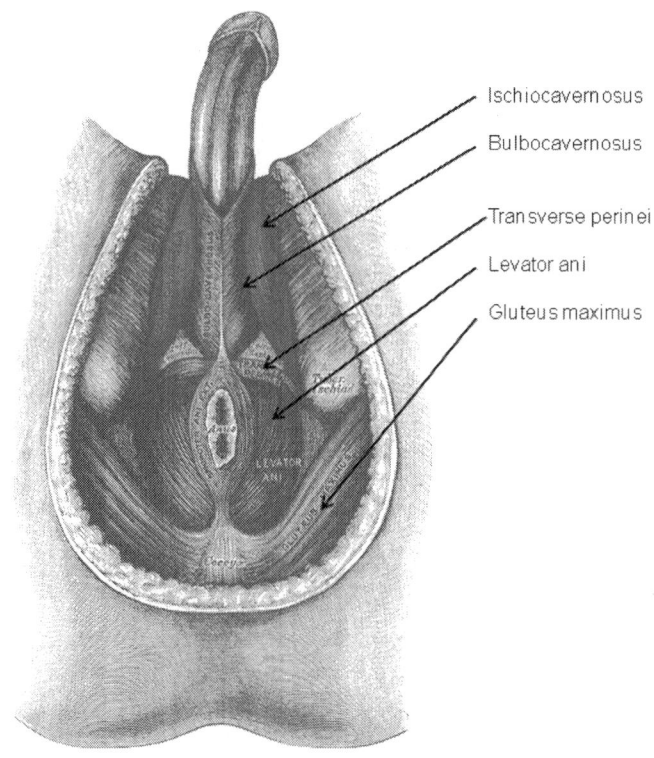

Figure 7.18. Pelvic outlet muscles (male).

Pelvic and Upper Thigh Muscles (fig. 7.19)

Muscles that move the thigh connect to the pelvis and femur. The anterior muscles include the psoas and iliacus and the posterior muscles include the gluteus maximus, gluteus medius, gluteus minimus, and tensor fascia latae.

The psoas muscle actually has two divisions. The psoas major attaches to the lower lumbar vertebra and extends to the lesser trochanter of the femur. The psoas minor muscle is smaller and inserts on the pubic bone. The Iliacus muscle attaches to the ilium and also extends to the lesser trochanter of the femur. Since both the psoas major and iliacus share a common insertion point they are often referred to as the iliopsoas. The iliopsoas works to flex the hip.

The gluteus maximus is one of the strongest and largest muscles of the body (fig. 7.19). It attaches to the iliac crest, sacrum, coccyx and the aponeurosis of the sacrospinalis. It extends to the linea aspera of the femur and the iliotibial band. It works to produce hip extension.

The gluteus medius lies deep to the gluteus maximus. It attaches to the ilium and extends to the greater trochanter of the femur. It works to produce hip abduction and extension.

The gluteus minimus lies deep to the gluteus medius. It is the smallest gluteal muscle.

The tensor fascia latae is located on the lateral aspect of the thigh. It attaches to the iliac crest and extends to a band of dense connective tissue called the iliotibial tract or band. The iliotibial band extends down the lateral aspect of the femur to the tibia. It is a flat tendon or aponeurosis. Tendonitis can develop in this tendon in a condition known as iliotibial band syndrome.

Deep muscles in the posterior pelvic area include the piriformis, obturator internus, obturator externus, superior and inferior gemellus and quadratus femoris muscles. All of these muscles work to externally rotate and abduct the hip.

To learn the external rotators of the hip think "Play Golf Or Go on Quaaludes:"

Piriformis
Gemellus superior
Obturator internus
Gemellus inferior
Quadratus femoris

Muscles on the proximal medial aspect of the thigh include the adductor longus, adductor brevis, adductor magnus, pectineus and gracilis. These muscles attach to the pubic bone and extend down the thigh to various insertion points on the femur. They work to adduct the hip.

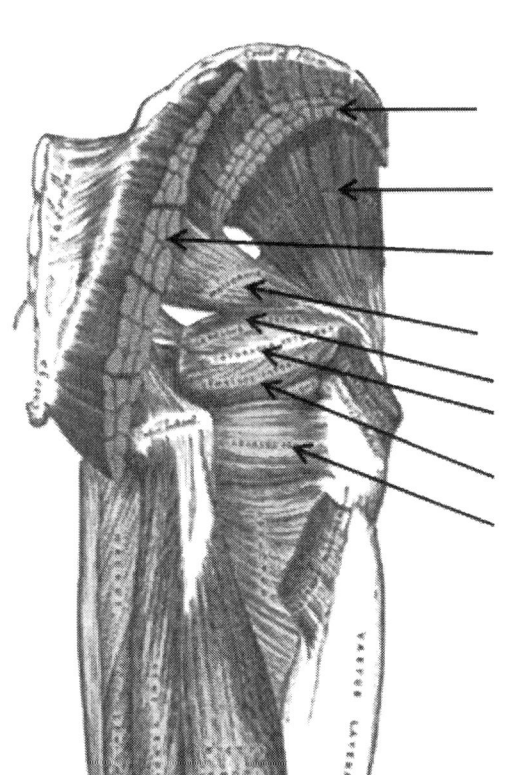

Figure 7.19. Deep muscles of the posterior pelvis.

Anterior Thigh Muscles (fig. 7.20)

The large muscles on the anterior portion of the thigh include the sartorius and quadriceps group.

The sartorius (tailor's muscle) attaches to the anterior superior iliac spine and extends from lateral to medial across the thigh to insert on the medial aspect of the upper tibia. This muscle has multiple actions including flexion, abduction and external rotation of the hip.

The quadriceps group consists of the rectus femoris, vastus medialis, vastus lateralis, and vastus intermedius. The quadriceps work together to produce knee extension.

The rectus femorus is located in the middle of the thigh. It attaches to the anterior superior iliac spine and extends inferiorly to the patella.

The vastus medialis is located in the medial aspect of the thigh. It attaches to the linea aspera of the femur and extends to the patella.

The vastus lateralis is located in the lateral aspect of the thigh. It attaches to the greater trochanter of the femur and extends to the patella.

The vastus intermedius lies deep to the rectus femorus. It attaches to the femur and extends to the patella.

All of the quadriceps have a common insertion point on the patellar ligament. The patellar ligament inserts on the tibial tuberosity.

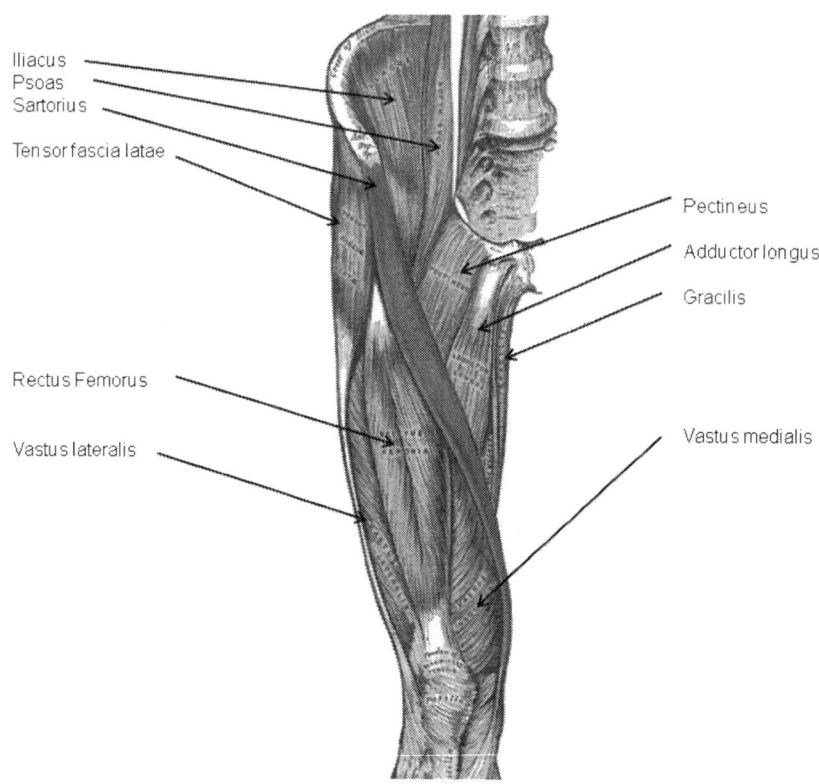

Figure 7.20. Muscles of the anterior thigh.

Posterior Thigh Muscles (fig. 7.21)

The posterior thigh muscles include the hamstring group. The hamstrings consist of three muscles which include the biceps femorus, semimembranosus, and semitendinosus. The hamstrings work to produce knee flexion.

The biceps femorus is a two-headed muscle. The long head attaches to the ischial tuberosity and the short head attaches to the linea aspera and lateral supracondylar line of the femur. The muscle then extends inferiorly to attach to the head of the fibula.

The semimembransous attaches to the ischial tuberosity and extends inferiorly to attach to the medial condyle of the tibia and lateral condyle of the femur.

The semitendinosus attaches to the ischial tuberosity and extends inferiorly to attach to the medial aspect of the upper tibia.

Posterior Knee

Located in the posterior portion of the knee is the popliteus muscle. If the femur is fixed the popliteus works to internally rotate the tibia. If the tibia is fixed it works to externally rotate the femur.

Muscles of the Anterior Leg (7.22)

The muscles of the anterior portion of the leg work to dorsiflex the foot. These include the tibialis anterior, extensor hallicus longus, extensor digitorum longus, and peroneus tertius.

The tibialis anterior is located just lateral to the tibia. It attaches to the lateral condyle of the tibia, the lateral aspect of the proximal portion of the tibia and the interosseous membrane that connects the tibia and fibula. It extends downward to attach to the medial cuneiform and first metatarsal. The tibialis anterior is involved in shin splints.

The extensor hallicus longus lies deep to the tibialis anterior. It attaches to the anterior aspect of the fibula and interosseous membrane and extends downward to attach to the first distal phalanx. Besides being a synergist for dorsiflexion of the foot it also extends the big toe.

The extensor digitorum longus also lies deep to the tibialis anterior. It attaches to the lateral condyle of the tibia, shaft of the fibula and interosseous membrane. It works as a synergist in dorsiflexion of the foot and extends the toes. It also works to tighten the plantar aponeurosis.

The peroneus tertius is part of the peroneal group that includes the peroneus longus and peroneus brevis. This muscle works to dorsiflex and evert the foot. It attaches to the medial surface of the lower portion of the fibula and extends to the fifth metatarsal. The peroneal group works together to evert the foot.

Managing Muscles 124

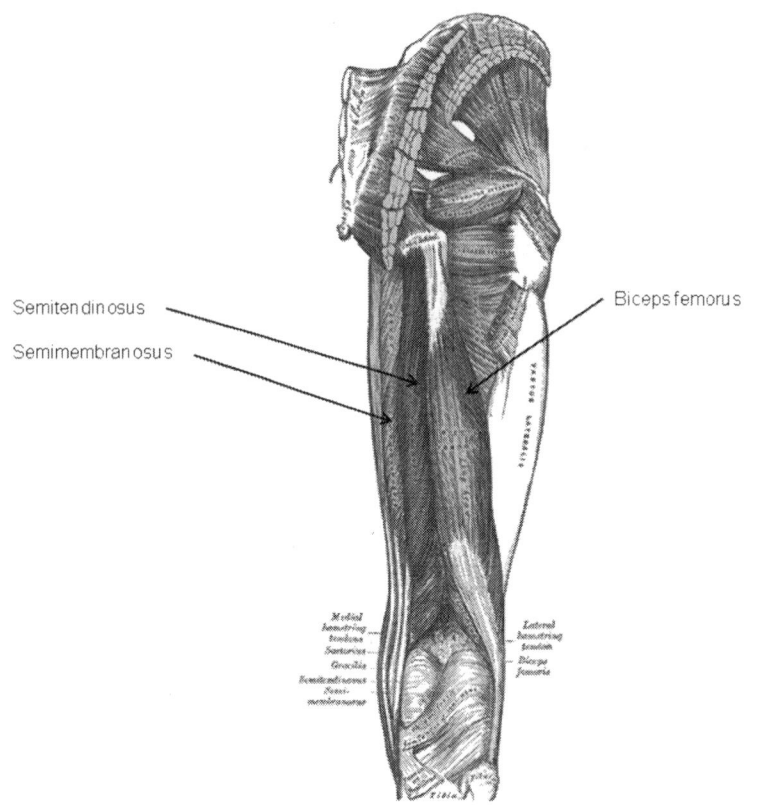

Figure 7.21. Muscles of the posterior thigh.

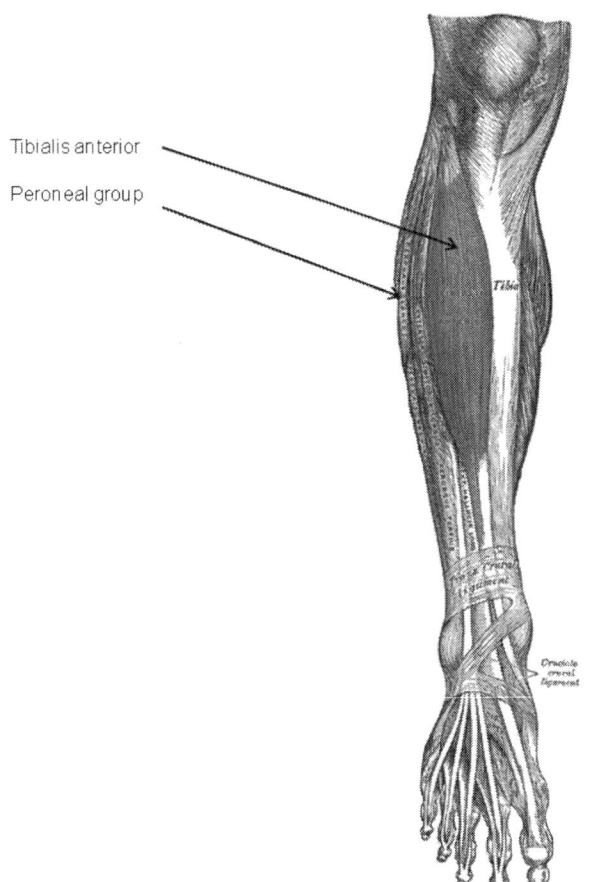

Figure 7.22. Anterior lower leg muscles.

Muscles of the Posterior and Lateral Leg (fig. 7.23)

The muscles of the posterior leg work to plantarflex the foot (point your toe). These include the gastrocnemius, soleus, flexor digitorum longus and tibialis posterior.

The gastrocnemius is a two-headed muscle that crosses both the knee and ankle joints. Its action in the knee is to help with knee flexion. It also works to produce ankle plantarflexion. It attaches to the femoral condyles and posterior surface of the distal femur and extends downward to attach to the calcaneus.

The soleus lies deep to the gastrocnemius. It attaches to the posterior aspect of the proximal fibula and tibia and extends downward to attach to the calcaneus. The soleus only crosses the ankle joint and produces ankle plantarflexion.

The gastrocnemius and soleus both insert on the large Achilles (calcaneal) tendon and are known collectively as the triceps surae.

The tibialis posterior is also a deep muscle of the posterior leg. It attaches to the posterior proximal surface of the tibia and fibula and extend downward to attach to the navicular, medial cuneiform and second to fourth metatarsals. It works to produce plantarflexion and also helps to control pronation of the foot while walking.

The flexor digitorum longus is a deep muscle of the posterior leg. It attaches to the posterior surface of the tibia and extends downward to attach to the second through fifth distal phalanges. It works to flex the toes and stabilizes the metatarsal heads.

The peroneus longus is located on the lateral aspect of the lower leg. It attaches to the tibia and fibula and extends to the medial cuneiform and first metatarsal.

The flexor hallicus longus is a deep muscle on the lateral aspect of the leg. It attaches to the distal portion of the fibula and interosseous membrane and extends to attach to the big toe. It works to flex the big toe.

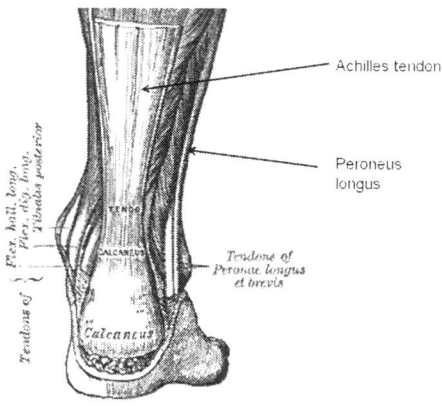

Figure 7.23. The gastrocnemius and soleus attach to the Achilles tendon. The peroneus longus is on the lateral aspect of the leg.

Muscles of the Foot (figs. 7.24, 7.25)

The top of the foot is known as the dorsum of the foot. The only muscle located exclusively on the dorsum of the foot is the extensor digitorum brevis. This muscle attaches to the calcaneus and extensor retinaculum of the ankle and extends to the big toe and tendons of the extensor digitorum longus. It works to produce extension of the toes.

The bottom or sole of the foot is known as the plantar region of the foot. This area contains four layers of muscles.

The layers from superficial to deep include:

Layer 1:

- Flexor digitorum brevis
- Abductor hallucis
- Abductor digiti minimi

Layer 2:

- Quadratus plantus
- Lumbricales

Layer 3:

- Adductor hallucis
- Flexor digiti minimi brevis
- Flexor hallucis brevis

Layer 4:

- Dorsal interossei
- Plantar interossei

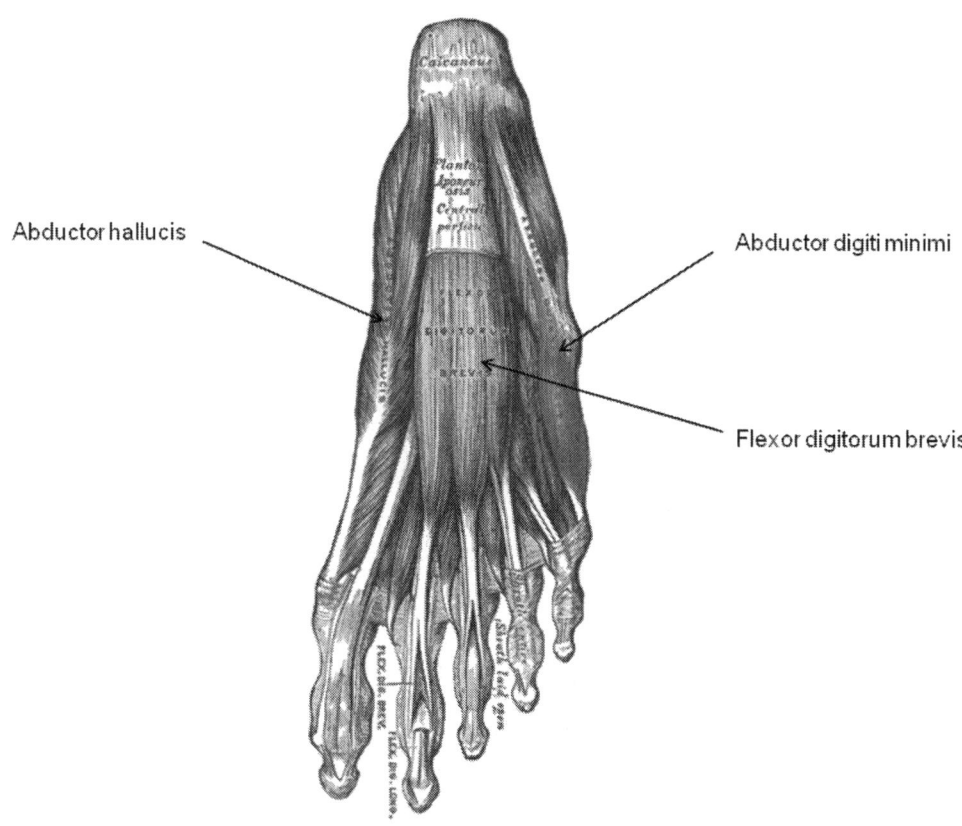

Figure 7.24. Dorsum of foot.

Managing Muscles 127

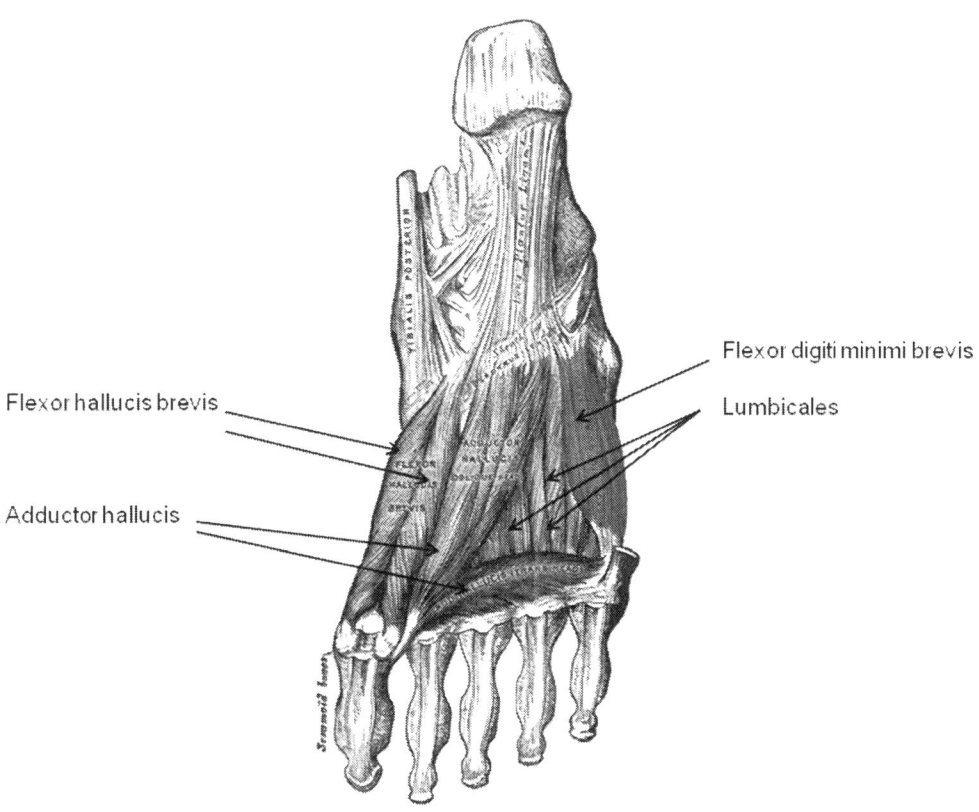

Figure 7.25. Deep muscles of dorsum of foot.

How Muscle Tissue is put Together

Big Picture: Muscle Tissue Structure

Muscle tissue consists of bundles within bundles within more bundles. The smallest bundles contain protein rods (filaments) that slide past each other when muscles contract.

Too bad the names for the bundles are not something simple like "big bundle," "smaller bundle," and so on. Of course this is anatomy so the names are a bit more complex. So here's how it goes.

The muscle is surrounded by a covering called the epimysium. If you look inside you will see bundles called fascicles. Each fascicle has a covering called the perimysium. Inside the fascicles are smaller bundles of muscle fibers covered by the endomysium. Protein filaments are located inside the fibers. See, that wasn't too bad! (fig. 7.26)

Now, there are two kinds of protein filaments. These are called actin and myosin. The actin and myosin overlap and can slide past each other when muscles contract. They connect by forming cross-bridges and are arranged in contractile units called sarcomeres. A sarcomere extends from one Z-disc protein to another. (fig. 7.27)

There is another important structure in muscle tissue we need to mention. The sarcoplasmic reticulum is network of tubules that wraps around the muscle fibers. At each end of the sarcoplasmic reticulum is an enlarged tube called the terminal cisterna. The outer membrane of the muscle fiber is called the sarcolemma. The sarcolemma also contains tubular structures called T-tubules. The T-tubules extend into the muscle fiber and wrap around the sarcomeres. They also extend to the extracellular fluid surrounding the muscle fiber. The T-tubules and terminal cistern form what is called a triad. The sarcoplasmic reticulum stores calcium needed for muscular contraction.

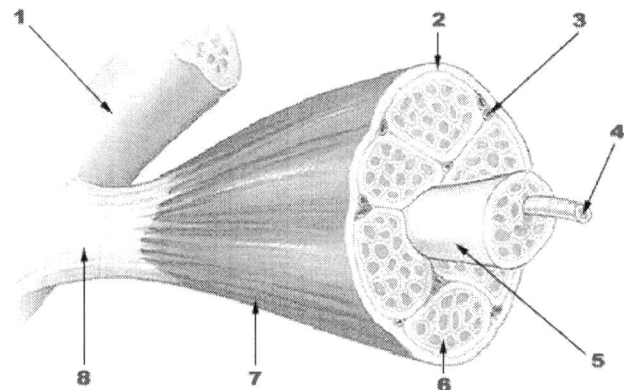

Fig. 7.26. Muscles consist of smaller and smaller bundles.

1. Bone
2. Perimysium
3. Blood vessel
4. Muscle fiber
5. Fascicle
6. Endomysium
7. Epimysium
8. Tendon

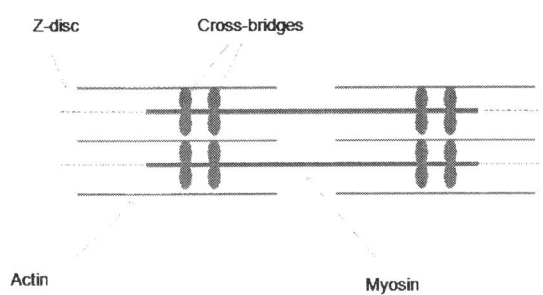

Fig. 7.27. Actin and myosin overlap and form cross bridges.

Muscle Contraction Physiology (figs. 7.28-7.34)

One of the more difficult concepts in the muscular system is the physiology of muscular contraction. Muscles contain a variety of microscopic structures that all work in harmony to produce muscular contraction.

A textbook or lecture might read something like this:

The action potential propagates down the axon and stimulates voltage-gated calcium gates in the axon terminal to open. The resultant inrush of calcium facilitates release of the neurotransmitter acetylcholine from the axon terminal. The acetylcholine moves across the synaptic cleft to the motor end plate of the muscle causing depolarization of the motor end plate. The depolarization facilitates the release of calcium from the sarcoplasmic reticulum. Once calcium is released it attaches to the troponin of the troponin-tropomyosin complex causing a conformational change resulting in the exposure of the myosin binding site on the actin myofilament. Once the binding site is exposed the myosin head binds to actin and releases its potential energy causing the myosin protein to pull the actin along. Adenosine triphosphate is converted to adenosine diphosphate releasing energy to power the myosin head...

The Big Picture: Muscle Contraction

Muscles are made of two different strands of protein.

Muscles contract when the strands of protein slide past each other.

In order for muscles to contract we need a source of energy.

Muscles respond to commands from the nervous system.

Once you understand this concept we can fill in the nitty gritty details:

Muscles are made of two different strands of protein.

Think of a mountain climber climbing a utility pole with a small pick. The climber is one type of protein (myosin) the pole is the other (actin). As the mountain climber climbs he reaches up, digs the pick into the pole and pulls himself higher.

Muscles contract when the strands of protein slide past each other.

Let's say that our pole climbing mountain climber must only dig his pick into certain parts of the pole. The pole is covered by a plastic sheet wrapped around it. A worker at the bottom pulls on the sheet to expose certain areas for the climber to dig into. The sheet is called the troponin-tropomyosin complex. The worker on the ground (calcium) pulls on the sheet to expose the myosin binding sites.

The worker on the ground must come from somewhere. He does. He comes from a construction trailer nearby (sarcoplasmic reticulum).

In order for muscles to contract we need a source of energy.

The climber needs energy in order to make his climb so he pops a small snack (ATP) in his mouth every time he drives the pick into the pole.

Energy is stored in a molecule called ATP (adenosine triphosphate). Think of ATP like a little energy ball in the body. If you need energy ATP releases it. You can also store energy in ATP.

Muscles respond to commands from the nervous system.

There must be some message to the worker in the trailer to go out and pull on the sheet. He receives this message from a phone inside the trailer. The message travels along wires much like the nerves that send messages to the muscles.

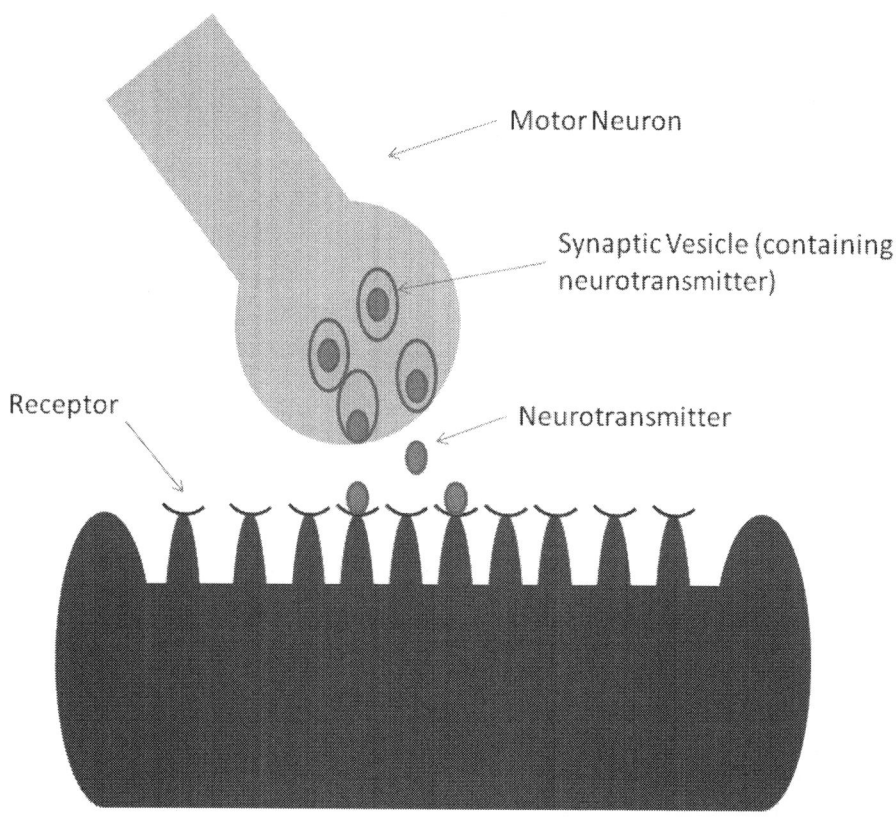

Figure 7.28. Motor Neuron and Muscle. The motor neuron sends information to the receptors on the motor end plate (on muscle) via neurotransmitters.

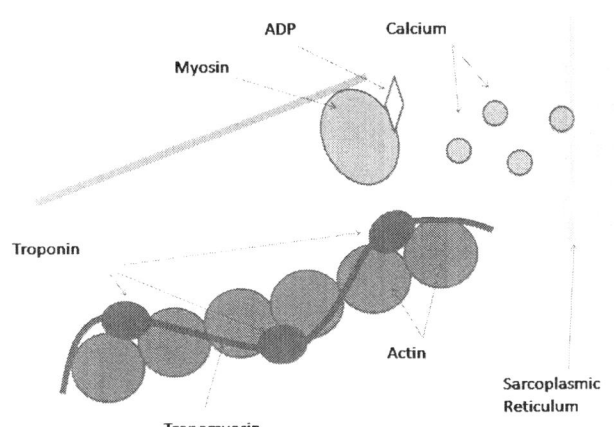

Figure 7.29. Muscle Contraction Physiology. The sarcoplasmic reticulum (trailer) responds to muscle fiber depolarization by releasing calcium (worker pulling sheet on pole).

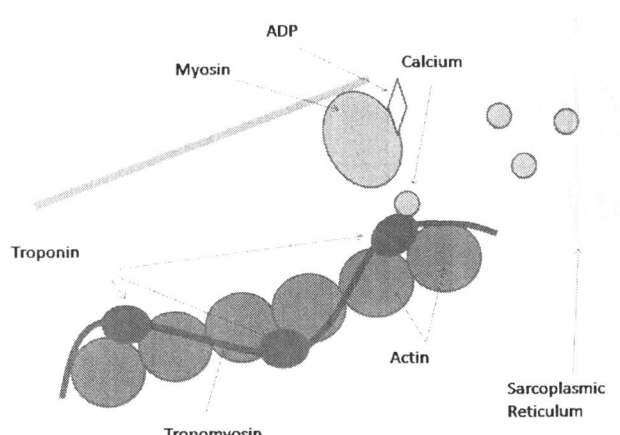

Figure 7.30. Muscle Contraction Physiology. Calcium (worker) attaches to troponin on the tropomyosin (plastic sheet on pole) surrounding the actin. ADP is attached to myosin.

Managing Muscles 132

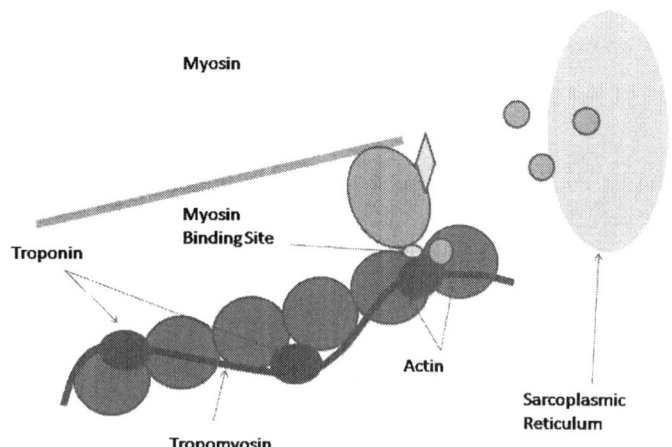

Figure 7.31. Muscle Contraction Physiology. Troponin-tropomyosin (sheet around pole) responds to the attachment of calcium (worker) by changing its shape and exposing myosin binding sites on actin (worker pulls on sheet).

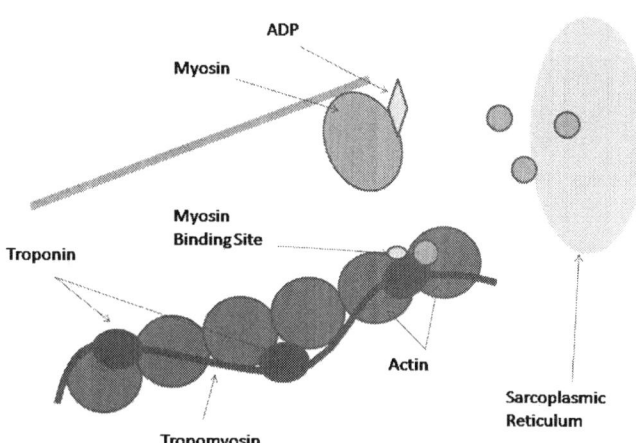

Figure 7.32. Muscle Contraction Physiology. Myosin binds to actin forming a cross-bridge (climber drives pick into pole).

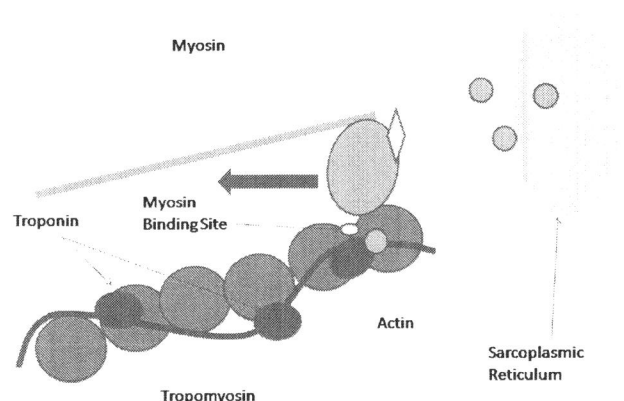

Figure 7.33. Muscle Contraction Physiology. Myosin (climber) can now bend and pull actin (climb pole) along causing muscle contraction and shortening the sarcomere.

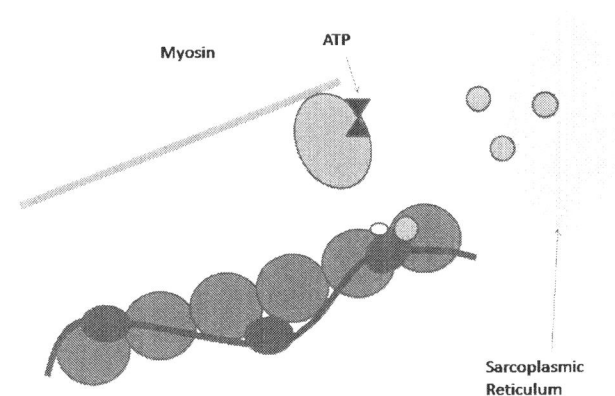

Figure 7.34. Muscle Contraction Physiology. Myosin releases from actin when a second ATP (energy snack) attaches to myosin. Myosin is now available for another cross-bridge formation.

Muscle Twitch

Big Picture: Muscle Twitch

When a muscle fiber receives a stimulus from the nervous system it contracts (twitches). The response occurs in 3 phases.

We can improve our understanding of muscle contraction by examining the contraction of one muscle fiber. A twitch occurs when one muscle fiber contracts in response to a command (stimulus) by the nervous system. The time between the activation of a motor neuron until the muscle contraction occurs is called the lag phase (sometimes called the latent phase). During the lag phase a signal called an action potential moves to the end of the motor neuron (axon terminal). This results in release of acetylcholine and depolarization of the motor end plate. The depolarization results in the release of calcium by the sarcoplasmic reticulum and subsequent binding of calcium to troponin which causes the myosin binding site to be exposed (fig. 7.35).

This is followed by the actual muscle contraction that develops tension in the muscle. This next phase is called the contraction phase. During the contraction phase, cross-bridges between actin and myosin form. Myosin moves actin, releases and reforms cross-bridges many times as the sarcomere shortens and the muscle contracts. ATP is used during this phase and energy is released as heat.

When the muscle relaxes the tension decreases. This phase is called the relaxation phase. During this phase calcium is actively transported back into the sarcoplasmic reticulum using ATP. The troponin moves back into position blocking the myosin binding site on the actin and the muscle passively lengthens.

Muscle Stimulus and Contraction Strength

Big Picture Muscle Stimulus and Contraction Strength

If the stimulus is strong enough (reaches threshold) the muscle fiber will contract with a given force (no more—no less).

A skeletal muscle fiber will produce a given amount of force if the stimulus is strong enough to reach the threshold for contraction. This is called the all or none law. Let's say that we are electrically stimulating a muscle fiber. We begin with a low amount of stimulation that does not reach the threshold to produce a contraction. The muscle fiber will respond by remaining relaxed, it will not contract. Now if we increase the stimulation so that enough is produced to reach the threshold the muscle fiber will respond by contracting. Finally if we continue to increase the stimulus so that it well exceeds the threshold the fiber will respond by contracting with the same force as when we just reached the stimulus. The muscle will not contract with greater force if the stimulus is greater. The muscle responds to stronger stimuli by producing the same force.

Motor Units

Big Picture: Motor Units

A motor unit is a motor neuron connected to a set of muscle fibers. As more force is needed, more motor units contract.

In skeletal muscles a motor neuron can innervate many muscle fibers. This is called a motor unit. There are numerous motor units throughout skeletal muscles. Motor units act in a coordinated fashion. One stimulus will affect all of the muscle fibers innervated by a given motor unit.

Whole muscles containing many motor units can contract with different amounts of force. More motor units are recruited to increase the force of contraction when needed. This phenomenon is called summation. In other words, increasing numbers of motor units are activated in order to increase the muscle's force of contraction.

stimulus. If the stimulus is strong enough to produce an action potential we say that the stimulus is a threshold stimulus. As the stimulus increases more motor units are recruited. We call this stimulus a submaximal stimulus.

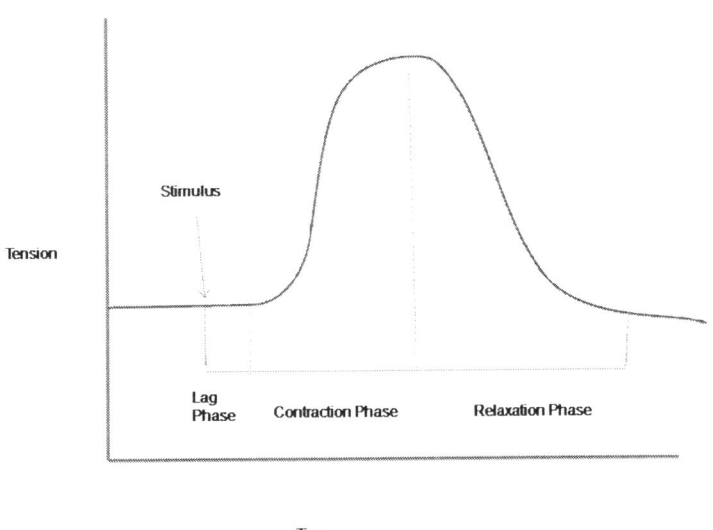

Figure 7.35. Muscle Twitch Phases.

Let's look at an example. Let's say that I am helping a friend move to a new house. I am holding an empty box while my friend fills it up with various items. The weight of the box or "load" is increasing. My biceps muscles must respond by increasing their force of contraction so that I will avoid dropping the box. As the load increases more motor units are recruited and the force of contraction increases to accommodate the load.

Nerves contain many axons of neurons that innervate many motor units. If a nerve is stimulated to produce a stimulus that is below the threshold no action potential is generated in the neurons and there is no muscle contraction. This is called a subthreshold

When the stimulus is strong enough to cause activation of all of the motor units associated with the nerve we say that the stimulus is a maximal stimulus. A stimulus greater than a maximal stimulus (supramaximal stimulus) will not have any additional affect on contraction of motor units.

The ratio of neurons to muscle fibers differs in various muscles. Muscles involved in more precise movements such as in the hands have a smaller ratio of neurons to muscle fibers, whereas muscles involved in gross movements such as the muscles in the thigh have a higher number of fibers innervated by one neuron.

Muscle Contraction Frequency

Big Picture: Muscle Contraction Frequency

There is a maximum frequency of nervous system impulses that will cause a complete contraction of a muscle (tetanus).

When a muscle is stimulated by an action potential it will contract. The time it takes for an action potential to occur is much shorter that the time it takes to contract a muscle. This means that another action potential can produce another contraction. As the frequency of action potentials increases the frequency of muscle contraction also increases. There is a maximal frequency of action potentials that will cause a sustained contraction of a muscle. We call this phenomenon tetanus. Muscles in tetanus will not produce even a partial relaxation. The tension produced by muscles increases along with the frequency of stimulation by action potentials. This phenomenon is known as multiple-wave summation.

During muscle contraction calcium is released by the sarcoplasmic reticulum in response to depolarization of the sarcolemma. If a high frequency of action potentials is administered to the muscle the calcium levels inside the cell remain high and the muscle responds by remaining in a contracted state. This allows for more cross-bridge formation and subsequent increase in tension of the muscle.

If a muscle is stimulated by an action potential and then allowed to relax, the next stimulus will produce a stronger contraction. This will continue for a few contractions then the strength of contraction will level out. This phenomenon is called treppe.

Muscle Length-Tension Relationship

Big Picture: Muscle Length-Tension Relationship

As a muscle is stretched it produces more tension to a point. If the muscle is stretched beyond this point the tension decreases.

The length of a muscle is related to the tension generated by the muscle. Muscles will generate more force when stretched beyond their resting length to a point. Muscles stretched beyond this point will produce less tension.

If the muscle is at its resting length it will not produce maximal tension because the actin and myosin filaments excessively overlap. Myosin filaments can extend into the Z-discs and both filaments interfere with each other limiting the number of cross-bridges that can form.

If the muscle is stretched to a point the tension will increase in the muscle. The actin and myosin filaments can now optimally overlap so that the greatest number of cross-bridges can form.

If the muscle is overstretched the tension will decrease. The actin and myosin filaments do not overlap causing a decrease in the number of cross-bridges that can form.

Types of Muscle Fibers

Big Picture: Muscle Fibers

There are 3 types of fibers:

Fast twitch = high force, low endurance (sprinter)

Slow twitch = low force, high endurance (marathoner)

Intermediate = kind a like both fast and slow twitch (average Joe)

There are three major types of skeletal muscle fibers. These are called fast twitch, slow twitch and intermediate.

Generally, fast twitch fibers generate high force for brief periods of time. Slow twitch fibers generate lower amounts of force but can do so for longer periods of time. Intermediate fibers have some characteristics of both fast and slow twitch fibers. Fast twitch fibers are also called Type II fibers.

Fast twitch fibers are the predominant fibers in the body. They respond quickly to stimuli and can generate a good deal of force. They have a large diameter due to the large amount of myofibrils. Their activity is fueled by ATP generated from anaerobic metabolism.

Slow twitch fibers respond much more slowly to stimuli than fast twitch fibers. They are smaller in diameter and contain a large number of mitochondria. They are capable of sustaining long contractions and obtain their ATP from aerobic metabolism.

Slow twitch fibers are surrounded by capillary networks that supply oxygenated blood for use in the aerobic energy systems. They also contain a red pigment called myoglobin. Myoglobin can bind oxygen (like hemoglobin) and provide a substantial oxygen reserve. Because of the reddish color of myoglobin these fibers are often called red muscle fibers. Slow twitch fibers are also called Type I fibers.

Intermediate fibers resemble fast twitch fibers because they contain small amounts of myoglobin. They also have a capillary network around them and do not fatigue as readily as fast twitch fibers. They contain more mitochondria than fast twitch but not as many as slow twitch fibers. The speed of contraction and endurance also lie between fast and slow twitch fibers. Intermediate fibers are also called Type IIa fibers.

Muscles that have a predominance of slow fibers are sometimes referred to as red muscles such as in the back and areas of the legs. Likewise muscles that have a predominance of fast fibers are referred to as white muscles. It is interesting to note that there are no slow twitch fibers in the eye muscles or muscles of the hands.

The ratio of fast-slow-intermediate fibers is determined genetically. However training can change the ratio of these fibers in skeletal muscles that contain all three types. For example training for endurance can cause some fast twitch fibers to become more like intermediate fibers.

Muscle Response to Exercise

Big Picture: Muscle Response to Exercise

Increases in strength come first from motor units contracting together (synchronous contraction) and then from increases in size (hypertrophy and hyperplasia).

There are three basic ways the muscular system responds to exercise. Let's look at this in the context of Sally who is beginning an exercise program.

Sally is starting an exercise program. She has never been in a gym before and is excited to see the results of her efforts. Part of her program is weight lifting. Her trainer tests her on the first day and finds that she can lift 45 lbs. in a biceps curl. She then begins exercising three times per week. After about two weeks she finds that she can now lift 50 lbs. She is excited about her improvement in just two weeks of training. Sally asks her trainer to measure her biceps and they find that there is no difference in size. If the muscle size has not changed, then what is responsible for Sally's increase in strength?

One of the first ways muscles respond to training is to increase synchronous contraction of motor units. When motor units contract at different points in time (asynchronous contraction) they cannot generate as much force as when they contract together. Training increases synchronous contraction so that the motor units work together to generate higher amounts of force.

Sally continues her program and finds that after about 8-10 weeks there is some increase in her biceps circumference. This is primarily due to hypertrophy or an increase in the cross-sectional diameter of muscles fibers. The number of muscle fibers does not change but the size of the fibers increases. The number of protein filaments, mitochondria, enzymes, and glycogen reserves increases.

Sally may also experience some small amount of hyperplasia. Hyperplasia is an increase in the number of muscle fibers resulting from mitosis. The increase is slight as most of the increase in size is attributed to hypertrophy.

Cardiac Muscle

Cardiac muscle is only found in the heart. Like skeletal muscle it has a high concentration of myofilaments and is striated. There are a number of structural differences between skeletal and cardiac muscles.

Cardiac muscles are smaller and generally contain one nucleus whereas skeletal muscles are multinucleated. They have a different arrangement of T-tubules and no triads. The sarcoplasmic reticulum does not have a terminal cisternae. Cardiac muscle fibers are powered by aerobic metabolism and contain energy reserves in the form of glycogen and lipids. Cardiac muscle cells contain large numbers of mitochondria to utilize aerobic energy systems.

Cardiac muscle cells also contain a specialized kind of cell junctions called intercalated discs that allow the flow of chemicals between cells and help to maintain the structure of the muscle. This allows for a greater transmission of electrical signals across large areas of cardiac muscles. The discs also allow adjacent fibers to pull together in a more coordinated contraction. Instead of motor units working separately in skeletal muscle, intercalated discs allow cardiac muscle to contract in large uniform segments.

Cardiac muscle can also contract without a stimulus from the nervous system. Cardiac muscle contains self-generating action potential cells called pacemaker cells or nodes. The pacemaker cells however can respond to the nervous system by changing the rate and force of contraction of cardiac muscle cells.

Cardiac muscle cannot undergo tetanic contractions due to the structure of the cell membrane.

Smooth Muscle

Smooth muscle cells are found throughout the body in organs, blood vessels and tubelike structures. Smooth muscles contain actin and myosin and are long spindle shaped cells. Actin and myosin are not arranged in sarcomeres so smooth muscle is not striated. Instead the actin and myosin are scattered about throughout the muscle. Smooth muscle has no T-tubules and the myosin has a larger number of globular protein heads.

Smooth muscle contraction differs from skeletal or cardiac contraction in that when calcium is released by the sarcoplasmic reticulum it binds with a calcium-binding protein called calmodulin that activates an enzyme called myosin light chain kinase. This enzyme allows for the formation of cross-bridges. Because of the structure of smooth muscle, length and tension are not related. When smooth muscle is stretched it adapts to its new resting length and can continue to contract.

Smooth muscle cells are classified as multiunit or visceral. Multiunit smooth muscle is organized into motor units that are innervated by the nervous system. However, each cell can be connected to more than one motor unit. Visceral cells do not connect directly with motor neurons and are arranged in layers. Gap junctions connect layers of smooth muscle so that one area can influence others when contracting. This can produce a wave-like contraction called peristalsis.

Review Questions

1. The temporomandibular joint is an example of which type of lever:
   a. First class
   b. Second class
   c. Third class
   d. Fourth class

2. Which best describes a first class lever:
   a. Fulcrum is between pull and weight
   b. Weight is between pull and fulcrum
   c. Pull is between weight and fulcrum
   d. Weight and pull are next to fulcrum

3. My car was stuck. I tried to push it but it wouldn't budge an inch. This is an example of which type of contraction:
   a. Isotonic
   b. Isometric
   c. Isokinetic
   d. Isoisonic

4. Which of the following is not an arm muscle:
   a. Brachioradialis
   b. Triceps
   c. Sartorius
   d. Flexor carpi radialis

5. This muscle is located on the side of the neck:
   a. Erector spinae
   b. Trapezius
   c. Sternocleidomastoid
   d. Latissumus dorsi

6. Which of the following is a muscle of facial expression:
   a. Trapezius
   b. Platysma
   c. Orbicularis occuli
   d. Levator scapula

7. Which of the following is not a hamstring muscle:
   a. Rectus femorus
   b. Semitendinosus
   c. Biceps femorus
   d. Semimembranosus

8. Which muscle works to flex the hip:
   a. Tibialis anterior
   b. Iliopsoas
   c. Rectus abdominus
   d. Soleus

9. Which of the following is the deepest abdominal muscle:
   a. Transverse abdominus
   b. External oblique
   c. Internal oblique
   d. Rectus abdominus

10. The latissumus dorsi muscle inserts:
    a. On the thoracic wall
    b. On the humerus
    c. On the scapula
    d. On the cervical spine

11. Which of the following consists of a connective tissue layer that covers the entire muscle:
    a. Fascicle
    b. Epimysium
    c. Endomysium
    d. Perimysium

12. The "thick" filament in muscle is known as:
    a. Actin
    b. Myosin
    c. Troponin
    d. Tropomyosin

13. The troponin-tropomyosin complex covers _____ on the actin.
    a. Sarcolemma
    b. Sarcoplasmic reticulum
    c. Calcium
    d. Myosin binding site

14. Which of the following binds to the troponin-tropomyosin complex causing a conformational change:
    a. Potassium
    b. Calcium
    c. Sodium
    d. Magnesium

15. Which neurotransmitter is released by the axon terminal and propagates to the motor end plate:

a. Dopamine
b. Acetylcholine
c. Norepinephrine
d. Serotonin

16. Which of the following consists of thin threads that hold the myosin in place:
a. Actin fibers
b. Troponin
c. Titin protein
d. Tropomyosin

17. A sarcomere extends from ____ to _____:
a. Z-disc, Z-disc
b. I-band, I band
c. A-band, I-Band
d. I-band, H-zone

18. Which of the following electrolytes is responsible for depolarization of the motor end plate:

a. Sodium
b. Potassium
c. Calcium
d. Magnesium

19. Increasing the stimulation to a muscle fiber until it contracts is known as:
a. Fiber contraction hypothesis
b. Sliding filament theory
c. All or none law
d. Invoked potential law

20. A motor neuron and all of the muscle fibers it innervates is called:
a. Motor unit
b. Contractile element
c. Sarcomere
d. A-band contraction

21. Which type of muscle fiber would be working harder in a marathon runner:

a. Slow twitch
b. Fast twitch
c. Intermediate
d. Secondary

22. Which of the following is the first muscular response to exercise:

a. Hypertrophy
b. Atrophy
c. Hyperplasia
d. Synchronous contraction of motor units

23. Which type of muscle can perform a contraction known as peristalsis:
a. Cardiac
b. Skeletal
c. Smooth
d. All can perform this contraction

Chapter 8

What's a Joint like This Doing in a Girl Like You?

Chapter 8

What's a Joint like This Doing in a Girl like You?

Joints can get a bit confusing especially when we read words like amphiarthrotic, fibrous, and degrees of freedom. Hopefully this chapter will make some sense out of the wonderful world of joints.

The Big Picture: Joints

Joints can be classified according to tissue or how they move.

Joints can be classified two ways. They can be classified according to their tissue and/or by their movements. Fortunately there are only three tissue classifications:

1. Fibrous
2. Cartilagenous
3. Synovial

There are also 3 movement classifications:

1. Synarthrotic--immovable
2. Amphiarthrotic—slightly movable
3. Diarthrotic—freely movable

Think of someone with arthritis. The word "arthrotic" sounds a lot like arthritis. You can also think of the statements "it is a sin (syn) to move" and "I would die to be able to sit still."

By and large most of the joints in the body are synovial (fig. 8.1). These are the most complex and common joints in the body. Examples of synovial joints include the shoulder, hip, knee and elbow.

Fibrous joints are simple structures because they just consist of a band of connective tissue holding the bones together. There are three basic types of fibrous joints:

1. Syndesmosis—interosseous ligament (band of connective tissue between bones like the radius and ulna).
2. Gomphosis—cone shaped process in cone shaped socket (like a tooth).
3. Suture—like those holding the skull bones together.

So the trick is in combining the movement classification with the tissue classification. For example:

A syndesmosis is slightly movable so it is amphiarthrotic.

A gomphosis is immovable so it is synarthrotic.

A suture is immovable so it is also synarthrotic.

Get the picture? It just takes a bit of practice.

Cartilagenous joints consist of, you guessed it, cartilage (which can be either hyaline or fibrocartilage). The cartilage unites the bones. These are all classified as slightly movable or amphiarthrotic. Examples include the discs in your spine and the pubic symphysis.

Like we said earlier, synovial joints are the most popular in the body. They consist of a complex structure and are freely movable or diarthrotic. One of the key things to know about synovial joints is that their shape dictates how they move. There are a number of shapes such as "ball and socket" and "hinge." Some are more exotic sounding like "condyloid." The key is to think of the shapes of the bones forming the joint.

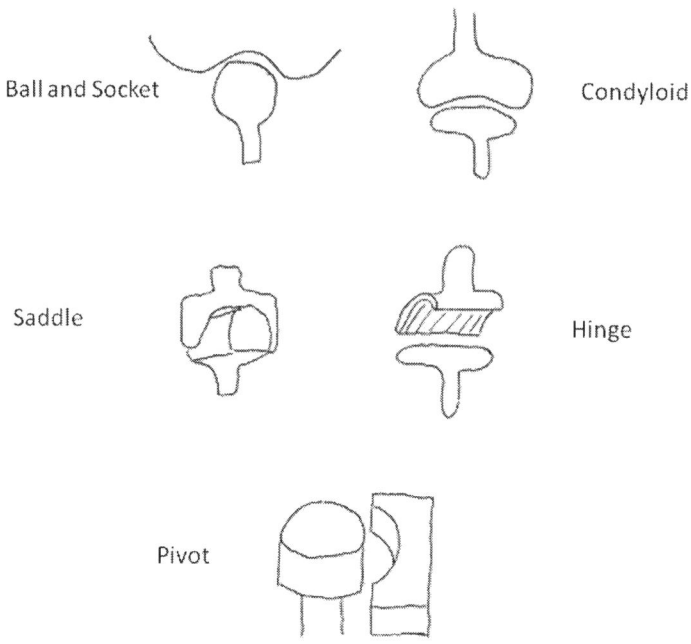

Figure 8.1 Types of synovial joints.

Joint Movements

You'll definitely want to learn the joint movements. The best way is to just do them while naming them out loud. Just be sure you are doing this in private. Also, most of the movements can be done in pairs.

Here is a list and short description of the movements:

Flexion—usually this means bending the joint from anatomical position.

Extension—straightening a joint.

Abduction—moving the joint so that the limb moves away from the midline of the body.

Adduction—moving the joint so that the limb moves toward the midline of the body (think of "adding" to the body).

Lateral flexion—side bending.

Rotation—twisting. This can either be internal/external rotation or right and left rotation (we'll explain this later).

Dorsiflexion—this is an ankle movement. Think of walking on your heels.

Plantarflexion—another ankle movement. Think of walking on your toes.

Inversion—more ankle movements. Think of bending your foot so the sole of your foot faces to the middle of the body.

Eversion—ankle movement where the sole of the foot faces outward.

Supination—forearm rotation so the palms face upward. Think "cup of soup."

Pronation—forearm rotation so the palms face downward.

Circumduction—tracing a circle with the distal end of a joint.

Ulnar/Radial Deviation—special movement of the wrist by bending toward the ulna or radius.

Protraction/Retraction—think of two flat surfaces stacked on top of each other. The top surface moves forward in protraction and backward in retraction. This can happen in the cervical spine.

Elevation/Depression—moving upward/downward.

Okay, now let's group these together according to the joints.

Spine—all of the movements are the same for the spine. You can just think of the spine in terms of the cervical, thoracic and lumbar areas.

Flexion/Extension—bending forward/bending backward

Right/Left Lateral Flexion—bending to the right and left

Right/Left Rotation—twisting to the right and left

Shoulder and Hip—these have the same movement because they are both ball and socket joints.

Flexion/Extension—moving the arm or leg forward and backward.

Abduction/Adduction—moving the limb out to the side and then back toward the midline of the body.

Internal/External Rotation—twisting the limb inward or outward.

Circumduction—tracing a circle with the end of the limb.

Knee—an easy one.

Flexion/Extension—bending and straightening the knee.

Elbow—not too bad.

Flexion/Extension—bending and straightening the elbow.

Supination/Pronation—twisting the forearm so the palms face upward/downward.

Ankle—these are special movements.

Dorsiflexion/Plantarflexion—bending the ankle as if to walk on heels/toes.

Inversion/Eversion—bending the ankle to point the sole of the foot inward/outward.

Circumduction—drawing a circle with your foot.

Wrist—more special movements.

Flexion/Extension—bending wrist forward from anatomical position/bending it backward.

Ulnar/Radial Deviation—bending to the ulnar side/radial side.

There are more joint movements but these should get you started.

Joints and Ligaments

Joints are held together by ligaments. Many ligaments tell you where they are located by virtue of their names. For example the coracohumeral ligament in the shoulder connects to the coracoid process of the scapula and (you guess it) the humerus.

Here is a listing of some popular ligaments:

Shoulder

The shoulder consists of the scapula, humerus and clavicle. The joint between the scapula and humerus is a synovial ball and socket joint. As in all joints the shoulder joint is held together by ligaments (fig. 8.2). Some of the important ligaments include:

Glenohumeral

These ligaments exist as three bands extending from the anterior wall of the glenoid fossa and attaching to the anatomical neck and lesser tubercle of the humerus.

Coracohumeral

This ligament extends from the coracoid process of the scapula to the greater tubercle of the humerus.

Transverse humeral

This ligament forms a band of connective tissue between the greater and lesser tubercles of the humerus. The long head of the biceps brachii is found in this groove.

Glenoid Labrum

This is a rim of fibrocartilage that attaches to the glenoid fossa.

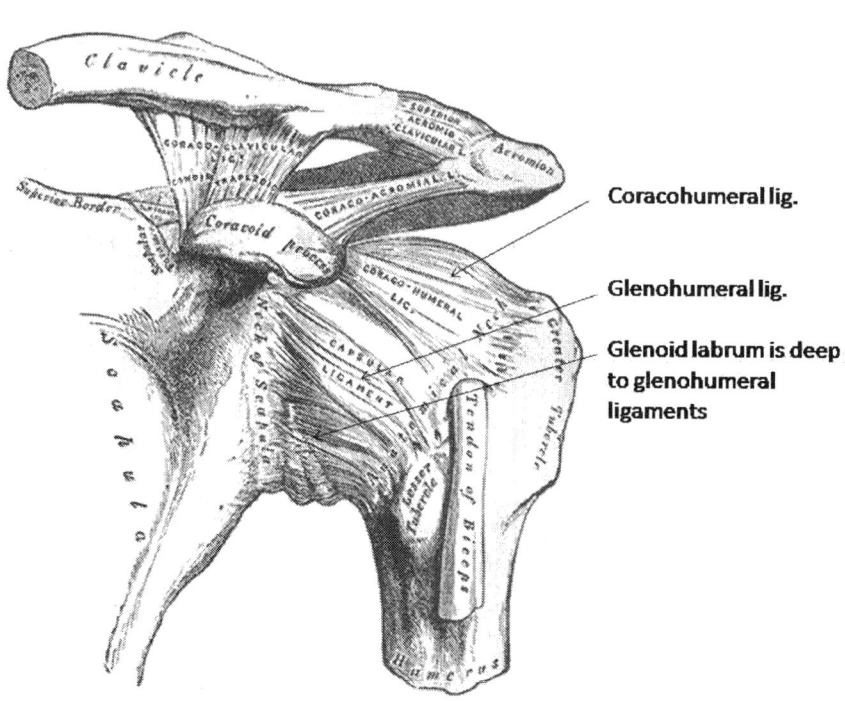

Figure 8.2. Shoulder

Elbow

The elbow contains two articulations (fig. 8.3). One involves the humerus and ulna. The other involves the humerus and radius. The humeralulnar joint is formed by the trochlea of the humerus and the proximal portion of the ulna. This joint can only flex and extend. The humeralradial joint is formed by the capitulum of the humerus and the radial head. This joint can rotate. Some of the important ligaments include:

Ulnar and Radial Collaterals

The ulnar collaterals connect the medial aspect of the medial epicondyle to the medial aspect of the coronoid process of the ulna.

The radial collaterals connect the lateral epicondyle to the annular ligament of the radius.

Annular

The annular ligament encircles the radial head and attaches to the trochlear notch of the ulna. This ligament can be prone to dislocation in children.

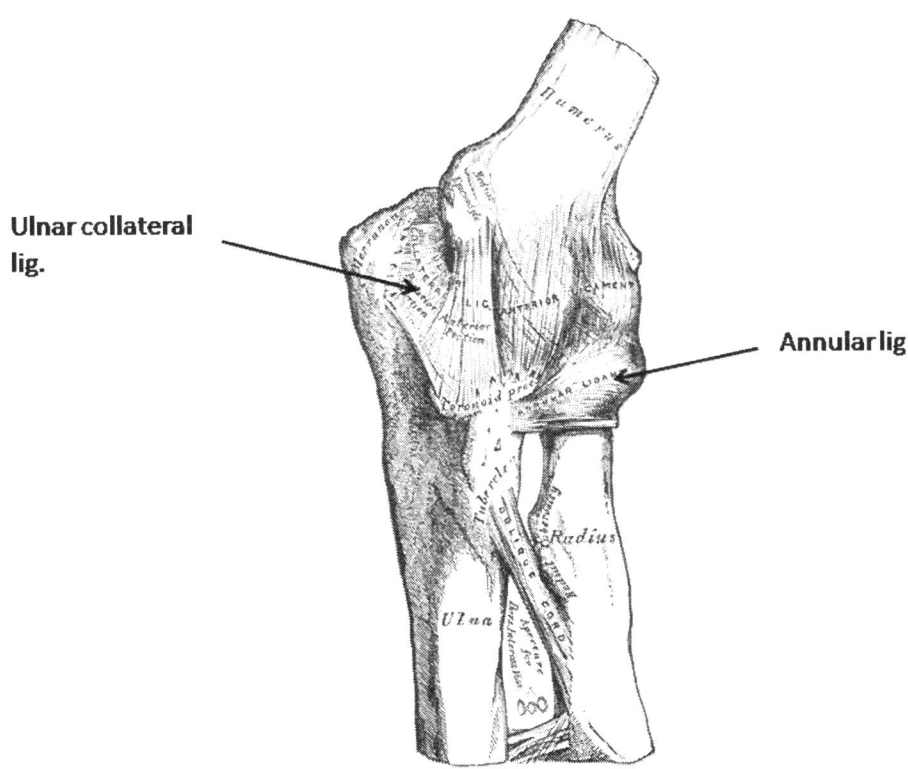

Figure 8.3. Elbow (medial projection). Notice how the annular ligament wraps around the head of the radius.

Hip

The hip joint consists of the femur and coxal bones. The head of the femur fits into the acetabulum of the coxal bone (figs. 8.4, 8.5). The hip joint has the same motions as the shoulder.

The ligaments include:

The iliofemoral ligament extends from the ilium to the greater and lesser trochanters of the femur. It is Y-shaped and is considered the strongest ligament in the body.

The ischiofemoral ligament extends from the ischium to the joint capsule of the femur.

The pubofemoral ligament extends from the pubis to the joint capsule of the femur.

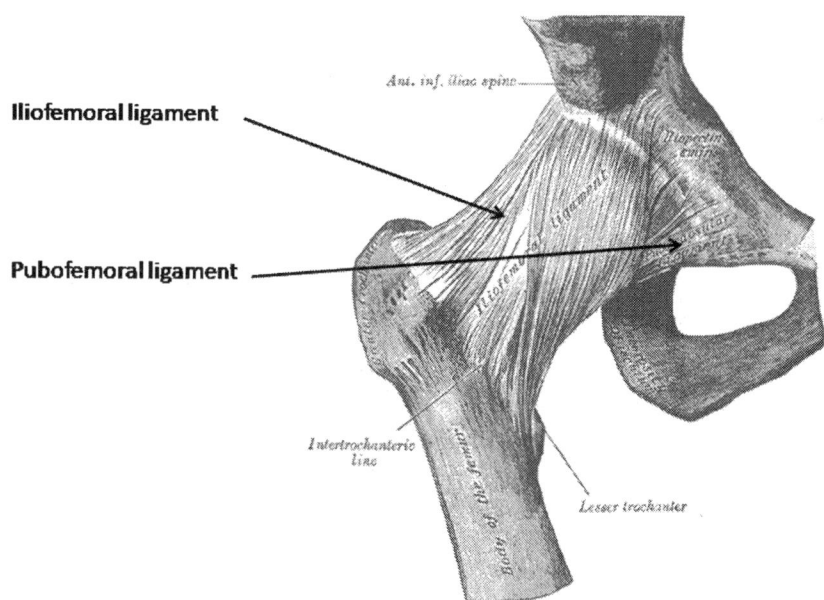

Figure 8.4. Hip (anterior view)

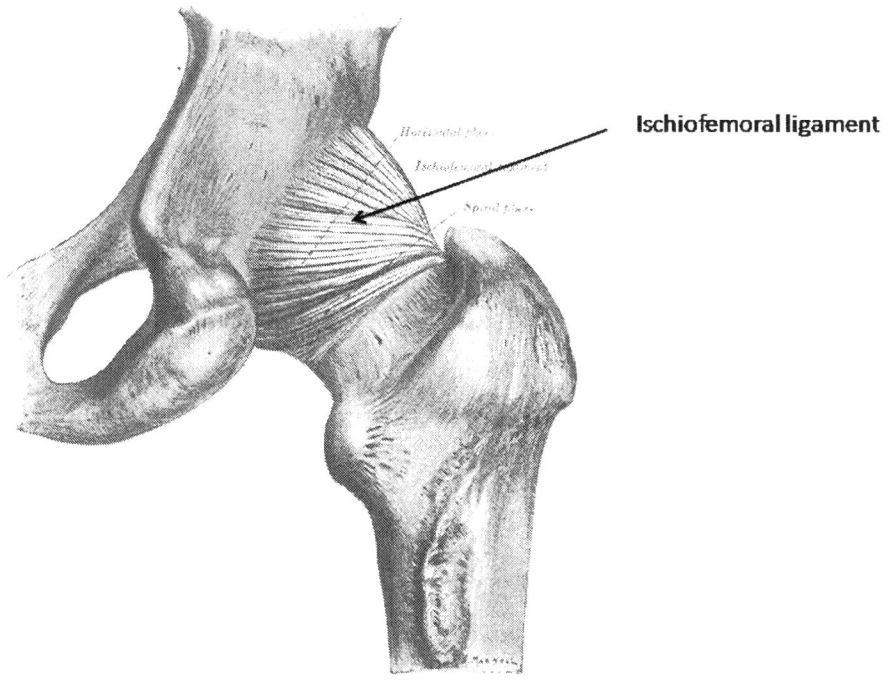

Figure 8.5. Hip (posterior view)

Knee

The knee is the most complex joint in the body. It is also the largest. It consists of the condyles of the femur articulating with the condyles of the tibia (fig. 8.6). The patella also articulates with the femur. The knee flexes and extends as well as rotates. It forms a locked position when extended. Some of the knee ligaments include:

The medial collateral ligament extends from the medial condyle of the femur to the medial condyle of the tibia.

The lateral collateral ligament extends from the lateral condyle of the femur to the head of the fibula.

The anterior cruciate ligament is inside the knee and extends from the posterior femur to the anterior tibia. The anterior cruciate (ACL) works to stop forward translation (movement) of the tibia on the femur.

The posterior cruciate ligament is also inside the knee and extends from the anterior femur to the posterior tibia. It works to stop backward translation of the tibia on the femur.

The arcuate popliteal ligament is on the posterior aspect of the knee. It is Y-shaped and extends from the lateral condyle of the femur to the fibular head.

The patellar ligament extends from the inferior aspect of the patella to the tibial tuberosity. It is an extension of the common quadriceps tendon.

The oblique popliteal is located in the posterior aspect of the knee and extends from the lateral condyle of the femur to the head of the fibula.

Joint Injuries

Sprains and Strains

When the force exceeds what the tissue can handle the tissue becomes damaged. Common injuries include tears to the muscles (strains) and tears to the ligaments (sprains).

Both sprains and strains are graded 1, 2, or 3. In a first degree injury 0 to 25% of the fibers are torn. These injuries typically take 1-2 weeks to heal.

A second degree injury is characterized by 25% to 50% of the fibers torn. These usually take from 2-4 weeks to heal.

Third degree injuries are the most severe with greater than 50% of the fibers torn. These injuries take at least 12 weeks to heal.

The body reacts to these injuries by producing inflammation. The joint will appear red and swollen and cause pain. Healing depends on the severity of the injury as well as the health of the subject. In severe sprains the joint will become unstable due to the torn ligaments. In some cases the joint must be stabilized with splints, supports or casts.

Osteoarthritis

Osteoarthritis is characterized by the breakdown of cartilage. It is the most common form of arthritis and tends to affect people in middle age and beyond. Osteoarthritis commonly affects the hands, knees, hips and spine. In osteoarthritis the normal cartilage repair mechanisms malfunction and the cartilage begins to wear out. The joint space will become smaller and may progress to the point of bone rubbing on bone. The joint surfaces become roughened and cause pain and inflammation. There is no cure for osteoarthritis however severe cases are treated with joint replacement.

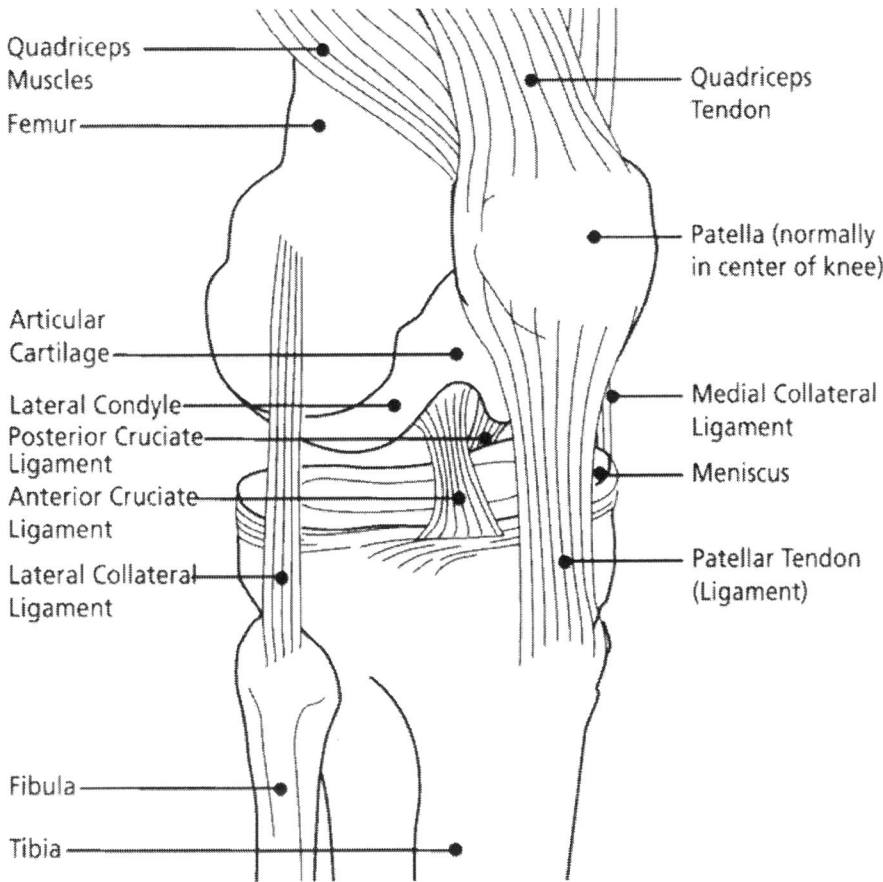

Figure 8.6. Knee ligaments.

Review Questions

1. Which of the following is not a joint category:
   a. Cartilaginous
   b. Fibrous
   c. Synovial
   d. Bony

2. A tooth is an example of which of the following types of joints:
   a. Cartilaginous
   b. Gomphosis
   c. Synchrondosis
   d. Amphiarthrosis

3. An epiphyseal plate is an example of which type of joint:
   a. Cartilaginous
   b. Synchondrosis
   c. Synovial
   d. Fibrous

4. Most of the joints in the body are which type:
   a. Fibrous
   b. Cartilaginous
   c. Synovial
   d. Amphiarthroses

5. Which of the following is not a synovial joint:
   a. Shoulder
   b. Knee
   c. Ankle
   d. Intervertebral disc

6. The shoulder joint is which type:
   a. Hinge
   b. Modified hinge
   c. Condylar
   d. Ball and socket

7. The structures that hold joints together are called:
   a. Loose connective tissue
   b. Tendons
   c. Fibers
   d. Ligaments

8. Which joint motion is not performed at the hip:
   a. Flexion
   b. Extension
   c. Abduction
   d. Supination

9. Rotating the forearm so the palm of the hand points upward is called:
   a. Internal rotation
   b. Supination
   c. Lateral flexion
   d. Pronation

10. When standing on your toes your ankle joint performs this motion:
    a. Extension
    b. Dorsiflexion
    c. Eversion
    d. Plantar flexion

Chapter 9

Don't Get Nervous About the Nervous System

Chapter 9

Don't Get Nervous About the Nervous System

One of the real grade killers in anatomy courses is the nervous system. Usually students know a few bones, muscles and even some ligaments but when it comes to structures like the corpora quadrigemina or pons they haven't a clue where to find them. Here's where you really have to study from general to specific and get the big picture first.

Let's begin by looking at the structures of the central nervous system.

The Big Picture: Nervous System

The central nervous system consists of the brain and spinal cord. That's it—see, it's not so hard after all!

Before we get into some nitty gritty anatomy, let's take a look at how the system works. We can understand a good deal of how the nervous system works by examining how information flows through it. Let's say that I have reached for a hot cup of coffee and have moved it slightly, spilling coffee on my hand. I must make the decision to let go of the cup and move my hand away.

The sensation of touch, heat and pain are first processed by sensory receptors located in my skin. All sensory receptors take information from the environment and convert it to a form that can be processed by the nervous system. The environment can be internal (inside the body) or external (outside the body). The information going into the receptor can be in many forms. For example, light rays enter the eye, sound waves enter the ear, and pressure is sensed by receptors that are deformed by either light or heavy pressure in the skin. Heat is also sensed by temperature receptors in the skin. The information coming out of the receptor is in the form of electrochemical impulses called action potentials (more about these later).

The impulses from the sensory receptor then travel to the central nervous system via afferent pathways (the word "afferent" means toward the central nervous system). These pathways generally consist of sensory nerves that attach to the receptors. The pathway continues to the spinal cord which is part of the central nervous system.

The impulse then travels upward toward the brain via a special pathway in the spinal cord called a spinal tract. Since the tract travels upward to the brain it is called an ascending tract. The impulse then travels to the brain where the sensations of pressure and heat are processed. A decision is made in the brain to move the muscles of my arm and hand to let go of the cup. The impulse is now a motor impulse and it travels down the spinal cord following a spinal tract (this time a descending tract) and moves along an efferent (away from) pathway consisting of a motor nerve(s) to the muscles of my hand and arm. My hand lets go of the cup and moves away.

We can think of the nervous system then in terms of sensory stimulus (hot coffee) and motor response (move hand). Many nervous system functions occur this way. But before we go deeper into how the nervous system works we need to examine some structure.

Here's an overview of what was just said:

Sensory Pathway

Information flows from sensory receptor (eyes, ears, taste buds, etc.) via an afferent pathway (nerve) to the spinal cord where it follows an ascending spinal tract (special pathway in spinal cord) to the brainstem and then is routed to the brain.

You could think of sensory information as riding on an "up" elevator (spinal tract) to the brain.

Motor Pathway

Information now flows from the brain to the brainstem to the spinal cord where it follows a descending spinal tract and exits the cord via an efferent pathway (nerve) to the effectors (muscle, organ or gland).

You could think of motor information riding on a "down" elevator (spinal tract) from the brain to where it needs to go (like a muscle, organ or gland).

That's how information flows through the nervous system (in a nutshell of course).

Try to keep this basic pathway in mind when we start going over the various nervous system parts.

Meninges

Let's begin with the coverings of the brain and spinal cord called the meninges.

There are three layers of tissue covering the brain and spinal cord. These are the dura mater, arachnoid mater and pia mater. This should help you remember:

Dura mater—tough **dura**ble outer covering.

Arachnoid mater—middle covering. This one looks like a spider web.

Pia mater—thin, inner covering. The word "pia" is the smallest word of the three and is the thinnest and innermost layer.

You can also think of the meninges as a nice **PAD** covering the brain. **PAD** of course stands for pia, arachnoid and dura maters.

The Big Picture: Spinal Cord Structure

The spinal cord consists of a core of grey matter surrounded by white matter surrounded by a 3-layer covering called the meninges.

Let's take a very basic look at the structure of the spinal cord. The cord begins at the foramen magnum of the occipital bone and ends at about the second lumbar vertebra in a cone shaped process called the conus medullaris. The remaining nerves come off the end of the cord as the cauda equina (horse's tail).

The basic structure is a core of grey matter surrounded by white matter surrounded by meninges. There are also two nerve "roots" exiting the cord. One root comes out of the front and is called the ventral root while the other exits the back and is called the dorsal root.

You might also notice that the spinal cord is symmetrical (like the body). The right and left sides are connected by a bridge of grey matter called the grey commissure. Each grey matter area is further divided into what are called horns (they kind of look like horns). There are anterior, posterior and lateral horns.

The white matter is also divided into areas called funiculi. There are anterior, posterior and lateral funiculi.

What's the difference between grey and white matter? White matter looks lighter in color than grey matter because it contains a fatty substance called myelin. We'll talk more about myelin later.

There are two spaces that are important as well. These are the subarachnoid space which is between the arachnoid and pia mater and the epidural space which is between the dura mater and the bony walls of the spine.

Spinal Cord Tracts

The Big Picture: Spinal Cord Tracts

Sensory information goes up. It follows ascending spinal tracts.

Fasciculus gracilis/cunneatus—carry touch and pressure information and cross in the medulla oblongata.

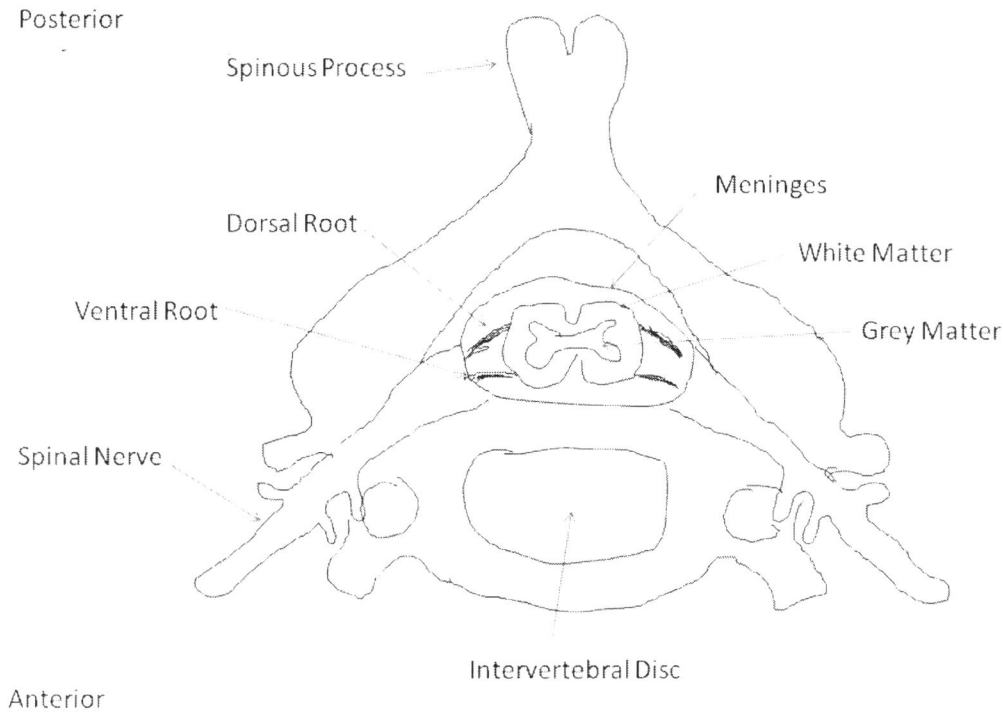

Figure 9.1. Spinal cord anatomy.

Lateral spinothalamic—carries pain and temperature information and crosses 1-2 segments above where the information enters the cord.

Anterior spinothalamic—carries light touch and pain information and crosses 1-2 segments above where the information enters.

Spinocerebellar—has 2 parts: anterior and posterior. Carries information for coordination and the fibers do not cross.

Motor information goes down. It follows descending spinal tracts.

Corticospinal—carries motor information to muscles.

Rubrospinal—carries motor information to muscles.

Reticulospinal—caries motor information for muscle tone and sweat glands.

One major function of the spinal cord is to carry information to and from the brain. This information is carried by areas in the white matter called spinal tracts. Tracts are like pathways that carry various types of information. Sensory information is carried to the brain by ascending tracts and motor information is carried from the brain by descending spinal tracts. Some tracts cross over (decussate—undergo decussation) to the contralateral side. The right side of the brain processes sensory information and sends motor information to the left side of the body and vice versa.

Important Ascending Tracts

Some important ascending tracts include the fasciculus gracilis, fasciculus cunneatus, spinothalamic and spinocerebellar. There are generally three neurons that carry information from the stimulus to the brain. The first-order neuron (first neuron in the series) carries information from the sensory receptor to the spinal cord. The second-order neuron carries the information to an area in the brain called the thalamus and the third-order neuron carries the information to the cortex of the brain.

The fasciculus gracilis is located in the posterior funiculus (that sounds catchy). This tract carries information related to discriminative touch (can you tell that I touched you in two places at the same time?), visceral pain (organ pain like from a heart attack), vibration, and proprioception (position of joints in space). The tract carries this information from the middle thoracic and lower areas of the body.

The fasciculus gracilis is part of the posterior spinal cord called the dorsal column. At the middle thoracic region (about T6) it combines with the fasciculus cunneatus. It contains first order neurons that travel up the ipsilateral side of the cord and cross over at the brainstem in an area known as the medulla oblongata (specifically in the graczile nucleus).

The fasciculus cunneatus is also located in the posterior funiculus (still sounds catchy). It carries the same type of information as the fasciculus gracilis from the middle to upper areas of the body (T6 and above). It is also part of the dorsal column and its fibers cross over in the medulla (cunneate nucleus) as well.

The second order fibers of the fasciculus gracilis and cunneatus combine to form an area known as the medial lemniscus from the medulla oblongata to the thalamus.

The spinothalamic tract consists of two portions. The anterior spinothalamic and lateral spinothalamics are located in the anterior and lateral funiculi. The spinothalamics are sometimes referred to as the anterolateral system.

The anterior spinothalamic tract carries information related to light touch and pain. Light touch is clinically defined as perceived sensation from stroking an area of the skin without hair. The fibers from the anterior spinothalamic tract cross at one to two segments above their entry point in the spine.

The lateral spinothalamic tract is an important clinical tract because it carries information related to pain and temperature. Its fibers also cross in a similar way to the anterior spinothalamic tract. Lesions of the lateral spinothalamic tract will result in loss of pain and temperature. For example in a Brown-Sequard lesion (sometimes called a hemisection of the spinal cord) there is a contralateral loss of pain and temperature below the level of the lesion as well as a bilateral loss of pain and temperature at the segmental level of the lesion.

The spinocerebellar tract also consists of two portions. The anterior and posterior spinocerebellar tracts are both located in the lateral funiculus. The fibers in the posterior tracts do not cross while the anterior fibers cross at the medulla oblongata. The spinocerebellar tracts carry information related to coordination of muscles from the lower limbs and trunk to the cerebellum.

Motor tracts carry information from the brain to the effectors. These consist of the corticospinal, rubrospinal and reticulospinal.

The corticospinal tract consists of anterior and lateral portions located in the anterior and lateral funiculi. These tracts are sometimes referred to as the pyramidal tracts. Fibers in the lateral tract cross over at the medulla oblongata. Fibers in the anterior portion cross at various levels in the spinal cord. Both tracts convey motor information to skeletal muscles.

The rubrospinal tracts are located in the lateral funiculi. The fibers from these tracts cross over in the brain and descend through the lateral funiculi. The rubrospinal tracts also carry motor information to skeletal muscles. They also carry information about posture and coordination.

The reticulospinal tracts consist of anterior and lateral tracts. They are located in the anterior and lateral funiculi. Some of the fibers cross while others do not. These tracts carry information related to muscular tone and activity of sweat glands.

Spinal Nerves

Remember those dorsal and ventral roots coming off of the spinal cord? Well they will combine to form spinal nerves. There are 31 pairs of spinal nerves. They are named after their attachment point in the spine. For example cervical nerves are named C1-C8, thoracic T1-T12, lumbar L1-5, and sacral S1-S5. All spinal nerves are mixed nerves and carry both sensory and motor information.

Once they exit the cord the spinal nerves branch. There are three branches:

Anterior. The ventral branch (ramus) innervates the sides and anterior trunk

Posterior. The posterior branch (sometimes called a ramus) innervates the back. It carries sensory information from the central region of the back as well as motor information to the muscles of the spine.

Visceral. The visceral branch becomes part of the autonomic nervous system.

There is also a meningeal branch that courses back into the spinal canal and innervates the vertebrae, meninges, and spinal ligaments.

Spinal nerves carry sensory information from the surface of the body and motor information to the muscles. Each nerve carries sensation from a specific area of the body called a dermatome (fig. 9.4).

Some spinal nerves combine to form complex networks called plexi. There are four major plexi in the human body. The cervical plexus (C1-C4) innervates the posterior head and skin of the neck. The brachial plexus (C5-T1) consists of the ventral rami (branches) from spinal nerves C5-T1. The rami form three trunks and the trunks become six divisions which again join to form three cords (yes I know it's complicated). Five branches emerge from the three cords which constitute the major nerves of the upper extremity. These include the axillary, radial, musculocutaneous, ulnar and median nerves.

The lumbar plexus consists of the ventral rami from spinal nerves L1-L4. The sacral plexus consists of the ventral rami from spinal nerves L4-S4. Sometimes both plexi are referred to as the lumbosacral plexus. The major nerves exiting the lumbosacral plexus include the obturator, femoral, and sciatic.

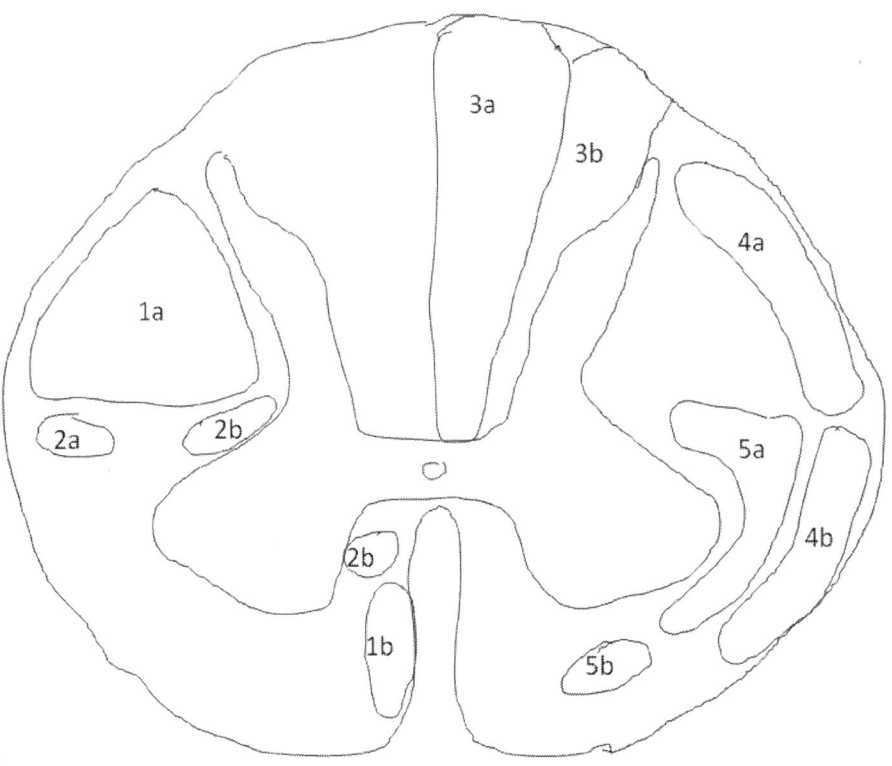

Figure 9.2. Spinal Tracts

1. **Pyramidal**
1a. Lateral Corticospinal
1b. Anterior Corticospinal
2. **Extrapyramidal**
2a. Rubrospinal
2b. Reticulospinal
3. **Dorsal Column**
3a. Fasciculus Gracilis
3b. Fasciculus Cunneatus

4. **Spinocerebellars**
4a. Posterior Spinocerebellar
4b. Anterior Spinocerebellar
5. **Spinothalamics**
5a. Lateral Spinothalamic
5b. Anterior Spinothalamic

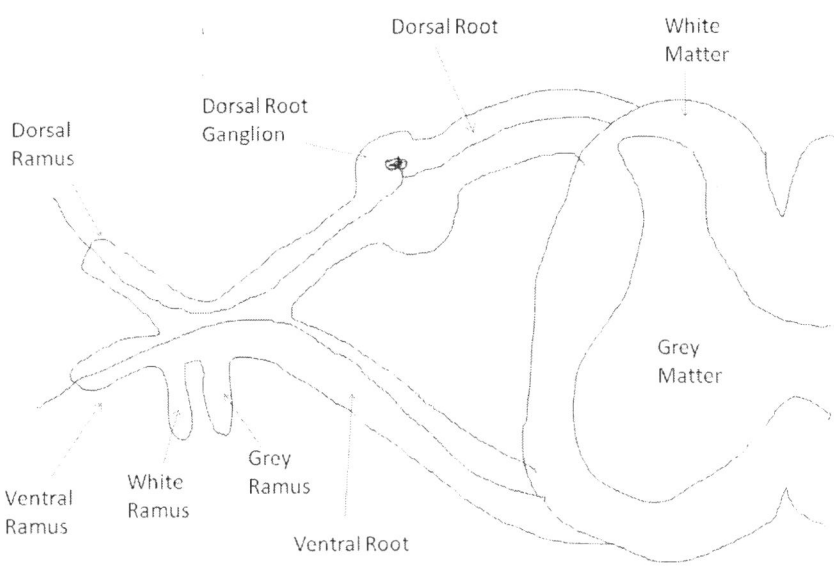

Figure 9.3. Spinal Nerve

Enough of this spinal cord stuff. Let's move on to the brain.

The Big Picture: The Brain

The brain consists of four major structures. These include the cerebral cortex, diencephalon, brainstem and cerebellum.

The Cerebrum

The cerebrum is the largest portion of the nervous system. The cerebrum consists of two hemispheres (right and left) connected by a white matter bridge called the corpus callosum. On the surface of the cerebrum are folds called gyri and grooves called sulci. Deep grooves are known as fissures. Each hemisphere is divided into lobes. The lobes are the frontal, parietal, temporal and occipital (fig. 9.5).

The frontal lobe processes information involving motor movements, concentration, planning and problem solving as well as the sense of smell and emotions. The parietal lobes process sensory information with the exception of hearing, smell and vision. The temporal lobes process information related to hearing, smell and memory as well as abstract thought and making judgments. The occipital lobe processes visual information.

Some lobes are divided by fissures. Along the superior aspect of the cerebrum lies the longitudinal fissure that divides the parietal lobes. The lateral fissure (Sylvian fissure) is located on the side and separates the temporal

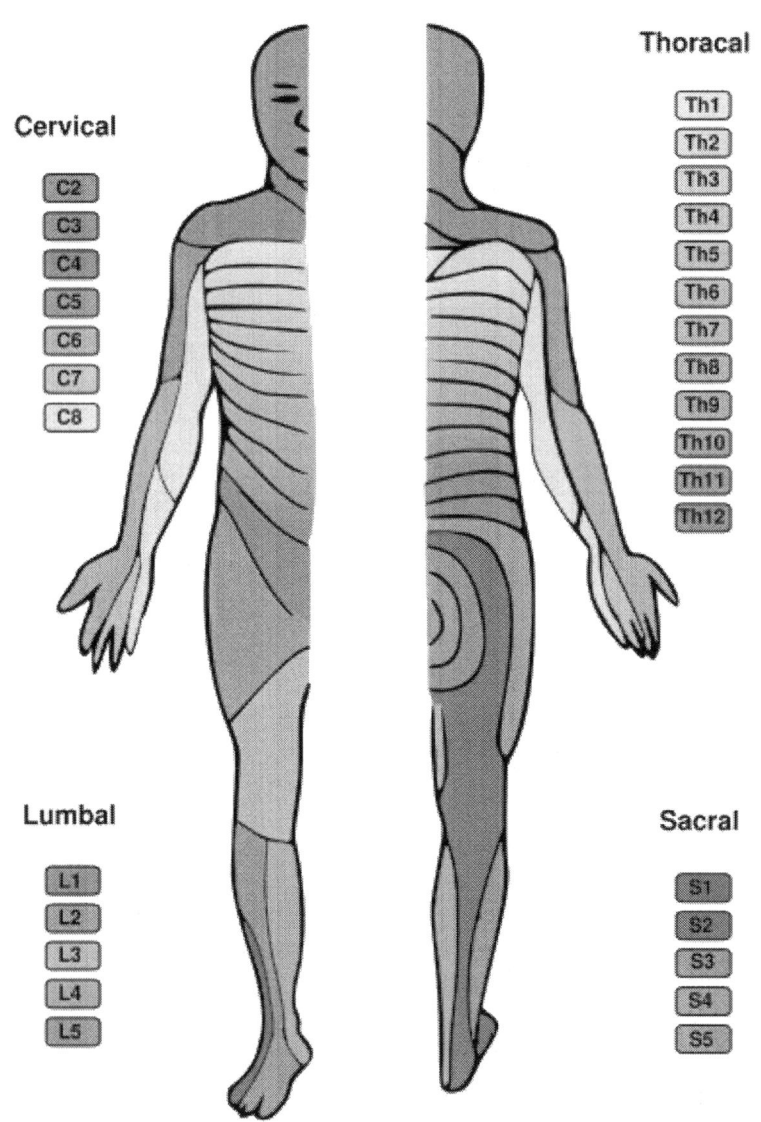

Figure 9.4. Dermatomes are specific areas of the body that carry sensation by spinal nerves.

from parietal lobes. One sulcus called the central sulcus is located midway on the side of the cerebrum and separates the frontal from parietal lobes.

Deep in the lateral fissure is the insula which is often referred to as a fifth lobe of the cerebrum.

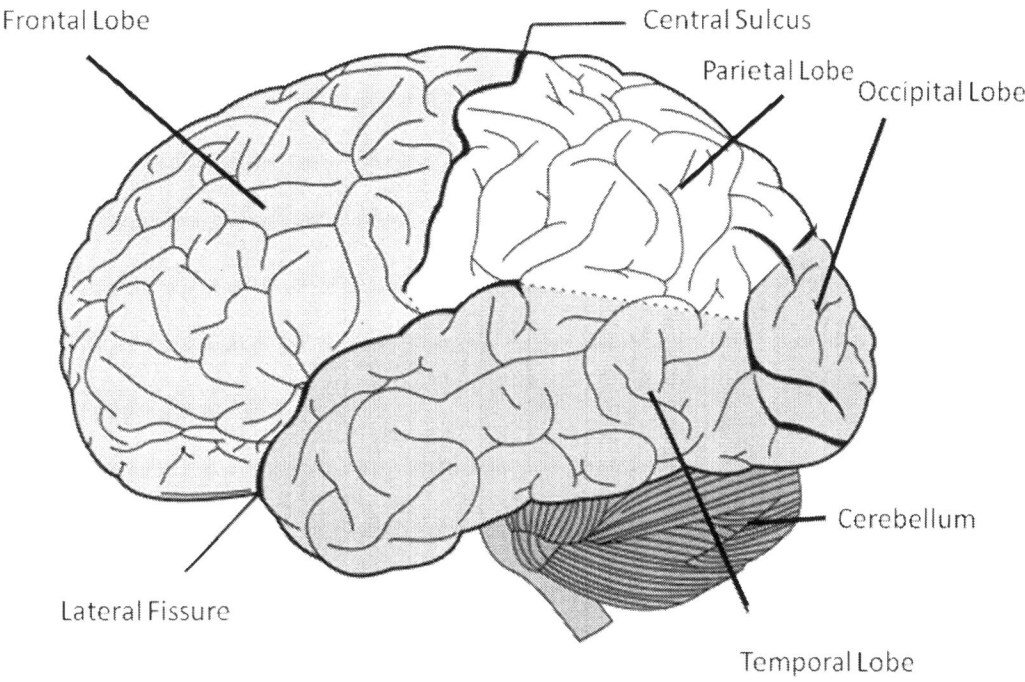

Figure 9.5. Lobes of cerebrum, lateral view.

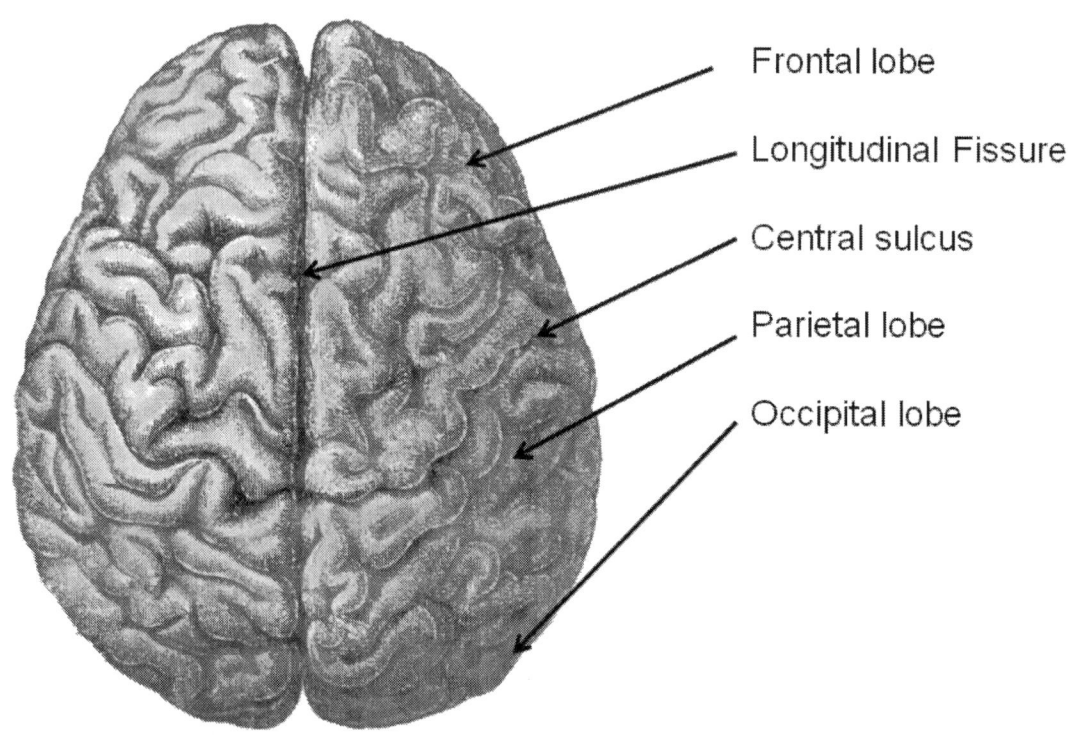

Figure 9.6. Cerebrum superior view.

Big Picture: The Diencephalon

The diencephalon consists of the thalamus and hypothalamus.

Some students struggle with the diencephalon. One way to think of it is that it lies between the cerebrum and the brain stem. It consists of some great nitty gritty structures including the thalamus, hypothalamus, subthalamus and epithalamus (how many thalami are there?). Most courses will just focus on the thalamus and hypothalamus.

The thalamus is like a relay station.

The thalamus is the largest part of the diencephalon. It consists of two lateral portions connected by a stalk called the interthalamic adhesion which is sometimes referred to as the intermediate mass. The thalamus carries all sensory information to the cerebral cortex with the exception of the sense of smell which is carried directly to the frontal lobe of the cerebral cortex by the olfactory nerves. The thalamus is sometimes referred to as a relay station for sensory information. The thalamus is also intimately involved in emotions due to its connections to the limbic system.

The hypothalamus is like a maintenance department.

The hypothalamus lies inferior and anterior to the thalamus. It contains the mamillary bodies on its anterior surface. The mamillary bodies process information associated with the sense of smell and emotions. A stalk-like projection called the infundibulum projects anterior and inferior and connects to the pituitary gland. The hypothalamus is intimately connected with the endocrine system and helps to regulate hormones. The hypothalamus also regulates body temperature, thirst, hunger and sexual drive and is involved in processing emotions, mood, and sleep along with the reticular activating system.

The epithalamus is located posterior and superior to the thalamus. It is a small area that works to process the sense of smell and emotional responses. The pineal body (gland) is also located in this area. It is a pine shaped structure that helps to regulate sleep-wake cycles by secreting the hormone melatonin.

The subthalamus is located inferior to the thalamus. It contains nuclei that are involved in controlling motor information.

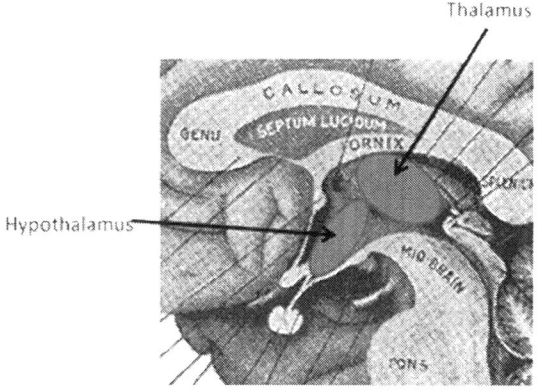

Figure 9.7. Diencephalon: thalamus and hypothalamus.

Big Picture: The Brainstem

The brainstem has 3 parts: the midbrain, pons and medulla oblongata.

The brainstem lies between the cerebral cortex and the spinal cord. It consists of the midbrain, pons and medulla oblongata. The medulla oblongata is the most inferior portion of the brainstem and contains a number of centers for controlling heart rate, respiration, swallowing, vomiting and blood vessel diameter. These centers consist of nuclei which are clusters of neuron cell bodies. The spinal tracts also continue through the medulla connecting the spinal cord with the brain. The medulla contains two rounded structures called olives (not the martini kind) which consist of nuclei that help to control balance, coordination and sound information. On the anterior surface of the medulla lie two enlargements called pyramids. The pyramids consist of the descending spinal cord tracts.

The pons is the middle section of the brainstem. The pons also contains spinal cord tracts as well as nuclei that help to control respiration and sleep.

The midbrain is the most superior portion of the brainstem. There is a roof (tectum) that contains four bumps (nuclei) called the corpora quadrigemina. The two superior nuclei (bumps on the top) are called the superior colliculi while the inferior are called the inferior colliculi (bumps on the bottom).

The superior colliculi help to control the movement of the head toward stimuli including visual, auditory, or touch. That means if you are a guy and a pretty girl walks by, your head turns and follows her by virtue of the superior colliculi. Next time you can tell your wife or girlfriend:

"Hey, I can't help it. It's just my superior colliculi!"

The superior colliculi receive input from the eyes. The inferior colliculi help to process hearing and also receive input from the skin and cerebrum. So again you can think of the function of the corpora quadrigemina as follows:

Pretty girl walks by causing gentleman's head to turn---superior colliculi.

Wife/girlfriend calls gentleman a jerk and hits him over the head—inferior colliculi.

The floor of the midbrain is called the tegmentum. It contains two reddish colored structures called the red nuclei that process information for unconscious motor movements. The midbrain also contains the cerebral peduncles (I like the term pee-duncle) that carry motor information from the cerebrum to the spinal cord. The substantia nigra resides in the midbrain and processes information relating to tone and coordination of muscles.

The reticular formation is located throughout the brainstem and is primarily concerned with regulating sleep-wake cycles.

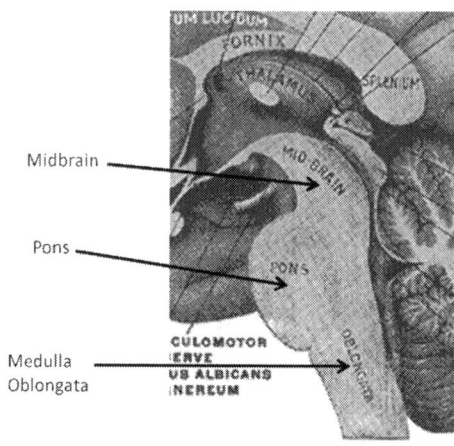

Figure 9.8. Brainstem.

Big Picture: The Cerebellum

The cerebellum processes information for balance, coordination, fine motor movements, and joint position.

The cerebellum is like a small version of the brain. It is located in the back of and below the cerebrum. It is connected to the brainstem via three cerebellar peduncles (there's that pee-duncle word again) (superior, middle and inferior peduncles). The cerebellum contains both gray and white matter. The white matter branches much like a tree and is called the arbor vitae.

The cerebellum contains a number of different types of neurons but one in particular; the Purkinjie cell is the largest cell in the brain. These cells can synapse with as many as 200,000 other fibers. Purkinjie cells are inhibitory cells and function in processing motor information.

The cerebellum can be divided into three parts. The flocculonodular lobe is the inferior portion. The vermis constitutes the middle portion and the two lateral hemispheres make up the remaining portion.

The cerebellum functions in processing information related to complex movements, coordination and unconscious proprioception.

Big Picture: Limbic System

Limbic system = emotions

Often called the seat of emotions, the limbic system consists of portions of both the cerebrum and diencephalon. The limbic system is also involved in reproduction and memory (hey emotions go with memory and reproduction). The limbic system contains the cingulate gyrus located just superior to the corpus callosum and the parahippocampal gyrus located on the medial aspect of the temporal lobe.

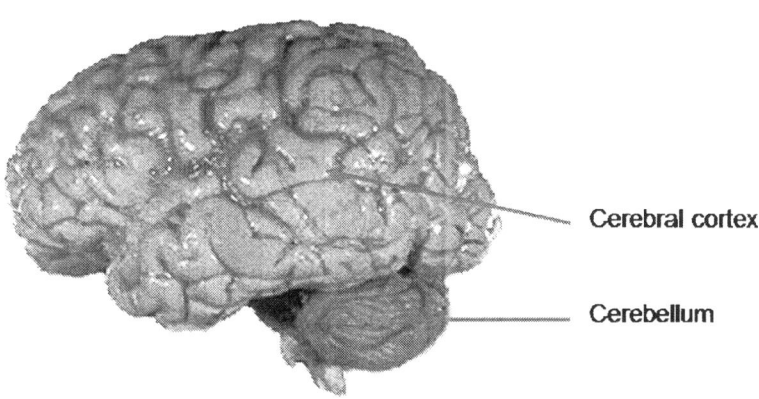

Figure 9.9. The cerebellum lies inferior to the cerebrum.

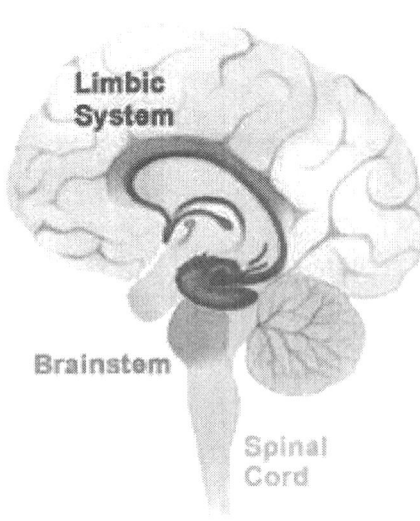

Figure 9.10. Limbic system.

Big Picture: The Cerebral Spinal Fluid System

Fluid circulates inside and outside of your brain and spinal cord.

Did you know that there are large hollow chambers inside of your brain? It's true and these chambers are called ventricles and contain a fluid known as cerebral spinal fluid or CSF.

Cerebral spinal fluid (CSF) has to come from somewhere, and lo and behold it comes from the good ole blood. It acts as a shock absorber and cushions the brain and spinal cord. CSF is produced by small vascular structures called choroid plexi (plexus).

The blood vessels in a choroid plexus form a blood-brain barrier between the blood and CSF. The capillaries inside of the brain also form a blood-brain barrier. Examples of substances that can pass through the blood-brain barrier include lipid soluble drugs and alcohol (hmm that explains why I have a hard time studying after a few drinks). Water soluble substances can also enter the brain via transport proteins.

CSF not only circulates in the subarachnoid space but also within the hollow ventricles. There are four total ventricles in the nervous system. There are two lateral ventricles separated by a fibrous membrane called the septum pellucidum, a third and fourth ventricle. The lateral ventricles are located within the cerebral hemispheres. The third ventricle lies between the two halves of the thalamus in the diencephalon. The fourth ventricle lies between the brainstem and the cerebellum.

The ventricles are all connected via foramen (holes) or tubular passages. The lateral ventricles connect to the third ventricle via the interventricular foramen. The third ventricle connects to the fourth via a tube passing through the midbrain called the cerebral aqueduct (aqueduct of Sylvius). The fourth ventricle connects with the central canal of the spinal cord. The fourth ventricle also connects with the subarachnoid space via lateral and medial apertures. The median aperture is called the foramen of Magendie and the two lateral apertures are called the foramen of Luschka.

CSF is produced by the choroid plexi that make about 500 ml/day. However some of the CSF is absorbed so there is only about 140 ml in the system at any one time. This is due to the CSF being absorbed by arachnoid granulations. Arachnoid granulations are masses of arachnoid tissue located in the dural venous sinuses. CSF can move into the blood at these locations.

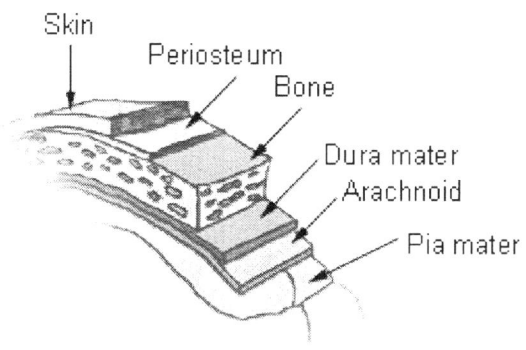

Figure 9.11. The brain has 3 layers of meninges.

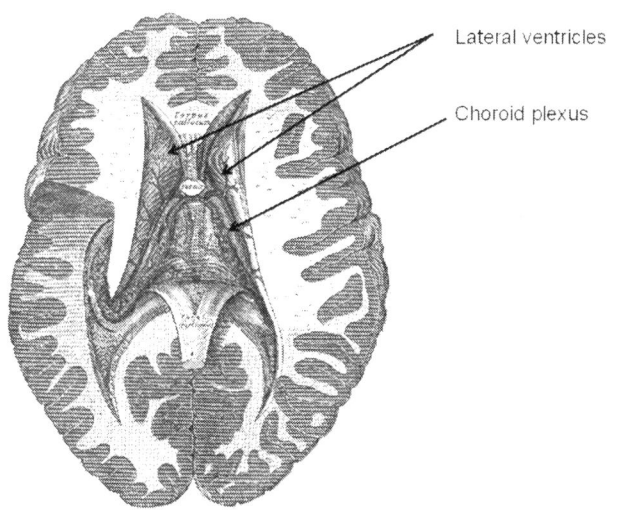

Figure 9.12. Lateral ventricles of the brain.

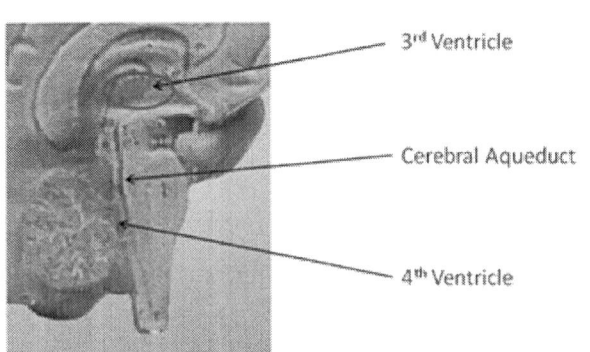

Figure 9.13. Third and fourth ventricles of brain.

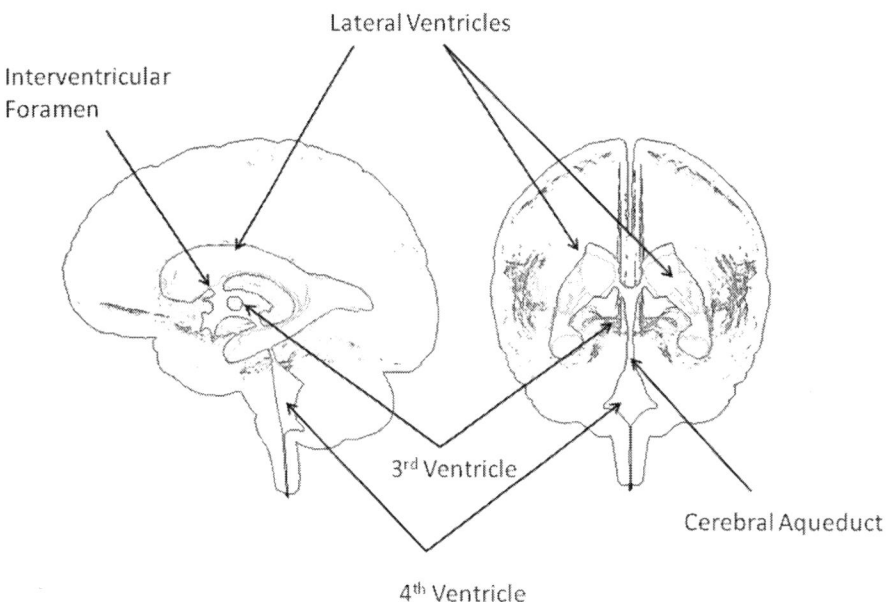

Figure 9.14. CSF circulatory structures.

Here is a summary of the flow of CSF in the brain:

CSF is produced by the choroid plexi and flows between the two lateral ventricles (in the cerebral hemispheres) through the interventricular foramen to the 3rd ventricle (near the thalamus) through the cerebral aqueduct to the 4th ventricle (between the brainstem and cerebellum) and on to the spinal cord (fig. 9.12-9.14).

Big Picture: The Cranial Nerves

Cranial nerves carry sensory and motor information mostly from the face.

There are 12 pair of cranial nerves. Eleven of these originate in the diencephalon or brainstem while one pair originates in the frontal lobe of the brain. The cranial nerves can carry sensory information, motor information or both. The sensory information consists of touch, pain, taste, hearing and vision. Motor information controls skeletal muscles, organs and glands. Some cranial nerves also carry information for the parasympathetic nervous system. The cranial nerves are usually designated as Roman numerals (I—XII).

Anatomy students really need to get the name, Roman numeral and at least some functions down cold. Here are a couple of mnemonics to help you remember the cranial nerves:

Cranial Nerves

This was the one I used to memorize them.

> I On (Olfactory)
> II Old (Optic)
> III Olympus' (Oculomotor)
> IV Towering (Trochlear)
> V Tops, (Trigeminal)
> VI A (Abducens)
> VII Finn (Facial)
> VIII And (Auditory)
> IX German (Glossopharyngeal)
> X Viewed (Vagus)
> XI Astounding (Accessory)
> XII Hops (Hypoglossal)

Here is one that's a bit more risqué that has to do with the type of cranial nerve.

> I Some (Sensory)
> II Say (Sensory)
> III Marry (primarily Motor)
> IV Money, (primarily Motor)
> V But (Both)
> VI My (primarily Motor)
> VII Brother (Both)
> VIII Says (Sensory)
> IX Big (Both)
> X Bras (Both)
> XI Matter (primarily Motor)
> XII More (primarily Motor)

Here is an overview of the function of the cranial nerves to help you get a handle on functions:

Cranial Nerve		Type	Function
I.	Olfactory	Sensory	Carries information for sense of smell to frontal lobe.
II.	Optic	Sensory	Carries information for sense of sight to occipital lobe.
III.	Occulomotor	Motor	Innervates eye muscles.
IV.	Trochlear	Motor	Innervates superior oblique muscles of eye.
V.	Trigeminal	Mixed	Carries sensory information from face. Motor to masseter and temporalis muscles (chewing muscles).
VI.	Abducens	Motor	Innervates lateral rectus muscles of eye.
VII.	Facial	Mixed	Carries sensory information from anterior 2/3 of tongue. Motor to muscles of facial expression.
VIII.	Auditory (Vestibulocochlear)	Sensory	Carries information for sound, static and dynamic equilibrium from inner ear.
IX.	Glossopharyngeal	Mixed	Carries sensory information for taste from posterior 1/3 of tongue. Motor to swallowing muscles.
X.	Vagus	Mixed	Sensory information from esophagus, respiratory and abdominal viscera. Motor to heart, stomach, intestines and gallbladder. Also helps to coordinate swallowing.
XI.	Spinal Accessory	Motor	Motor to trapezius and sternocleidomastoid muscles.
XII.	Hypoglossal	Motor	Motor to tongue.

The Big Picture: Autonomic Nervous System

The autonomic nervous system has two parts. The sympathetic division pumps you up (fight or flight) while the parasympathetic division calms you down.

The autonomic nervous system can be thought of as an "automatic" system because it works to maintain homeostasis in the body even when it is in an unconscious state. The autonomic nervous system (ANS) can control respiratory, cardiovascular, urinary, digestive and reproductive functions. It works to maintain balance of fluids, electrolytes, blood pressure, nutrients, and blood gasses. The ANS does this by sending motor impulses to viscera, cardiac and smooth muscle. Since it sends motor impulses to viscera, the ANS is also known as a visceral motor system.

The ANS is divided into two subdivisions. The sympathetic is often referred to as the "fight or flight" system. The sympathetic division emerges from the thoracic spine. The parasympathetic division begins in the cervical and lower thoracic spine and sends fibers to the same organs as the sympathetic. The sympathetic and parasympathetic divisions typically have the opposite effect on organs and thus work to maintain balance based on the body's needs. For example the sympathetic system can increase heart rate while the parasympathetic system decreases it.

Here is a table that summarizes the function of the autonomic nervous system.

Organ	Autonomic Innervation	Type of Receptor	Action
Eye : pupil	sympathetic	alpha	dilation of the pupil
	parasympathetic	muscarinic	constriction of the pupil
Eye : ciliary muscle	sympathetic	beta	allows far vision
	parasympathetic	muscarinic	allows near vision
Tear glands	sympathetic	beta	vasoconstriction
	parasympathetic	muscarinic	secretion of tears
Salivary glands	sympathetic	alpha	vasoconstriction and secretion of mucous with a low enzyme count
	parasympathetic	muscarinic	secretion of watery saliva with a high enzyme count
Heart	sympathetic	beta	dilation of coronary arteries, increased heart rate, increased force of contraction, increased rate of pacemaker conduction
		alpha	coronary artery constriction

	parasympathetic	muscarinic	slows, heart rate, reduces contraction and conduction, constricts coronary arteries
Bronchii	sympathetic	beta	dilation
	parasympathetic	muscarinic	constriction and mucous secretion
Esophagus	sympathetic	alpha	vasoconstriction
	parasympathetic	muscarinic	peristalsis, secretion of mucous
Stomach and Intestines	sympathetic	beta	inhibition of peristalsis and secretion
		alpha	vasoconstriction, spinctre contraction
	parasympathetic	muscarinic	peristalsis and secretion
Spleen	sympathetic	alpha	contraction
Adrenal medulla	sympathetic	-	adrenaline and noradrenaline secreted into the bloodstream
Liver	sympathetic	beta	break down of glycogen (glyogenolysis)
Gall Bladder	sympathetic	beta	relaxation
	parasympathetic	muscarinic	contraction
Pancreas	sympathetic	alpha	inhibition of insulin secretion
		beta	stimulation of insulin secretion
Descending colon	sympathetic	alpha	vasoconstriction
		beta	inhibition of peristalsis and secretion
	parasympathetic	muscarinic	peristalsis and secretion
Sigmoid colon, rectum and anus	sympathetic	alpha	constriction of sphincter muscles
		beta	inhibition of peristalsis and secretion
	parasympathetic	muscarinic	peristalsis and secretion
Bladder	sympathetic	alpha	contraction of sphincter
		beta	relaxation of detrusor muscle
	parasympathetic	muscarinic	contraction of detrusor muscle
Penis	sympathetic	-	ejaculation
	parasympathetic	muscarinic	erection
Clitoris	parasympathetic	muscarinic	erection
Uterus	sympathetic	alpha	contraction

		beta	relaxation
Blood vessels in:			
Skin	sympathetic	alpha	constriction
Muscle	sympathetic	cholinergic	dilation
Sweat glands except palm of hands	sympathetic	muscarinic	sweating
sweat glands on palms of hands	sympathetic	alpha	sweating
Arector pili muscles at root of body hair	sympathetic	alpha	piloerection (making hair "stand on end")
Adipose tissue	sympathetic	beta	lipolysis (break down of fat to release energy)

Big Picture: Neurons

Neurons are nervous system cells that talk to each other.

We've been looking at the large structures of the nervous system up until now. However the real fun is in learning how the bits and pieces work. We will now take a look at the small structures of the nervous system, namely the neurons.

The neuron is the carrier of information in the nervous system. The neuron has a number of parts but there are three basic parts you should learn:

Cell body—sometimes called the perikaryon contains many of the cell organelles we described in chapter two. These include mitochondria, microtubules, Golgi apparatus, and a granular cytoplasm. The cell body also contains Nissl Bodies which are membranous packets of chromatophilic substance consisting of rough endoplasmic reticulum (remember, this makes proteins).

Dendrites—extensions of the cell body that typically receive information from other neurons.

Axon—a long process that transmits information to other neurons. Some axons contain a wrapping called the myelin sheath. The axon connects to the cell body via a structure called the **axon hillock.**

There are a number of different types of neurons but the one we will be learning about is called the multipolar neuron.

Neuroglia

The nervous system also contains cells that support neurons called neuroglia. There are a few different types of neuroglia that have a number of important functions.

Astrocytes--provide structural support and may also help in regulating electrolytes. They are star shaped and can be found between neurons and blood vessels. Astrocytes help to maintain the blood-brain barrier and help to repair damaged areas in the central nervous system by forming scar tissue.

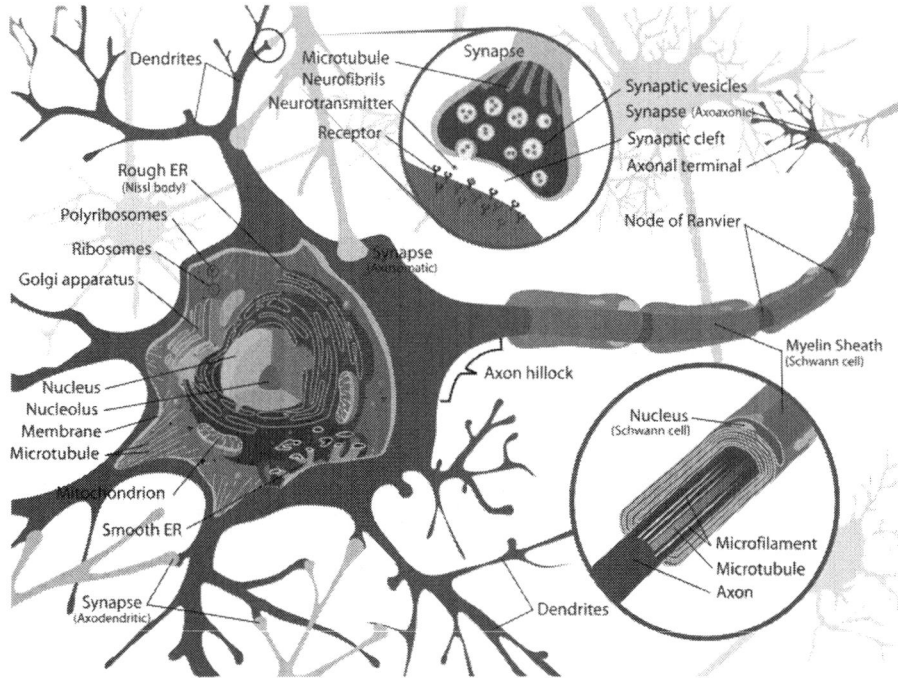

Figure 9.15. Multipolar neuron.

Oligodendrocytes-- produce the myelin that surrounds white matter axons in the brain and spinal cord.

Empendymal cells-- form the lining of the central canal and ventricles in the spinal cord and brain. They are also found in the choroid plexi of the brain. They help to produce CSF by providing a porous membrane for blood plasma to pass through.

Microglia-- are very small cells that are located throughout the central nervous system. They provide support and help to clean up debri through phagocytosis.

Connecting Neurons

There are literally billions and billions of connections in your nervous system. One neuron can have up to 10,000 connections with other neurons. The axon of one neuron can connect with the dendrites of another neuron. The connection is called a **synapse**.

The first neuron in a 2-neuron system is called the **pre-synaptic neuron** and will secrete a substance called a neurotransmitter that floats across what is called the synaptic cleft to the second neuron called the **post-synaptic neuron**. The neurotransmitter is then picked up by the dendrites of the post-synaptic neuron.

It is important to know that neurons can only send one of two different messages:

Either a neuron can send a message to move the message forward or to hold the message back.

More about this later…

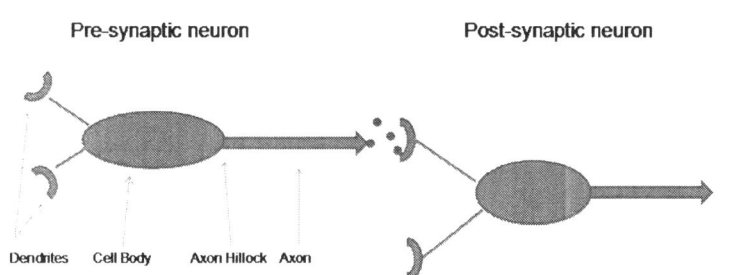

Figure 9.16. Neurons can connect with other neurons to send messages in the form of neurotransmitters.

Big Picture: How Neurons Work

Neurons talk to each other by generating electrical messages called action potentials that result in sending messages called neurotransmitters to one another.

The Resting Membrane Potential

This can be a difficult concept to learn completely so I will attempt to simplify it. Remember in the big picture approach it is necessary to see the big picture first before seeing all of the details.

Let's begin by saying that neurons don't like to exist in a state of equilibrium. They are a bit unbalanced. They get unbalanced because their membranes are permeable to some substances but not to others. It just so happens that some of those substances can move out of the cell while others are trapped in. Since these substances have charges the cell is left with a negative charge on the inside of its membrane.

The substances we are talking about are mostly sodium (Na+) and potassium (K+). This brings us to two important rules in human physiology:

There is always more sodium outside the cell than inside.

There is always more potassium inside the cell than outside.

Now let's say there are some negative ions trapped inside the cell. And let's say that potassium can move out of the cell (remember there's more potassium inside than out). Both of these conditions work to provide more positive charges outside the cell than inside.

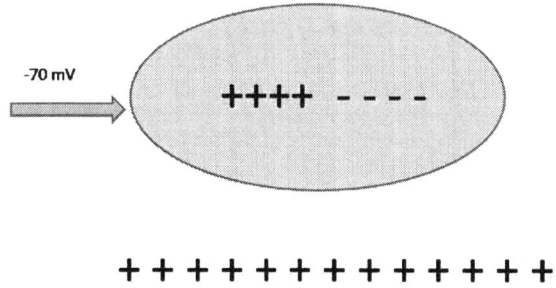

Figure 9.17. The resting membrane potential exists because the cell membrane is negatively charged as compared to the fluid surrounding it. There are more positive charges outside the cell than inside.

What happens next is that something stimulates the cell so that sodium gates open. Following our rule that there is more sodium outside the cell than inside; sodium rushes into the cell as it moves down its concentration gradient. This produces the condition in our next diagram.

This process is called **depolarization.** It is best to associate sodium with depolarization. Okay so say it with me three times:

Sodium = Depolarization--- Sodium = Depolarization--- Sodium = Depolarization

Why is it called depolarization? Well when the cell exists in its unbalanced state it is said to be polarized. Perhaps you have a relative that takes an opposite stance on an issue than you.

You could say that with regard to that issue you and your relative are at polar opposites or polarized. The movement of sodium into the cell tends to reduce this difference in charge from outside to inside. This would be like you and your relative agreeing or finding some middle ground on the issue. We say the cell is becoming less polar or **depolarizing.**

The cell will continue to depolarize all the way to + 30 millivolts.

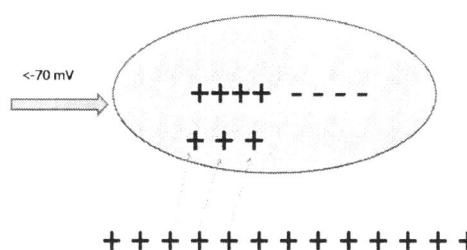

Figure 9.18. Positively charged sodium ions move into the cell causing the cell to become less negative.

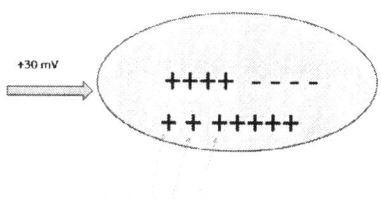

Figure 9.19. The movement of sodium inside the cell causes it to depolarize all the way to +30 mV.

Next, the sodium gates close and potassium gates take over causing potassium to move out of the cell (actually the potassium gates were open all of the time but the net effect of potassium moving out was negligible as compared to sodium moving in).

We say the cell is repolarizing.

The cell repolarizes all the way back to -70mV.

So where is the action potential?

As the cell depolarizes from -70mV it becomes less negative. At some point it will reach -55mV. This is an important number because something special happens at -55mV.

-55mV is called the Threshold.

The something special that happens at the threshold is that a large number of sodium gates open causing the cell to rapidly depolarize all the way to +30 mV. Once the cell depolarizes to -55mV it can't be stopped.

The action potential is the rapid change in voltage from -55mV to +30 mV.

What happens if the cell depolarizes but doesn't reach the threshold?

Nothing. The cell just repolarizes back to -70mV.

Where are all these sodium gates that open when the threshold is reached?

They are located in the axon hillock—where the axon connects to the cell body.

What happens if potassium gates are opened when the cell is at resting membrane potential?

The cell goes into what is called hyperpolarization and no action potential is produced.

What causes the cell to depolarize or hyperpolarize in the first place?

Neurons respond to substances known as neurotransmitters. Keep in mind our physiology rule:

Neurotransmitters can do one of two things. They either move the message forward or hold the message back.

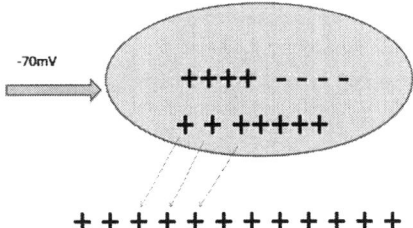

Figure 9.20. Positively charged potassium moves out of the cell causing it to become more negative once again. The cell repolarizes.

In other words there are **excitatory** and **inhibitory** neurotransmitters. Excitatory neurotransmitters move the message forward by depolarizing the neuron and producing and action potential. Inhibitory neurotransmitters hold the message back by hyperpolarizing the neuron and keeping it from producing an action potential. Remember that depolarization has to do with opening sodium gates and hyperpolarization has to do with opening potassium gates.

How the Action Potential Moves Down the Axon

There are two different types of axons. Some consist of grey matter while others have a fatty substance called myelin surrounding them. You could think of the difference as being similar to bare wires as compared to insulated wires.

In the grey matter axons (bare wire) the action potential moves down in a wave-like motion. Sections of the axon depolarize causing subsequent sections to depolarize. We say the action potential propagates down the axon.

In the white matter axons (insulated wire) action potentials move in a different way but before we can discuss this we need to look at some other differences of white matter axons. The myelin substance in white matter axons is produced by special cells called Schwann cells. Since there are a number of Schwann cells that make up the myelin sheath there are gaps in the myelin. These gaps are very special and even have a special name. They are called "Nodes of Ranvier (Rhan vee ay)."

Not only do these nodes have a special name, they also have a special structure. There are large numbers of sodium gates located at these nodes and they open in response to an action potential. When the action potential reaches a node, the sodium gates open causing sodium to rush into the axon. The end result is the impulse appears to jump from node to node. This also has a special name. This type of movement of an action potential along a myelinated axon is called **saltatory conduction**. One way to help remember this is to think of the *salt* in saltatory which has to do with sodium gates.

What is the purpose of myelinated axons?

It turns out that saltatory conduction in myelinated axons is much faster than unmyelinated axons. Nerve impulses have to travel a long distance in some neurons. Think of an impulse going from the spinal cord to your big toe. That's a lot of distance to cover. Myelinated axons are more numerous in areas that require this fast conduction.

We can illustrate the difference in conduction speed of myelinated versus unmyelinated axons with an analogy. We have two rows of students with 12 students in each row. The student at the end of each row has a ball and has to get it to the student to the other end of the row as quickly as possible. The first row is given instructions to pass the ball from one student to the next until reaching the last student. The second row is told to throw the ball to the 4th student who then throws it to the 8th student and so on until reaching the 12th student at the end of the row. The instructor tells the students to begin at the same time. Which row will win the race? Obviously the second row wins because time is lost with the handling of the ball by every single student in the row versus every 4th student. The first row then represents an unmyelinated axon while the second row represents a myelinated axon.

Likewise when the myelin becomes damaged (demyelination) the impulse does not conduct as rapidly. This can result in numbness, tingling, pain and even loss of muscle function. One example of this problem is seen with carpal tunnel syndrome.

Myelinated Axon

Action potential appears to jump from node to node.

Figure 9.21. In saltatory conduction the impulse appears to jump from node to node.

Release of the Neurotransmitter

Once the impulse reaches the end of the axon it enters a structure called the axon terminal. This is where the neurotransmitter is located that will carry the message to the next neuron. Upon reaching the axon terminal the action potential causes another ion gate to open. This time the ion is calcium. Since these calcium gates respond to the action potential they are often referred to as voltage gated calcium channels.

Once these calcium channels open calcium rushes into the axon terminal causing the release of the neurotransmitter from a package called a synaptic vesicle. The neurotransmitter then moves across the synaptic cleft to the next neurotransmitter.

Examples of Neurotransmitters

Here are a few examples of neurotransmitters.

Name of Neurotransmitter	Effects
Acetylcholine	Decreases heart rate Increases secretions of sweat and saliva Facilitates muscle contractions
Norepinephrine	Increases heart rate Decreases circulation of blood
Dopamine	Increases happiness and alertness
GABA	Decreases anxiety, alertness, muscle tension and memory
Serotonin	Increases well being and happiness Decreases pain

Big Picture: Reflexes

Reflexes are involuntary responses to sensory stimuli.

One of the simplest communication structures in the nervous system is the 2-neuron reflex arc. Think of going to the doctor and having her tap your leg just below your knee. Your usual response is to briefly straighten your leg in a jerking motion. Sometimes this is referred to as a monosynaptic (one synapse) reflex.

In this reflex we have only two neurons. There is a sensory neuron that carries the message that the muscle length has changed to the spinal cord. There is also a motor neuron that carries the message from the cord back to the muscle.

When you strike the tendon of a muscle (say the patellar tendon just below the knee) it causes a brief lengthening of the muscle. This is sensed by a special receptor in the muscle called a muscle spindle. The receptor takes the information about stretch to the spinal cord where it relays it to another neuron. This neuron is a motor neuron that carries the message to contract the muscle (to take up the slack) back to the muscle. The result is the familiar knee jerk.

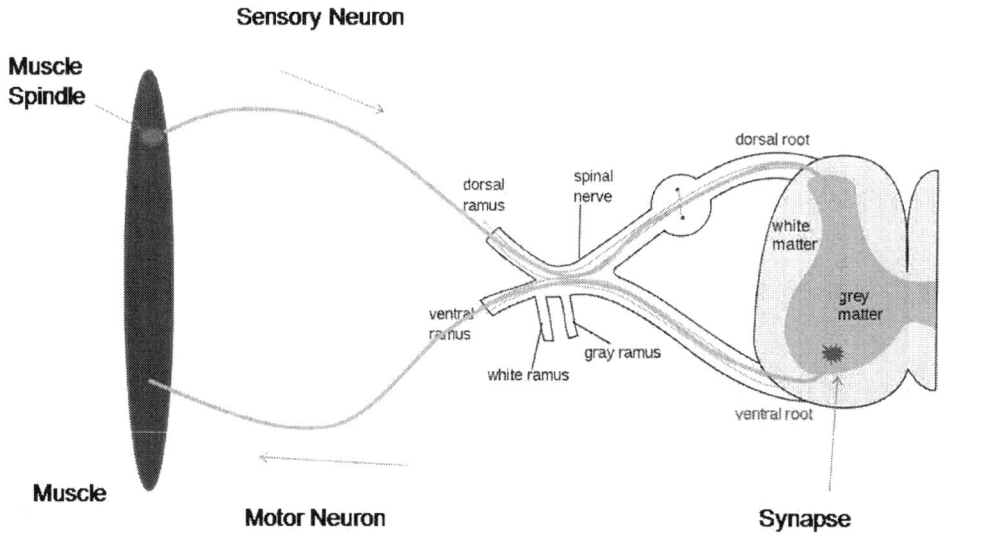

Figure 9.22. Reflex Arc.

Spilled Coffee Revisited

Now that you know a lot more about the nervous system we can take a detailed look at how information flows through the nervous system when spilling a cup of hot coffee on my hand and deciding to put it down.

I pull up to the drive through and reach for the cup of hot coffee. As I grab the cup the lid pops off and hot coffee spills on my hand. The sensory information was picked up by temperature and pressure receptors in my hand. The information was converted to electrochemical information called action potentials by the sensory receptors and sent along afferent pathways called nerve to the spinal cord. The spinal nerves transmit the sensory information via the dorsal roots to the posterior portion of the spinal cord so that it can travel via ascending tracts to the brain.

The first neuron to carry the information from sensory receptor to the spinal cord is called a primary neuron. This neuron synapses with a secondary neuron that carries the information from the spinal cord to the brain.

The temperature information travels via the lateral spinothalamic tract in a secondary neuron. This tract is located in the lateral funiculus of the spinal cord and crosses at cord level to the contralateral side. So if I am holding the cup in my left had (which I frequently do) then the information travels via a cervical spinal nerve (or nerves) to the spinal cord via the dorsal roots to the spinothalamic tract that crosses over to the right side of the cord.

The information then ascends to the brainstem (medulla oblongata, pons, midbrain) and synapses with a tertiary neuron in the thalamus. The information then goes to the post-central gyrus of the parietal lobe for processing. The parietal lobe sends the information to other areas of the cortex for interpretation and decision making.

The pressure information is picked up by sensory receptors in the skin (Meissner's and Pacinian corpuscles) and sent via a primary neuron to the spinal cord as well. There it synapses with a secondary neuron in the cord and is carried to the thalamus via the dorsal column pathway consisting of the fasiculus gracilis/cunneatus located in the posterior funiculus of the spinal cord. These tracts cross over (decussate) in the medulla oblongata and synapse in the thalamus with a tertiary neuron. This neuron carries the information to the post-central gyrus as well.

Now aside from any reflex activity (withdrawal or crossed extensor) I must interpret the sensory information using association areas in the cortex and make a decision using my frontal lobe to set the cup down. The premotor areas will process the decision and work to coordinate the actions of moving the appropriate muscles of my arm to set the cup down. They will send the information to the precentral gyrus of the frontal lobe.

The information then reaches the prefrontal gyrus (on the right side) of the frontal lobe and travels toward the brain stem bypassing the thalamus. The primary or upper motor neuron here begins in the precentral gyrus and travels via descending spinal tracts after crossing in the medulla oblongata. The descending tracts include the corticospinal tract. The neurons synapse with lower motor neurons in the anterior horn of the spinal cord.

The lower motor neurons carry the information to the skeletal muscles via the ventral root of spinal nerves to spinal nerves to the brachial plexus to terminal nerves to the muscles of my arm and hand. I then can set the cup down.

Memory

Big Picture: Memory

There are basically 2 types of memory. Short-term memory allows you to remember small bits of information for brief periods. Long-term memory allows for storing lots of information for long periods of time.

The brain is capable of storing vast amounts of information in its memory. There are two basic types of memory. Short-term memory stores 6-8 pieces of information for brief periods. For example a telephone number is 7 pieces of information long and can be stored for a short amount of time until the person is asked to remember something else. Long-term memory as its name implies allows for storage of information for much longer periods of time (as long as a lifetime). Types of long-term memory include declarative and procedural. Declarative is sometimes referred to as explicit and procedural is referred to as implicit.

Declarative memory occurs in part of the temporal lobes and the hippocampus and amygdala. The hippocampus is involved in retrieving stored memories whereby the amygdala stores emotions associated with memories. Declarative memory is also stored in various parts of the cerebrum. Memories are grouped together as well. For example, faces may be stored in a different location than names. Retrieving a memory involves accessing various components and assembling them. Over time memories decay and can lead to false memories.

Procedural memory involves storing skills such as playing an instrument or driving a car. Procedural memories are stored in the premotor area of the cerebrum and cerebellum.

Information to be remembered moves from short-term to long-term memory. Neurons in long-term memory actually change in response to storing information. The phenomenon of long-term potentiation occurs when memories are stored. This involves changes in neurotransmitter storage and release as well as protein synthesis. New connections are made and maintained between neurons. This flexible and adaptive characteristic of the brain is known as neural plasticity.

Review Questions

1. The middle layer of meninges is called:
   a. Pia mater
   b. Dura mater
   c. Arachnoid mater
   d. Visceral mater

2. This structure lies between the arachnoid and pia mater:
   a. Epidural space
   b. Spinal cord
   c. Subarachnoid space
   d. Choroid plexus

3. White matter in the spinal cord is divided into:
   a. Triangles
   b. Funiculi
   c. Horns
   d. Dendrites

4. Which spinal tract carries pain and temperature information:
   a. Spinocerebellar
   b. Fasciculus gracilis
   c. Lateral spinothalamic
   d. Rubrospinal

5. Which of the following tracts carry fibers that do not cross in the spinal cord:
   a. Spinothalamic
   b. Fasciculus cunneatus
   c. Spinocerebellar
   d. Fasciculus gracilis

6. Which of the following tracts carries motor information for posture and coordination:
   a. Corticospinal
   b. Spinothalamic
   c. Rubrospinal
   d. Reticulospinal

7. Which branch of a spinal nerve is considered the autonomic nervous system branch:
   a. Dorsal
   b. Visceral
   c. Ventral
   d. Recurrent menigeal

8. Which of the following brain structures helps to regulate sleep/wake cycles:
   a. Superior colliculus
   b. Pons
   c. Reticular formation
   d. Medulla oblongata

9. Which is the most superior portion of the brainstem:
   a. Pons
   b. Midbrain
   c. Medulla oblongata
   d. Cerebral aqueduct

10. Which of the following is considered the middle portion of the cerebellum:
    a. Vermis
    b. Flocculonodular lobe
    c. Lateral hemispheres
    d. Arbor vitae

11. This structure consists of a tree-like arrangement of white matter:
    a. Vermis
    b. Flocculonodular lobe
    c. Lateral hemispheres
    d. Arbor vitae

12. This structure is a stalk-like projection that connects the hypothalamus with the pituitary gland:
    a. Mamillary body
    b. Superior colliculi
    c. Infundibulum
    d. Inferior colliculi

13. Folds on the surface of the cerebrum are known as:
    a. Sulci
    b. Gyri
    c. Rugae
    d. Cerebri

14. Which of the following lobes primarily processes information related to concentration, planning and problem solving:
    a. Temporal
    b. Occipital
    c. Frontal
    d. Parietal

15. Which lobe primarily processes information related to vision:
    a. Temporal
    b. Occipital
    c. Frontal
    d. Parietal

16. The central sulcus divides which 2 lobes:
    a. Frontal, occipital
    b. Parietal, temporal
    c. Both parietal
    d. Frontal, parietal

17. Which structure connects the 3rd and 4th ventricles of the brain:
    a. Cerebral aqueduct
    b. Interventricular foramen
    c. Choroid plexus
    d. Arachnoid villi

18. Which structure reabsorbs CSF:
    a. Choroid plexus
    b. Cerebral aqueduct
    c. Arachnoid villi
    d. Pia mater

19. Which cranial nerve is responsible for moving facial muscles:
    a. Trigeminal
    b. Facial
    c. Spinal accessory
    d. Hypoglossal

20. Which cranial nerve carries information for taste from the anterior 2/3 of the tongue:
    a. Vagus
    b. Facial
    c. Hypoglossal
    d. Trigeminal

21. Which cranial nerve carries information regarding balance and equilibrium:
    a. Vagus
    b. Trigeminal
    c. Vestibulocochlear
    d. Spinal accessory

22. Which cranial nerve is motor to the trapezius muscle:
    a. Trigeminal
    b. Spinal accessory
    c. Facial
    d. Vagus

23. Which cranial nerve is motor to the tongue:
    a. Spinal accessory
    b. Facial
    c. Hypoglossal
    d. Vagus

24. Most post-ganglionic parasympathetic neurons secrete the neurotransmitter:
    a. Norepinephrine
    b. Acetylcholine
    c. Epinephrine
    d. Dopamine

25. Which of the following is not an effect of the sympathetic nervous system:
    a. Pupils dilate
    b. Heart rate increases
    c. Respiration increases
    d. Digestion increases

26. Which type of neuron secretes myelin in the central nervous system:
    a. Bipolar neuron
    b. Schwann cell
    c. Oligodendroctye
    d. Multipolar neuron

27. The resting membrane potential of a neuron is typically:
    a. +30 mV
    b. -70 mV
    c. -55 mV
    d. Neutral

28. Which of the following best describes depolarization of a neuron:
    a. Sodium gates open and sodium enters the cell
    b. Potassium gates open an potassium enters the cell
    c. Sodium gates open and sodium leaves the cell
    d. Calcium gates open and calcium enters the cell

29. The afterpotential is caused by:
    a. Sodium gates remaining open
    b. Potassium gates remaining open
    c. Calcium gates remaining open
    d. Potassium gates closing

30. Which best describes saltatory conduction:
    a. Action potential moves down the axon in a wavelike fashion.
    b. Action potential appears to jump from node to node
    c. Action potential moves down a section then stops for a brief period
    d. Action potential resets midway down an axon

31. Which gate is responsible for releasing the neurotransmitter:
a. Sodium
b. Calcium
c. Potassium
d. Chloride

32. An inhibitory post-synaptic potential is characterized by the opening of:

a. Potassium gates
b. Sodium gates
c. Calcium gates
d. Acetylcholine gates

33. The withdrawal reflex incorporates the use of neurons called:

a. Neuroglia
b. Oligodendrocytes
c. Astrocytes
d. Interneurons

34. The sense of smell is processed in which part of the brain:
a. Temporal lobe
b. Parietal lobe
c. Occipital lobe
d. Frontal lobe

35. Which specialized area of the brain has to do with speech recognition:
a. Parietal lobe
b. Broca's area
c. Occipital lobe
d. Wernicke's area

36. Short-term memory can handle about ____ pieces of information:
a. 4-6
b. 6-8
c. 8-10
d. 10-12

Chapter 10

Making Sense of the Sensory System

Chapter 10

Making Sense of the Sensory System

It would be a miserably bleak existence if it wasn't for our senses. Our senses take in information from outside (and inside too) of our bodies and convert it to a form our nervous systems can understand. Our sensory systems work kind of like computers. In a computer system there is the CPU or central processing unit, you know the big box thing that you connect everything to.

Let's say all you have is a CPU and you plugged it in and turned it on—eureka—nothing happens. There is no monitor, no keyboard, no mouse. It just sits there, loads the software and waits for something to happen. The CPU is a lot like the CNS (central nervous system) and the peripherals are like our senses.

Okay so now we hook up a keyboard, mouse and monitor to our CPU. We start to tap on the keys and click away on the mouse. What is happening is the information from pressing buttons and keys is converted to electrical impulses that are recognized by the CPU. The information has been converted from movements of buttons to the movement of electrons. This is just what happens in the sensory system.

Our senses sense changes in the environment by the actions of a number of structures called sensory receptors. There are of course the obvious eyes, ears, smell and taste receptors. In fact these are called the special senses. There are a number of other receptors as well.

The Big Picture: Sensory System

All sensors do the same thing. They take information in many forms and convert it to electrochemical impulses so it can be sent to the nervous system.

Here is a list of some of the sensory receptors:

Chemoreceptors sense changes in chemical concentration.

Pain receptors (nociceptors) sense tissue damage.

Thermoreceptors sense changes in temperature.

Mechanoreceptors sense mechanical deformation of tissue.

Proprioceptors sense changes in position of joints.

Stretch receptors sense changes in tissue length.

Photoreceptors sense changes in light intensity.

Somatic Sensory System

We can divide the sensory system into the senses that sense the skin, muscles, joints and organs and the special senses we mentioned above. The somatic senses include touch, pressure, temperature, pain, and stretch.

Here is a list of specific receptors:

Free nerve endings sense touch and pressure and produce pain.

Merkel's discs sense fine touch and pressure.

Pacinian corpuscles sense heavy pressure.

Ruffini corpuscles sense skin movement and pressure.

Two Important Receptors in Muscles

There are two important sensory receptors located in muscles. These are the muscle spindles and Golgi Tendon Organs.

Muscle spindles are located more centrally in muscles. These are the receptors involved in the reflexes we described in the nervous system chapter. Muscle spindles sense stretch or the change in muscle length.

Golgi tendon organs (GTOs) are located at the ends of the muscle near the origins and insertions. These act as a protective mechanism to help keep muscles from pulling off the bone with extreme forces. For example, think of helping a friend move to a new house. You are holding an empty box while your friend fills it full of heavy objects. Eventually the box becomes so heavy that it triggers the GTOs and your arm muscles relax. The box drops to the floor and injures your foot (see joint section for information on sprains and strains).

Big Picture: The Special Senses

The special senses include taste, smell, hearing and vision.

As we mentioned previously, the special senses include taste, smell, hearing and vision. Taste and smell are very similar in that they are both sensed by chemoreceptors. This is why taste and smell are so closely related. Have you ever had a bad cold or sinus infection where you couldn't taste food? Smell actually accounts for a large percentage of taste. Your sense of smell can distinguish between 2000-4000 different smells.

The sense of smell is also closely related to the limbic system (the seat of emotions). This is why certain smells can trigger memories (remember grandma's apple pie and how content you were after eating it?).

The organs for taste reside on the tongue (where else?). There are also taste buds on the walls of the cheek and the throat. You have about 3000 taste buds.

A number of textbooks say there are four primary taste sensations. Tastes are combinations of these four primary sensations.

1. Sweet
2. Sour
3. Salty
4. Bitter

However there is a 5th taste sensation called umami. This one is a hearty, meaty taste that is produced by L-glutamate (think monosodium glutamate). In fact, you can experience all of the tastes by visiting your local Chinese buffet.

Here is an example:

1. Sweet---sweet and sour pork
2. Sour—sweet and sour pork
3. Salty—hey the eggdrop soup is too salty!
4. Bitter—the tea is bitter
5. Umami—just about everything ladened with MSG in some restaurants!

Taste receptors are more sensitive to unpleasant stimuli. For example we are about a thousand times more sensitive to acids than sweet tastes (although my 10 year old daughter the candy expert is very sensitive to sweets).

Vision

To take our food analogy a bit further we now come to the meat and potatoes section of the sensory system. This includes vision and hearing and we will spend a good deal of time here. Let's begin with vision.

Vision is sensed by the eyes (of course). So we will spend some time getting the big picture of how the eye is put together.

The Big Picture: The Eye

The eye consists of three layer called tunics.

The eye is put together in three layers called tunics (doesn't everyone own at least one tunic?). There is an outer, inner and middle tunic in each eye. In order to understand how the eye is put together let's look at an old fashioned camera.

Here we have a very basic camera (fig. 10.1). There is a frame to hold things together, a lens, an aperture and a piece of film. Light rays from the subject pass through the aperture, lens and a space in the frame to reach the film. The aperture can change its diameter to let in more or less light. Notice that the image is upside down on the film.

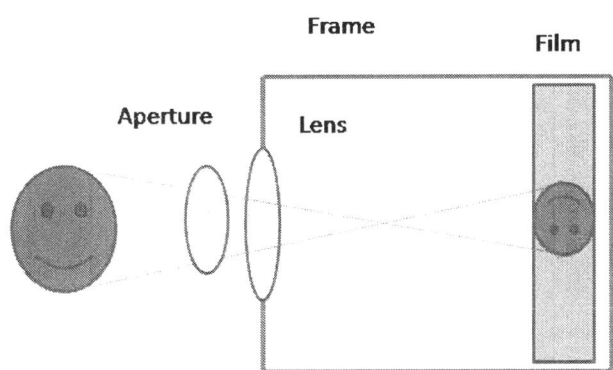

Figure. 10.1 A basic camera works much like the eye.

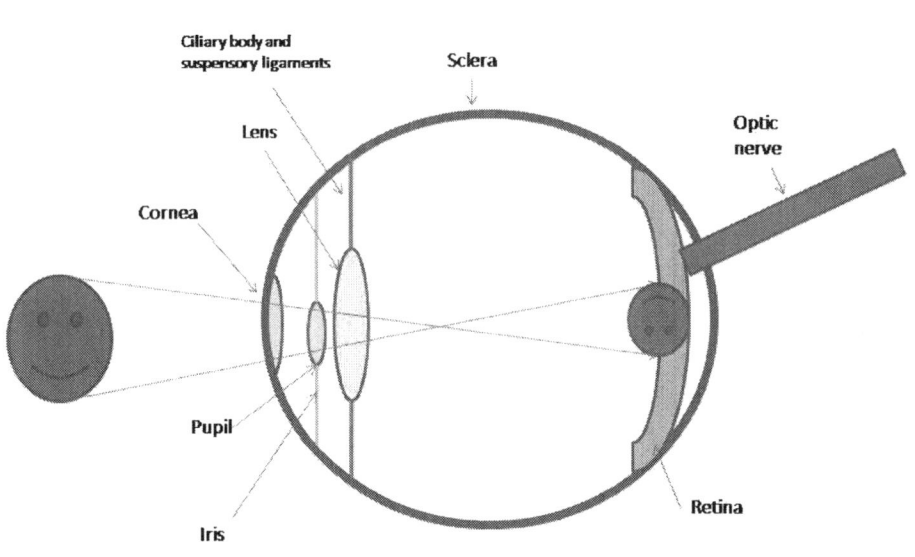

Figure. 10.2. Basic eye structures.

Notice the similarities between the camera and the eye (fig. 10.2). The "frame" of the eye is the sclera (white portion of the eye). Cameras are usually painted black on the inside. This provides a good contrast for the image on the film. The eye has a similar structure called the choroid coat. It is in the middle layer of the eye and is a dark colored membrane.

The aperture is the pupil, the lens is still the lens, and the film is now the retina. There are some additional structures such as the cornea which is a transparent portion of the outer tunic to allow light into the eye, the iris or colored part of the eye, and the ciliary body that holds the lens in place (fig. 10.3, 10.4).

The large space between the lens and the retina is filled with a jelly like fluid called vitreous humor. The ciliary body also secretes a watery fluid called aqueous humor which is in the front part of the eye. The optic nerve carries the visual information from the retina to the brain (remember the occipital lobe processes vision).

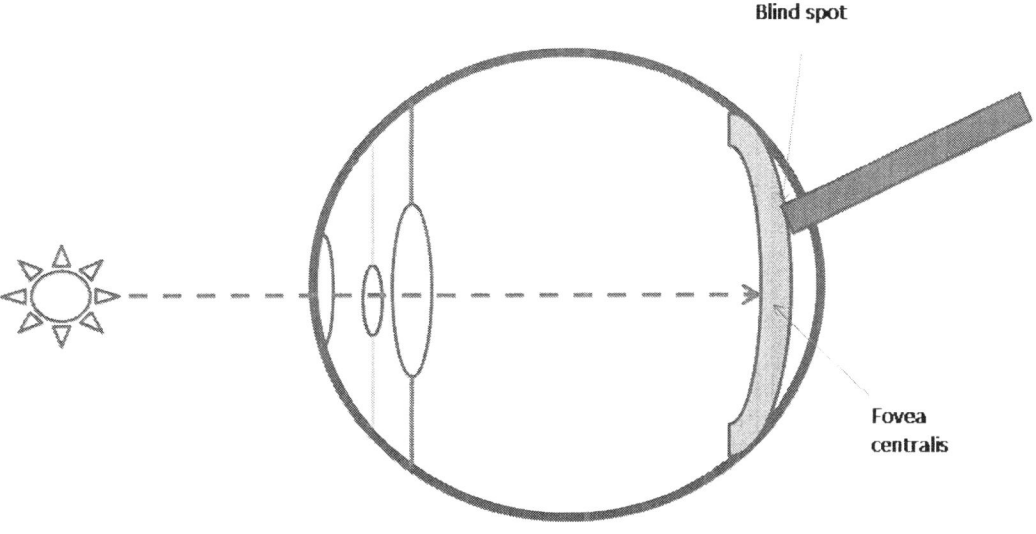

Figure 10.3. The fovea centralis is the area of sharpest vision. The blind spot is where the optic nerve connects to the back of the eyeball.

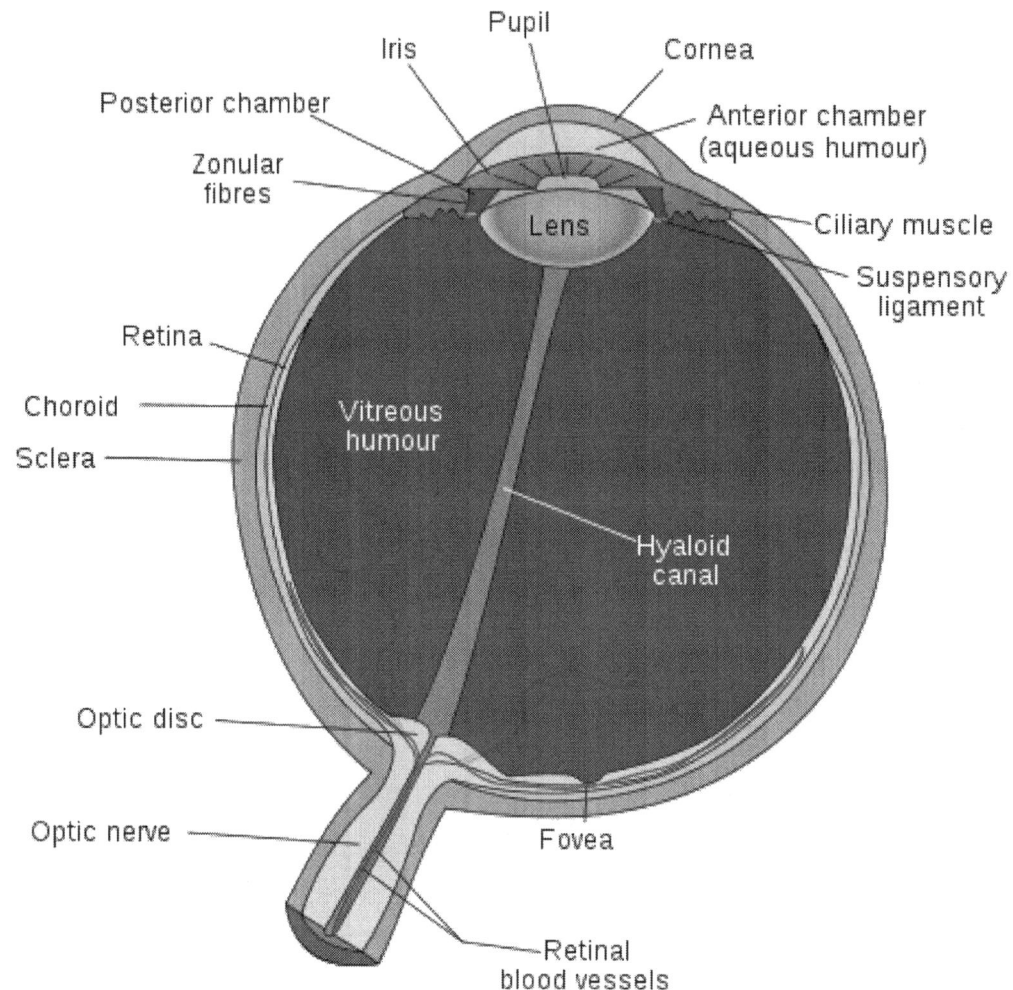

Figure 10.4. Anatomy of the Eye.

Light rays that enter the eye and follow a straight path to the back of the eye reach an area called the fovea centralis. This is also known as the area of sharpest vision and is conveniently located in this straight path. This area has the largest concentration of photoreceptors. A bit off to the side is the blind spot where the optic nerve connects to the eyeball. There are no photoreceptors here and the brain "fills in" the area so you don't even realize it.

Moving the Eye

The eye can perform complex movements. Just try to follow a ping pong game. It can do this because there are six muscles on each eye. Remember that these muscles are controlled by the cranial nerves. Think of holding a beach ball. Your arms are the eye muscles:

1. Stand behind the ball. Hold the ball with one hand on top and the other on the bottom. These are the superior and inferior rectus muscles. Bending either arm will simulate the action of the muscle.

2. Stand behind the ball. Hold the ball with one hand on each side. Bending either arm will simulate the action of the medial and lateral rectus muscles.

3. Instead of standing behind the ball stand to the side of it. Reach across your body and place on hand on top and the other on the bottom of the ball. This simulates the action of the superior and inferior oblique muscles.

Rods and Cones

There are two different types of photoreceptors in the eye. One is for color and the other for black and white vision. The photoreceptors for color are called cones. The others are the rods for black and white vision. Cones work better in daylight and rods work better at night. Cones are also more concentrated at the central area of the retina (fovea centralis) while rods are more concentrated in the periphery of the retina (peripheral vision).

These receptors work by virtue of chemical reactions. In the rods the key substance is rhodopsin. Rhodopsin consists of two parts which include opsin (a protein) and retinal. Rhodopsin is imbedded in the rod's cell membrane. Rhodopsin breaks down in light causing the retinal to separate from it. The separation of retinal causes changes in the membrane potential that travels to the nervous system.

One important thing about retinal is that it contains vitamin A. It was actually found that soldiers in World War II who did not get enough vitamin A developed night blindness. Carrots are high in vitamin A, so eat your carrots and you will see better in the dark.

Cones require a higher intensity of light in order to trigger action potentials. Cones use different opsins corresponding to the primary colors they absorb (blue, green, red). The reaction of retinal and opsins is essentially the same as in rods. The end result is a change in membrane potential.

If you don't have the right cones or a small amount of a certain type of cone you will be color blind. The most common type is red-green color blindness in which red and green are seen as the same color. Up to 8-10% of the male population may have some degree of color blindness. I have a similar issue with picking out clothes to wear. It's not color blindness but more like style blindness. I solved this problem by buying lots of clothes that all match. Now when I reach in my closet I know that my outfit will at least be color coordinated. I learned that one from Einstein who had similar problems with fashion.

Making Sense of the Sensory System 196

The Ear

The last special sense we will cover is the ear. I like to tell my students that when it comes to the ear it's all about hair cells. Not the hair growing out of your ear but special receptors that contain tiny hairs. You'll understand this a bit later. Hang in there.

The Big Picture: The Ear

The ear consists of 3 parts: outer, middle and inner ear.

The eye had three layer or tunics and the ear has three divisions (fig. 10.5). These are the outer, middle and inner ear. The structures aren't too complex until you get to the inner ear. So we will summarize the structures with a diagram and brief explanation so you get the big picture.

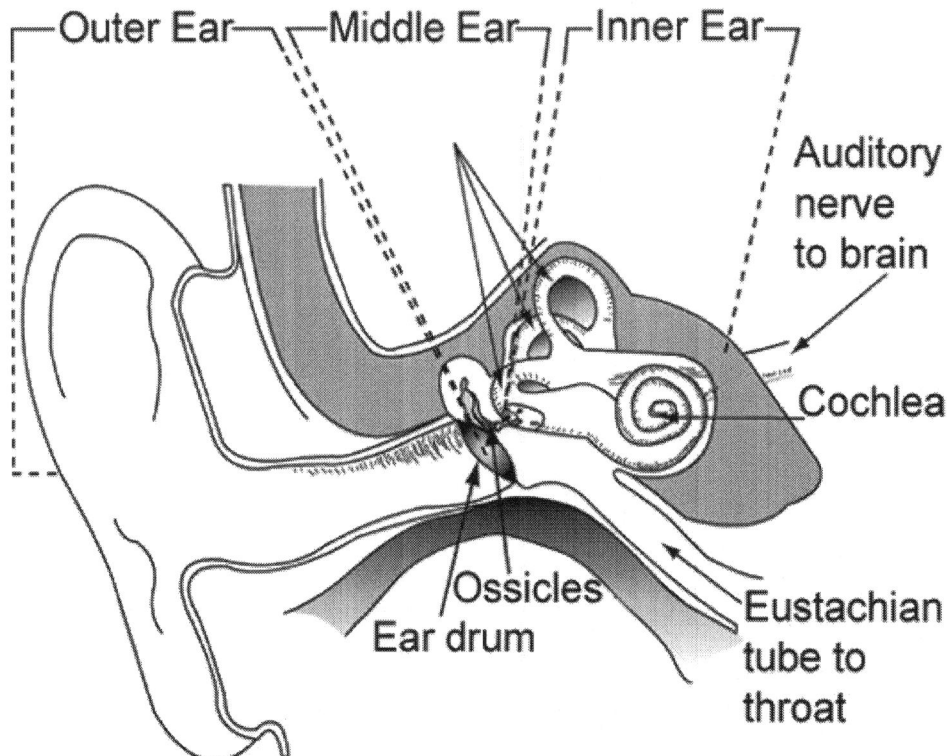

Figure 1: The Outer, Middle, and Inner Ear

Figure 10.5. Ear anatomy

Outer Ear

Auricle (or pinna)—the outer portion of the ear.

External auditory meatus—tubelike structure that runs in the temporal bone.

Middle Ear

Tympanic membrane—eardrum.

Malleus, Incus and Stapes—tiny bones called ossicles. Think of a hammer, anvil and stirrup.

Eustachian tube—passageway between middle ear and upper nasal passages to equalize pressure.

Inner Ear

Cochlea—looks like a snail's shell. This is where hearing is sensed.

Semicircular canals—look like loops. This is where dynamic equilibrium is sensed.

Vestibule—area between cochlea and semicircular canals. This is where static equilibrium is sensed.

Oval window—where the stapes connects to the inner ear.

Round window—membrane that helps to equalize fluid pressure.

Vestibulocochlear nerve-good ole cranial nerve VIII.

Some Nitty Gritty Structures

The inner ear is a bit more complex. Hey, not only does it sense hearing but also static equilibrium (am I standing or bending over?) and dynamic equilibrium (am I sitting on a chair or riding a roller coaster?). Also, remember what I said about hair cells? We'll be seeing these little guys soon.

Inside the Cochlea

Remember how the cochlea looks like a snail's shell? Well, if we were to "unroll" it we would see a tubelike structure that looks something like a windsock. If we were to slice the tube we would get something like the next picture (fig. 10.6).

Inside the cochlea we see three chambers filled with fluid. There is the scala vestibuli, scala tympani and scala media or cochlear duct. The cochlear duct contains the organ that senses hearing called the organ of Corti.

The next diagram gets into a bit more detail (fig. 10.7).

Making Sense of the Sensory System 198

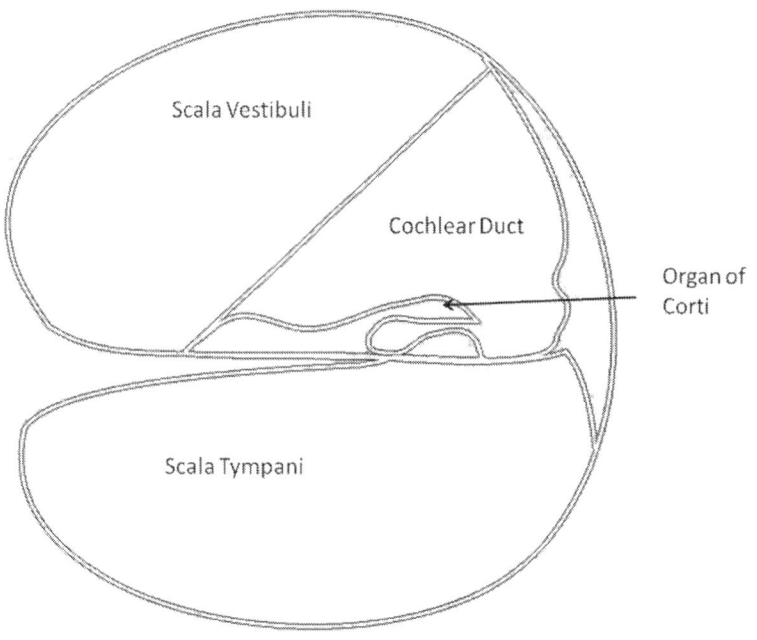

Figure 10.6. Chambers of the cochlea.

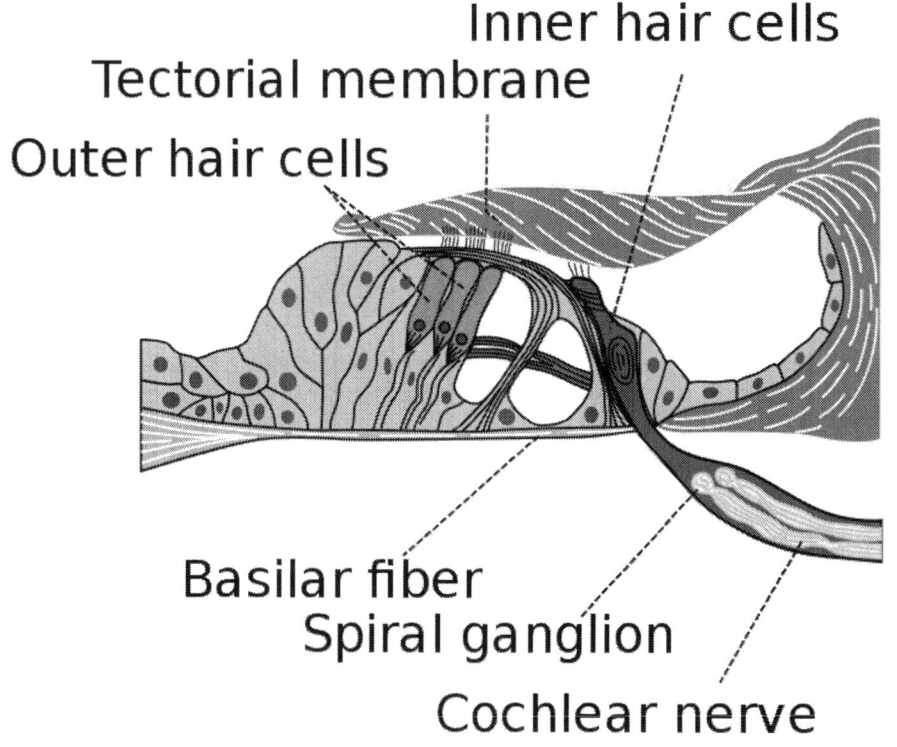

Figure 10.7. Inner ear structures.

If you look closely at the organ of Corti you will see little hairs coming out of the top of it. These are the hair cells. Also notice that the organ of Corti is anchored to a membrane called the basilar membrane. The hair cells stick up and connect to another membrane called the tectorial membrane. These will be important when we explain how hearing is sensed by the ear.

The Big Picture: Ear Physiology

When learning ear physiology you should remember that it's all about hair cells. Not the hair growing out of your ear but the tiny hairs called cilia emerging from tiny cells. Read on…

It all starts with sound waves. Sound is nothing more than changes in air pressure (or fluid pressure if you're underwater). In other words there must be a medium to carry sound. I am always a bit disappointed when watching science fiction movies that portray spacecraft screaming though space. In space since there is a vacuum. Literally no one can hear you scream and spacecraft would be silent.

These sound waves reach the ear and are collected by the pinna and the external auditory meatus much like taking a piece of paper and rolling it into a cone with both ends open then sticking the narrow end into your ear.

The sound waves reach the tympanic membrane and cause it to move in and out. This movement of the tympanic membrane is carried via vibrations of the auditory ossicles (malleus, incus and stapes) to the inner ear (at the oval window). The vibrations are then carried to the fluid filled chambers of the inner ear.

Think of striking a tuning fork and dipping the vibrating end into a bucket of water. You could see the waves produced by the vibrations. The sound waves have been converted to mechanical fluid waves.

The fluid waves reach the organ of Corti and cause the membranes (basilar and tectorial) to resonate and move. The movement is picked up by the hair cells and converted to good ole action potentials carried by cranial nerve VIII (vestibulocochlear nerve). The impulses basically end up in the temporal lobe where sound is perceived.

The Tympanic Reflex (Quieting the beating drum)

There are a couple of tiny muscles attached to the ossicles in the middle ear that help to either transmit the vibrations from the tympanic membrane of to dampen (inhibit) them. These are the tensor tympani and stapedius muscles.

Think of going to a rock concert. You end up sitting in front of some very large speakers and are subsequently exposed to some very loud sounds. At the end of the concert as you walk out of the arena you may notice your ears ringing. What happened was that the tiny muscles contracted pulling on the ossicles to dampen the sound to protect your ear.

Another analogy would be if your young child was given a noisy toy drum. He delights in beating the heck out of it. You walk over and press down on the head of the drum decreasing the irritating noise. This is much like how these muscles work.

Static Equilibrium (More hair cells and rocks in your head)

The ear also let's your brain know whether you are reclining on your sofa or standing up. In other words it senses changes in position. This is called static equilibrium.

Static equilibrium is sensed in the vestibule. Inside the vestibule are two structures called the utricle and saccule. Guess what they contain? You guessed it—hair cells! The hair cells are located in structures called macula and connect to a membrane called the otolithic membrane.

Making Sense of the Sensory System 200

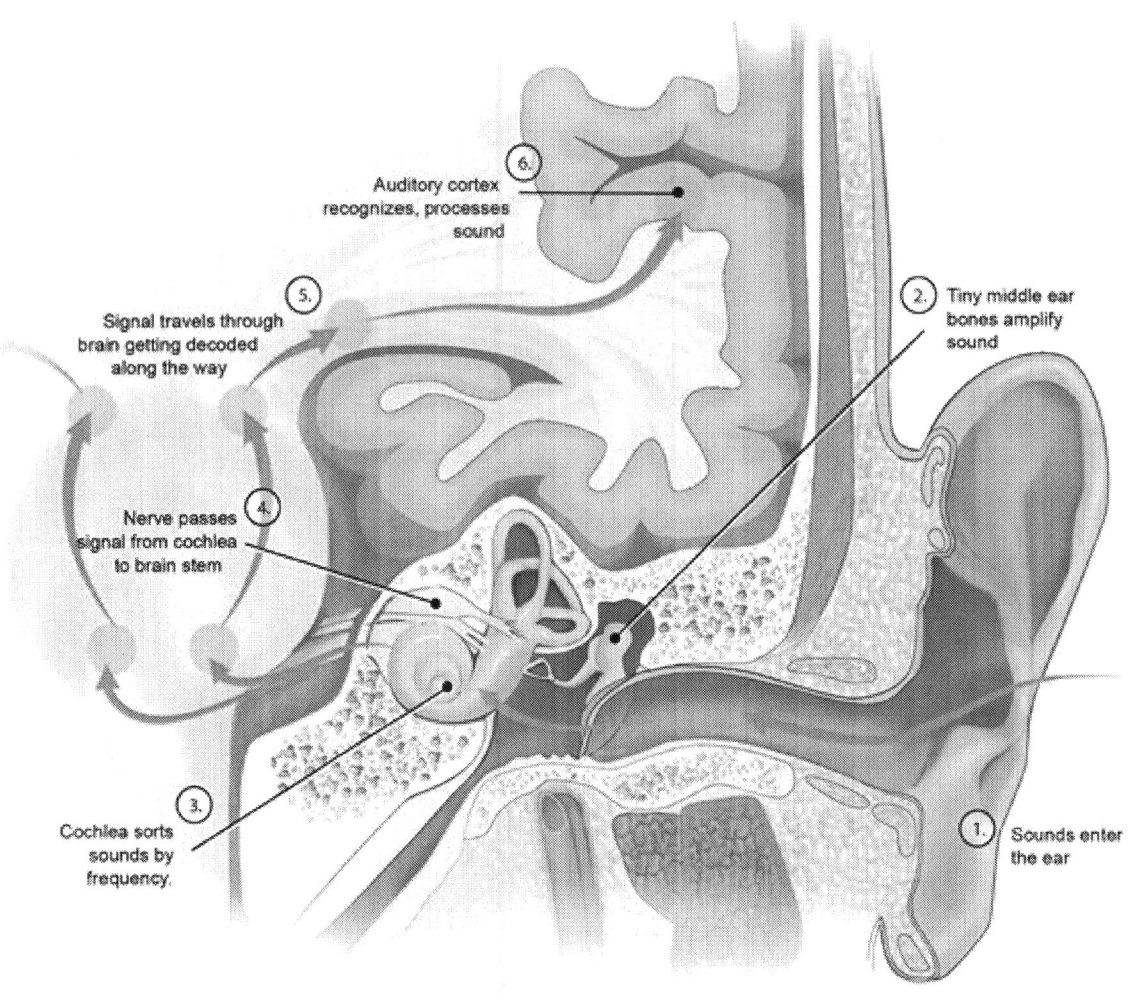

Figure 10.8 How the ear processes hearing.

The membrane contains small stonelike structures called otoliths (rocks in your head).

Okay, to simplify without using the anatomical terms. What you basically have are hair cells imbedded in a membrane that has rocks on it. When you change position the rocks pull on the membrane and bend the hair cells. This motion is converted to action potentials and sent off to the brain.

Once in awhile the rocks get out of position causing a condition called benign positional vertigo. Not to worry though because they can be put back into position with a relatively simple maneuver called otolithic repositioning.

Dynamic Equilibrium (Still more hair cells!)

Remember those loopy semicircular canals? Well those suckers are filled with fluid as well. At the base of each one is a bulge called the ampulla. Inside the ampulla is a structure called the crista ampullaris. Guess what's inside the crista ampularis—hair cells!

Also inside the semicircular canals is a fluid. The fluid moves when you move causing the hair cells to bend. The bending is once again translated to action potentials and sent off to the brain. Sometimes the fluid keeps moving even when you are stopped. I remember a recent trip to an amusement park where I rode one of those roller coasters that does loops and barrel rolls. It took me a few moments to stabilize after the ride was over.

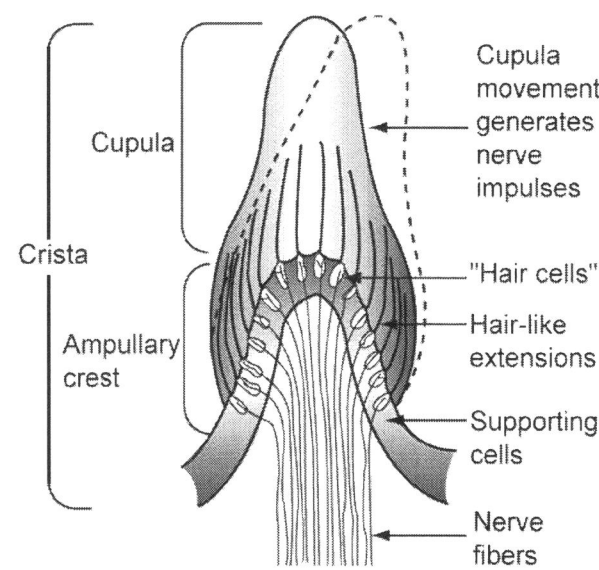

Figure 10.10. The crista ampullaris is located inside a semicircular canal. It contains a cupula that bends in response to fluid movement inside of the canals.

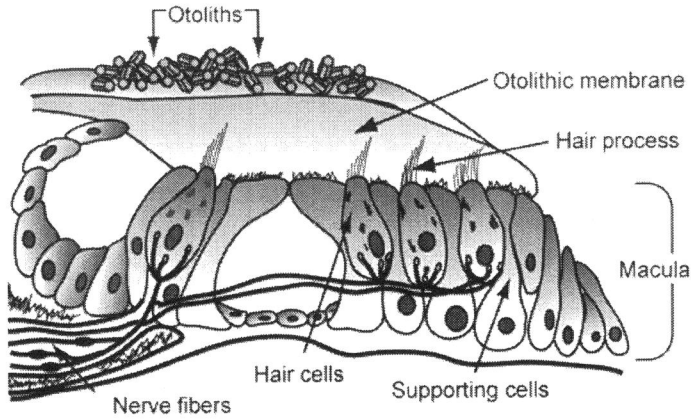

Figure 10.9 A change in the body's position results in the otoliths pulling on the otolithic membrane. The causes bending of hair cells and transmission of impulses to brain that are interpreted as changes in static equilibrium.

Review Questions

1. Which sensory receptor senses changes in joint position:
   a. Chemoreceptor
   b. Osmoreceptor
   c. Baroreceptor
   d. Proprioceptor

2. Which of the following receptors senses heavy pressure:
   a. Meissner's corpuscles
   b. Merkel's discs
   c. Pacinian corpuscles
   d. Ruffini corpuscles

3. Which of the following produces a protective mechanism in muscles:
   a. Golgi tendon organs
   b. Meissner's corpuscles
   c. Muscle spindles
   d. Ruffini corpuscles

4. How many different types of chemical substances can we sense with our sense of smell:
   a. 500-1000
   b. 1000-2000
   c. 2000-4000
   d. 4000-6000

5. Which of the following is not a primary taste:
   a. Bitter
   b. Water
   c. Sweet
   d. Salty

6. Which of the following is not an eye muscle:
   a. Superior rectus
   b. Inferior oblique
   c. Lateral rectus
   d. Medial oblique

7. As light passes through the eye which structure will it not pass through:
   a. Cornea
   b. Sclera
   c. Pupil
   d. Vitreous humor

8. Which structure contains the photoreceptors in the eye:
   a. Choroid coat
   b. Ciliary body
   c. Retina
   d. Optic nerve

9. In myopia the images focuses:
   a. In front of the retina
   b. On the retina
   c. Behind the retina
   d. None of the above

10. Which best describes the function of rods:
    a. Work better in dim light
    b. Sense color
    c. Work better in daylight
    d. Produce sharp central vision

11. Which is the most common form of color blindness:
    a. Red-orange
    b. Red-green
    c. Blue-green
    d. Green-yellow

12. Which structure forms the boundary between the outer and middle ear:
    a. Pinna
    b. Tympanic membrane
    c. Oval window
    d. Round window

13. Which structure is in the inner ear:
    a. Vestibule
    b. Stapes
    c. Tensor tympani
    d. Eustachian tube

14. Which part of the ear senses static equilibrium:
    a. Cochlea
    b. Semicircular canals
    c. Vestiblule
    d. Tympanic membrane

15. Which of the following is sensed by otoliths pulling on a membrane:
    a. Sound
    b. Static equilibrium
    c. Proprioception
    d. Dynamic equilibrium

16. Which cranial nerve carries the sensation of hearing:
    a. Hypoglossal
    b. Spinal accessory
    c. Vestibulocochlear
    d. Trigeminal

Chapter 11

Learning the Hell Out of Hormones

Chapter 11

Learning the Hell Out of Hormones

Yup, we're talkin about the endocrine system here. Another complex system with lots of feedback mechanisms, but don't let it scare you. The trick (as always) is to keep the big picture in mind.

Big Picture: Endocrine System

The endocrine system consists of hormones secreted by glands traveling through the blood to target tissues.

Let's take a simplified look at the endocrine system. When you think of the endocrine system you should think of hormones. Hormones are secreted by glands and travel via the bloodstream to what are called target tissues. Substances secreted by glands that don't travel via the bloodstream are called exocrine substances.

Okay, we know that glands secrete hormones, so how do the hormones know where to go? Do they just float around the blood looking for any old place to land? Of course they don't. They look for special receptors on the target tissue. Hormones work by virtue of their shape. They just fit certain receptors.

Here is a diagram showing important endocrine glands (fig. 11.1).

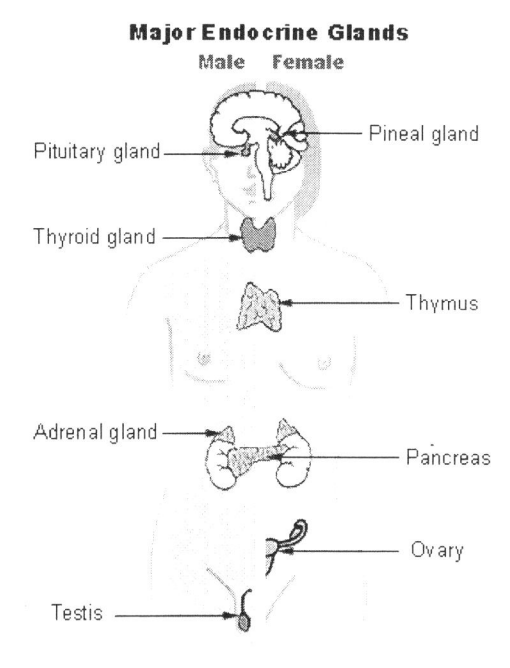

Figure 11.1. Endocrine System

Big Picture: Types of Hormones

There are three basic types of hormones: prostaglandins, steroids and non-steroids.

Simplest case: Prostaglandins

Prostaglandins are secreted by cells and have a local effect. This means that they are secreted by cells and only travel to nearby cells. This is known as a paracrine secretion. Once the hormone reaches the target cell, it can use what is known as the second messenger system (more about this later). Prostaglandins help to control smooth muscle contraction and relaxation. Prostaglandins also help to promote inflammation and pain.

More Complicated: Steroid Hormones

Steroid hormones are transported in the blood. They connect with a special transport protein known as a carrier protein. Once reaching the target cell, the hormone splits from the carrier protein.

Remember that lipid soluble substances can diffuse through a cell membrane. Since steroid hormones are considered lipids, they can diffuse through the cell membrane and enter the cell. Once inside the cell, steroid hormones combine with specialized receptors located within the cytoplasm of the cell. The hormone combines with the receptor and the receptor-hormone complex moves into the nucleus of the cell. There it causes changes in DNA transcription that cause changes in the metabolism and functioning of the cell.

Most Complex: Non-steroid Hormones

Non-steroidal hormones enter the cell differently than steroids. Non-steroidal hormones are not lipid soluble, since they cannot diffuse directly into the cell they must enter via a different process. Non-steroid hormones enter the cell by using what are known as second messengers.

Receptors for non-steroidal hormones are located in the cell walls of the target cells. When the hormone connects with the receptor on the outside of a cell membrane, a protein known as a G-protein activates and moves down the membrane into the cell. The G-protein binds to an enzyme known as adenylate cyclase and activates it. Adenylate cyclase then becomes involved in the reaction:

$$ATP \xrightarrow{\text{Adenylate cyclase}} cAMP + 2P$$

cAMP is known as cyclic adenylate monophosphate and is considered the second messenger in the system. cAMP in turn activates another inactive enzyme called protein kinase. Protein kinase facilitates the phosphorylation of various proteins. Phosphorylation occurs when phosphates are attached to a molecule. The phosphorylated proteins then activate some enzymes and inactivate others inside the cell. This alters the metabolic activity of the cell and the cell responds in accordance with the intended action of the hormone.

Results of second messenger activation include altered membrane permeability, activation of enzymes, protein synthesis, modulation of metabolic pathways, promoting movement of cells and causing secretion of other hormones.

cAmp works with a variety of hormones including those from:

- Hypothalamus
- Anterior pituitary
- Posterior pituitary
- Parathyroid
- Adrenals
- Thyroid
- Pancreas

There are other second messengers besides cAMP. These include:

— Diacyglycerol (DAG)
— Inositol triphosphate (IP3)
— Cyclic guanosine monophosphate (cGMP)

Hormones operating via 2nd messengers have a much greater response. Many 2nd messengers can be activated by one hormone.

Whoa! Too much information here. Maybe we can explain this second messenger stuff a little better.

Okay, so the idea is to transfer information from the hormone to the cell. This is much like instructions coming down from the administration in a large company. There's the president or CEO, the vice president (VP), the managers, and the workers. The information is in the form of a message beginning with the CEO (aka hormone).

The CEO (hormone) sends a message to the VP (G-protein). The VP in turn tells the managers (cAMP) who tell the workers (protein kinases). The workers carry out the instructions (alter the cell's functions). The message then has a powerful effect on the company (cell) because of the one to many process of communication.

Hypothalamus and Pituitary Gland

The hypothalamus and pituitary gland are both very important endocrine structures because they both contain a lot of hormones.

The pituitary gland can be divided into two parts. There is a part in the front (anterior pituitary) and the back (posterior pituitary). If you want to sound like you're getting your money's worth out of your education you can throw around words like adenohypophysis (anterior pituitary) and neurohypophysis (posterior pituitary). Parents love to hear words like this. They feel that their tuition money is going to a good cause.

Mom to student

Hi Frank, how's college? What have you been studying lately?

Student to mom

Well, I am in the process of studying the anatomical location and physiological functioning of the adenohypophysis and neurohypophysis.

Mom to student

Wow, that's sounds like you are really learning something. Would you like some spending money?

See, works every time.

Also, these words can help to impress girls.

Girl to Frank

Hi Frank, what are you reading there? Looks like a big book.

Frank to Girl

It's my A&P book.

Girl to Frank

Wow it sounds complicated.

Frank to girl

Yes, it is...We are learning the hormonal implications of the adenohypophysis...

Girl to Frank

Wow, I didn't know you were so smart--hey, wanna go out this Saturday night?

Or

Jon to girlfriend Mary

I can't understand why it takes you so long to study A&P. What could be so hard? I could probably learn it much faster than you.

Mary to Jon

Okay—tell me about the anatomical and physiological aspects of the adenohypophysis and neurohypophysis.

Jon to Mary

Never mind what I said. Do you want me to cook dinner for you?

See I told you it works...

The anterior pituitary has a different connection to the hypothalamus than the posterior pituitary. The anterior pituitary connects to the hypothalamus by way of a capillary network. This allows for a direct communication pathway between hypothalamus and anterior pituitary.

There is always a two step process with regard to secretions of the anterior pituitary:

Step 1: Hypothalamus secretes special hormones called releasing factors that target the anterior pituitary.

Step 2: Anterior pituitary responds by secreting hormones.

We will always see this two step process when it comes to anterior pituitary secretions.

Let's see an example of the 2-step process (figs. 11.2-11.6). We will be examining how growth hormone is secreted by the anterior pituitary.

Step 1: Hypothalamus secretes 2 hormones that control the secretion of growth hormone by the anterior pituitary.

Growth Hormone Releasing Hormone— facilitates secretion of growth hormone from anterior pituitary.

Somatostatin—inhibits secretion of growth hormone from anterior pituitary.

Step 2: Growth hormone is either secreted or inhibited depending on which releasing factor is secreted by hypothalamus.

Get the picture? We will always see the 2-step process for secreting anterior pituitary hormones.

The posterior pituitary has a different connection to the hypothalamus. It connects via special cells called neurosecretory cells. The cell body resides in the hypothalamus and the cell extends into the posterior pituitary. We don't have a 2-step process here.

Overview of Hormones

In order to get the big picture we will present a brief overview of the important hormones.

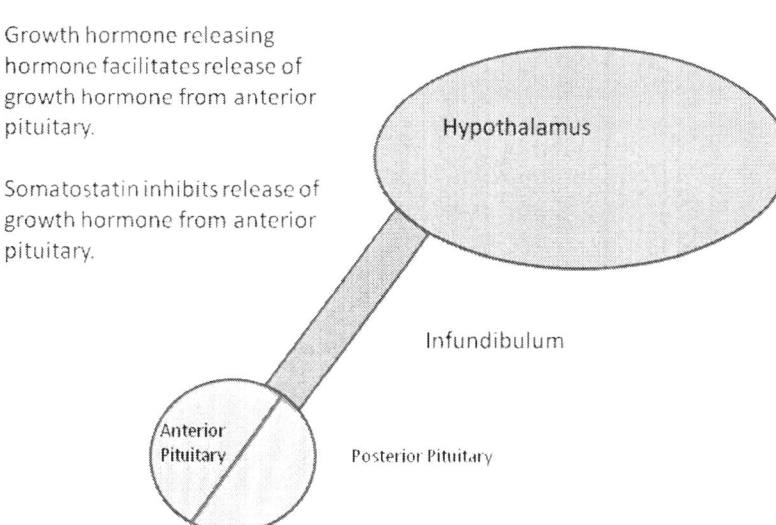

Figure 11.2. Growth hormone.

Growth hormone is secreted in response to two secretions by the hypothalamus: Growth hormone releasing hormone (GHRH) and somatostatin (SS). Growth hormone affects cellular metabolism by promoting the movement of amino acids into cells for protein synthesis which affects the overall growth of the organism. Growth hormone releasing hormone secreted by the hypothalamus stimulates the release of growth hormone by the anterior pituitary. Somatostatin inhibits the release of growth hormone.

Growth hormone stimulates cells to enlarge and undergo mitosis as well as increasing the rate of protein synthesis and increasing the cellular use of carbohydrates and fats.

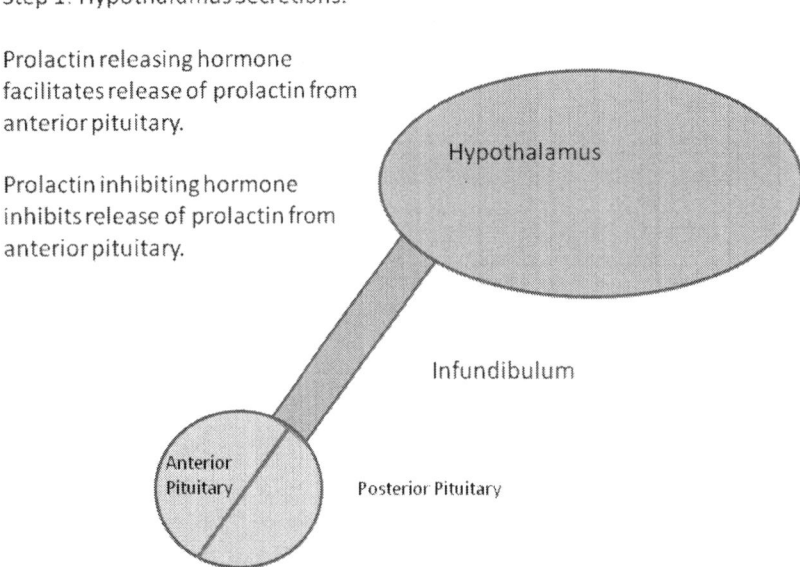

Figure 11.3 Prolactin.

Prolactin is secreted in response to two secretions by the hypothalamus. Prolactin releasing factor (PRF) stimulates secretion of prolactin by the anterior pituitary. Prolactin inhibiting hormone (PIH) from the hypothalamus inhibits secretion of prolactin by the anterior pituitary.

The function of prolactin is to stimulate milk production in females. In males, prolactin decreases the secretion of leutenizing hormone which facilitates production of the primary male sex hormones or androgens. Too much prolactin secretion in males can cause infertility.

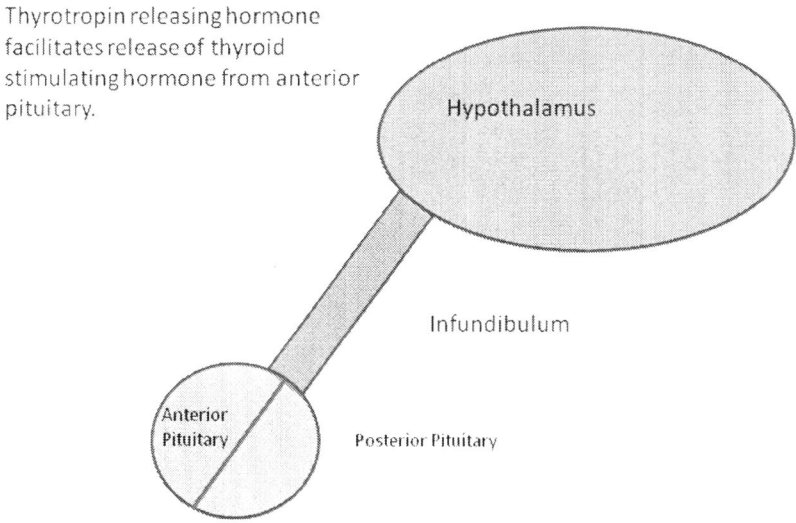

Figure 11.4 Thyroid stimulating hormone.

Thyroid stimulating hormone is released by the anterior pituitary in response to the release of thyrotropin releasing hormone from the hypothalamus. Thyroid stimulating hormone causes the thyroid gland to release the thyroid hormones triiodothyronine and tetraiodothyronine (T3 and T4). The blood concentration of thyroid hormones provides a negative feedback mechanism to the hypothalamus to help control the release of thyroid stimulating hormone. Secretion of T3 and T4 is also affected by stress.

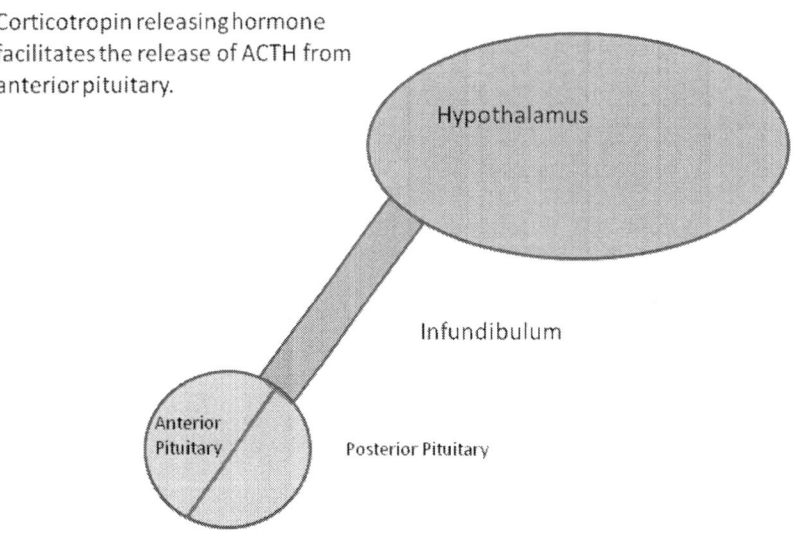

Figure 11.5 ACTH.

Adrenocorticotropic hormone is secreted by the anterior pituitary in response to secretion of corticotropic releasing hormone (CRH) by the hypothalamus. ACTH is picked up by the adrenal cortex and stimulates secretion of hormones by the adrenal cortex. Adrenal cortex hormones then provide feedback to the hypothalamus and anterior pituitary to help regulate secretion of ACTH. Stress also affects secretion of ACTH.

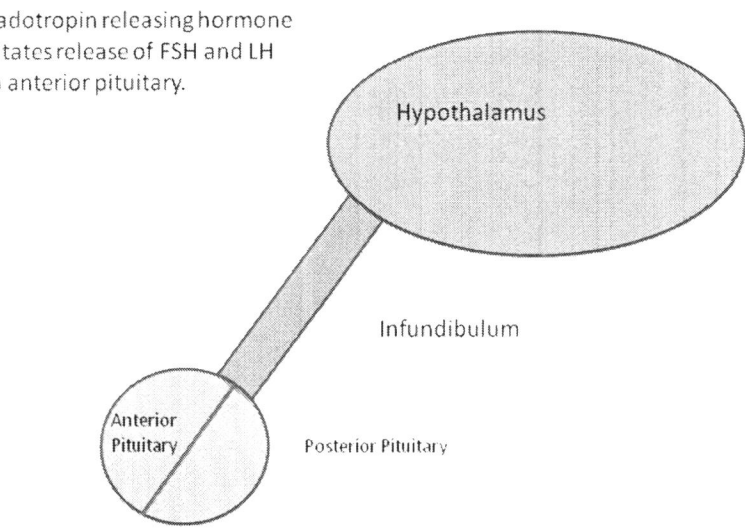

Figure 11.6. Follicle stimulating hormone and leutenizing hormone.

Follicle stimulating hormone is secreted by the anterior pituitary partly in response to the secretion of a releasing factor known as gonadotropin releasing hormone (GnRH). In females, FSH stimulates growth and development of egg-cell containing follicles in ovaries and stimulates follicular cells to produce estrogen. In males, FSH stimulates the production of sperm cells in the testes when the male reaches puberty.

Leutenizing hormone secretion is also partly controlled by the release of gonatotropin releasing hormone by the hypothalamus. Leutenizing hormone stimulates the glands of the reproductive system to produce sex hormones.

Posterior Pituitary

The posterior pituitary only secretes two hormones (thankfully). These are antidiuretic hormone and prolactin. Antidiuretic hormone does what its name implies. You might know what a diuretic does. If you don't, a diuretic makes you pee. In physiological terms it reduces fluid volume. So an antidiuretic does the opposite, it stops you from peeing. Or as we say in physiology, it facilitates the retention of fluid volume.

So how is ADH regulated? Well there are osmoreceptors that sense changes in chemical concentration of the blood in the hypothalamus. When the blood gets too concentrated the receptors tell the posterior pituitary to release ADH. This causes the body to retain fluid so the blood becomes less concentrated. ADH targets the kidneys to do this. I know you want to know just how this works but you'll have to be patient and wait until we get to the urinary system to explain this.

The other posterior pituitary hormone is oxytocin. Oxytocin stimulates uterine contractions during labor and delivery and milk ejection. Sometimes doctors give pitocin (oxytocin) to women in labor to help stimulate the uterine contractions and speed up the blessed event.

The secretion of oxytocin is one of those few positive feedback mechanisms. If you have ever experienced childbirth you'll probably remember that the contractions don't get milder over time, they get stronger.

Feedback—Feedback

The endocrine system is a great system for learning about feedback. All of those hormones have to be regulated in one way or another. This reminds us of our original definition of homeostasis which was that homeostasis works to maintain a range of values of substances.

Let's take a look at one system in particular, that of the thyroid hormones.

The thyroid gland is located in your neck just below the Adam's apple (aka thyroid cartilage) Fig. 11.7. There are two types of cells in the thyroid. Follicular cells secrete the hormones T3 (triiodothyronine) and T4 (tetraiodothyronine). Parafollicular cells secrete the hormone calcitonin (we'll talk about this one later).

Here's the way it works. The hypothalamus secretes TRH that travels to the anterior pituitary and tells it to secrete TSH (see diagrams). TSH then targets the thyroid gland causing the secretion of T3 and T4. These hormones then work to stimulate metabolism.

Negative feedback works to control the levels of T3 and T4 in the blood. For example, if the level of T4 gets too high then the secretions of TSH and TRH decrease which work to bring the level of T4 back into range. Likewise if the level of T4 gets too low then the secretions of TSH and TRH increase which raises the level of T4.

Calcitonin is the other thyroid hormone. It works with another hormone from the parathyroid glands called parathyroid hormone (PTH). The parathyroid glands are small masses of tissue located on the backside of the thyroid.

Both PTH and calcitonin work to regulate blood calcium levels. To help you remember think of calcitonin toning down the calcium in blood. PTH does the opposite and raises blood calcium levels. Both are controlled by the actual calcium in blood. For example if blood calcium levels get too high then calcitonin is secreted and PTH is inhibited.

More Hormones

Aldosterone is secreted by the adrenal gland fig. 11.9. It is produced by cells of the zona glomerulosa of the adrenal cortex. Aldosterone acts to regulate electrolytes such as magnesium and potassium. These are known as mineral electrolytes, thus aldosterone is known as a mineralcorticoid.

Aldosterone causes the kidney to conserve sodium and secrete potassium. The release of aldosterone is more strongly facilitated by the increase in plasma potassium concentration. The decrease in plasma sodium concentration does not affect the secretion of aldosterone as strongly. However, the decrease in sodium concentration can affect the renin-angiotensin system in the kidneys (see urinary system) and stimulates the release of aldosterone. Both aldosterone and the renin-angiotensin system work together to conserve blood volume and sodium. Aldosterone works by inhibiting the release of sodium by the kidney and the renin-angiotensin system works by causing vasoconstriction.

Aldosterone is also released via stimulation of the adrenal cortex by ACTH.

Cortisol

Cortisol is also secreted by the adrenal cortex, specifically by the cells of the zona fasciculata. Cortisol has an effect on glucose metabolism, thus it is called a glucocorticoid. Cortisol secretion increases glucose levels in the blood. It does this by stimulating the liver to convert non-carbohydrates into glucose. This process is called gluconeogenesis. It also stimulates the release of fatty acids for use as an energy source. These processes help to regulate the level of blood glucose between meals.

Cortisol is released in response to the release of ACTH by the anterior pituitary gland. Remember that ACTH is released in response to release of CRH by the hypothalamus. This system provides a negative feedback mechanism to help control the level of cortisol in the blood.

Norepinephrine and Epinephrine

The adrenal medulla, or inner portion of the adrenal gland is closely connected to the sympathetic nervous system. The adrenal medulla contains specialized cells called chromaffin cells that secrete chemicals called catecholamines. The catecholamines that are produced are norepinephrine and epinephrine. These chemicals should sound familiar because they were introduced as neurotransmitters in the sympathetic nervous system. Therefore, norepinephrine (NE) and epinephrine (E) have both neurotransmitter and hormonal actions.

NE and E are secreted by the adrenal medulla in response to impulses produced by the sympathetic nervous system (SNS). The SNS is connected via nerve fibers to the adrenal medulla. The actions of NE and E from the adrenal medullar are similar to the actions of the (SNS). Thus secretion of NE and E will cause an increase in heart rate, blood pressure, respiration, and a decrease in digestion. The hormonal action of NE and E lasts longer than neurotransmitter action because it takes longer to remove NE and E from the endocrine system. Both the adrenal glands and the SNS work together to provide the sympathetic response.

Insulin and Glucagon

Glucagon (secreted by alpha cells in the pancreas) works to increase the level of glucose in the blood. It does this by stimulating the liver to convert the storage form of glucose (glycogen) into glucose via a process known as glycogenolysis. Glucagon also stimulates the process of gluconeogenesis, which converts non-carbohydrates substances into glucose in the liver and breaks down fats into fatty acids and glycerol.

Glucagon is secreted when glucose levels are diminished in the blood. Secretion of glucagon is inhibited by high glucose blood levels.

Insulin (secreted by beta cells in the pancreas) works to decrease the levels of glucose in the blood. It does this by reversing the processes stimulated by glucagon. Insulin facilitates the storage of glucose in the liver by stimulating the production of glycogen from glucose. Insulin also inhibits the process of gluconeogenesis, stimulates protein synthesis and increases the storage of lipid in adipose tissue.

Insulin also facilitates the release of glucose into body tissues by stimulating facilitative diffusion of glucose carriers in cell membranes.

Insulin is secreted when blood glucose levels are high and inhibited when blood glucose levels are low.

Somatostatin (secreted by the delta cells) inhibits both glucagon and insulin secretion. Thus it also works to control glucose levels in the blood.

Melatonin

Melatonin is secreted by a small pinecone shaped gland located between the cerebral hemispheres called the pineal gland fig. 11.8. It attaches to the posterior portion of the thalamus. The pineal gland secretes melatonin. Melatonin is synthesized from the neurotransmitter serotonin and is involved in the regulation of sleep-wake cycles known as circadian rhythms. Melatonin secretion increases with a decrease in light. Melatonin also helps to regulate the menstrual cycle.

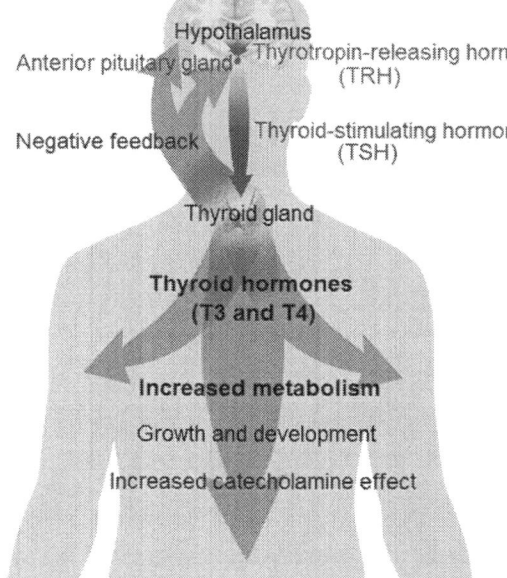

Fig. 11.7 Thyroid System

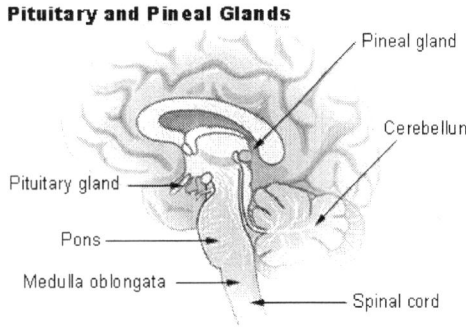

Fig. 11.8 Pineal Gland

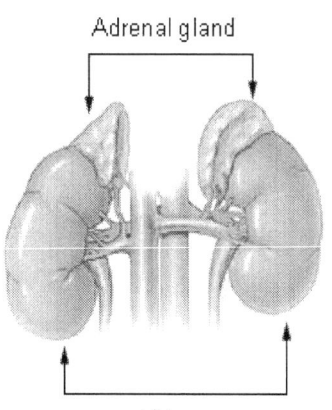

Fig. 11.9 Adrenal Glands

Review Questions

1. Which of the following is not an endocrine gland:

   a. Hypothalamus
   b. Pineal
   c. Thyroid
   d. Ileum

2. Which of the following class of hormones is lipid soluble and derived from cholesterol:

   a. Amines
   b. Steroids
   c. Peptides
   d. Carbohydrates

3. Cyclic adenylate monophosphate is commonly known as:
   a. Peptide
   b. Second messenger
   c. Neurotransmitter
   d. Hormone

4. Somatostatin elicits the following effect:
   a. Facilitates secretion of growth hormone
   b. Facilitates secretion of prolactin
   c. Inhibits secretion of growth hormone
   d. Inhibits secretion of prolactin

5. Which of the following hormones is not secreted by the hypothalamus:

   a. Thyrotropin releasing hormone
   b. Growth hormone releasing hormone
   c. Prolactin inhibiting hormone
   d. Adrenocorticotropin hormone

6. Which hormone is secreted in response to an increase in solute concentration of the blood:

   a. ACTH
   b. ADH
   c. Insulin
   d. Oxytocin

7. If blood calcium levels increase, which hormone will be secreted:

   a. Oxytocin
   b. Parathyroid hormone
   c. Calcitonin
   d. ACTH

8. Which of the following is a hormone of the adrenal cortex:

   a. Aldosterone
   b. Norepinephrine
   c. Insulin
   d. Corticotropin releasing hormone

9. Which hormone is secreted by the beta cells in the pancreas:

   a. Somatostatin
   b. ACTH
   c. Insulin
   d. Glucagon

10. Which hormone helps to regulate sleep wake cycles:

    a. Aldosterone
    b. Melatonin
    c. ACTH
    d. Insulin

Chapter 12

I Want to Drink Your Blood But I First Want to Know What's In It

Chapter 12

I Want to Drink Your Blood But I First Want to Know What's In It

This chapter will be of interest to you A&P students as well as the vampires lurking among us. We had a brief look at blood back in the tissue section. Now it's time to get into a bit more nitty gritty. But before we do let's look at an overview of what blood consists of.

The Big Picture: Blood

Blood consists of cells suspended in a fluid matrix called plasma.

Hey that's pretty simple, just cells and plasma. Okay, so what kinds of cells are there?

Blood contains red cells, white cells and cell fragments called platelets.

Let's begin our exploration into the nitty gritty of blood by learning more about red blood cells.

Big Picture: Red Blood Cells

Red blood cells are simple structures that carry oxygen.

First of all red blood cells (RBCs aka erythrocytes) are red (see how easy this is). They are red for a good reason too. The red comes from a pigment called hemoglobin (Hb). This stuff is important because it carries oxygen. In fact that's the main function of RBCs; to carry oxygen to our cells and tissues. The cells and tissues need oxygen in order to stay alive (I'd say that's important!).

When oxygen combines with hemoglobin we get:

Oxyhemoglobin

When oxygen breaks away from hemoglobin we get:

Deoxyhemoglobin

Since RBCs carry oxygen it is important that they have as large a surface as possible to do their job. They have a special shape. It's called a biconcave disc (fig. 12.1). I think it looks a bit like a donut where the hole did not go all the way through. The special shape works to increase the surface area to carry more oxygen.

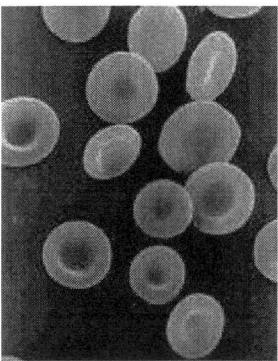

Figure 12.1. Red blood cells have a unique biconcave disc shape.

We have lots of RBCs in our blood. In fact our bodies make something like 2.5 million every second. We can count the number of RBCs in blood. Here are some ballpark normal values:

Red blood cell count = number of RBCs in cubic millimeter of blood

- 4,600,000 – 6,200,000 in males
- 4,200,000 – 5,400,000 in females
- 4,500,000 – 5,100,000 in children

Red blood cells are great workers. They happily do their jobs of carrying oxygen throughout their lives. They have relatively short lives that run about 120 days. You would think that after doing such a great job they would be able to enjoy a nice retirement. Not so, RBCs are recycled as soon as they retire.

Here's how it works. The old RBC actually becomes worn out. A bit frayed around the edges. As he wears out he gets smaller. When he gets small enough, he pops through the capillaries in the liver or spleen where the process of recycling begins.

Once inside the organ (let's say he is in the liver) he is attacked by white blood cells called macrophages. The macrophages attack and liberate the hemoglobin. The hemoglobin is further broken down into heme and globin portions. The heme gets broken down into iron and biliverdin. The iron moves into the blood and is carried off into storage by a plasma protein called transferrin. The biliverdin gets converted to bilirubin and is secreted by the bile. The globin portion gets recycled as plasma proteins.

A shortage of RBCs is known as anemia. There are a number of different types of anemias. Some of the more popular are pernicious, iron deficiency and blood loss anemias.

Pernicious anemia results from a deficiency of vitamin B12. Usually the body cannot absorb the vitamin because of problems with the stomach lining. Vitamin B12 is needed for the maturation of RBCs. Without it the cells are immature and large.

Iron is needed for the heme portion of hemoglobin. Without iron there may not be adequate hemoglobin. This affects the oxygen carrying capabilities of the cell.

Blood loss anemia is probably the simplest to understand. As blood is lost so are RBCs. The RBC count decreases and so do blood oxygen levels.

Big Picture: White Blood Cells

There are two kinds of white blood cells: granulocytes and agranulocytes.

White blood cells are called leukocytes. There are two categories of leukocytes. Those with granules in their cytoplasm (granulocytes) and those without (agranulocytes).

Granulocytes

Granulocytes (cells that have granules full of important stuff) tend to be significantly larger than red blood cells. There are three main types of granulocytes. These are the neutrophils, basophils and eosinophils. Granulocytes also have a significantly shorter lifespan (measured in hours) than red blood cells (120 days).

Neutrophils are larger than red blood cells and contain a segmented nucleus (fig. 12.2). They are the majority of leukocytes. Their function is primarily phagocytosis of bacteria and viruses. They are the first cells to arrive at an infection.

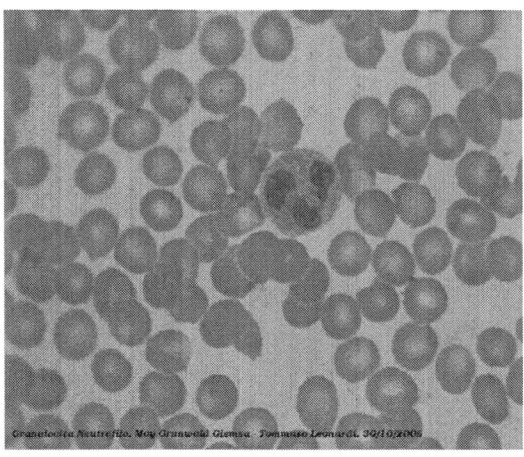

Figure 12.2. Neutrophil

Eosinophils

Eosinophils have a bilobed nucleus and granular cytoplasm (fig. 12.3.). They are relatively rare cells that only make up 1% to 3% of the leukocyte population. Their function is to moderate allergic reactions and defend against parasites.

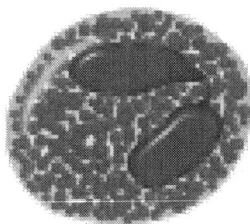

Figure 12.3. Eosinophil

Basophils

Basophils have the same size and shape of nuclei as eosinophils (fig. 12.4). They have fewer and much larger granules. The granules release histamine (a vasodilator) and heparin (an anticoagulant). Basophils function in inflammation. Think of the cardinal signs of inflammation; heat, redness, pain and swelling. The basophils release heparin and histamine that function to bring more blood to the area. The additional blood produces the signs of inflammation. Basophils are also relatively rare cells and represent only 1% of the leukocyte population.

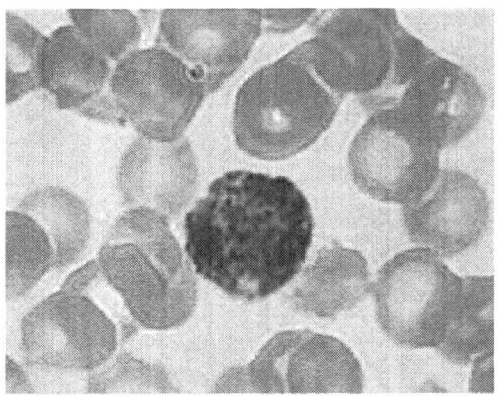

Figure 12.4. Basophil

Agranular Leukocytes

There are two agranular leukocytes. These include monocytes and lymphocytes.

Monocytes are the largest blood cells (fig. 12.5). Their nuclei can be spherical, kidney-shaped, oval or lobed. Monocytes have a function very similar to neutrophils. Monocytes work to clean up debris and phagocytize bacteria. They constitute 3% to 9% of the leukocyte population.

Lymphocytes are about the same size as red blood cells (fig. 12.6). They constitute 25% to 30% of the leukocyte population. There are two main types of lymphocytes. These include T-lymphocytes and B-lymphocytes. Both function in immunity. T-lymphocytes attack pathogens and help activate B-lymphocytes. B-lymphocytes produce antibodies when activated that attack pathogens.

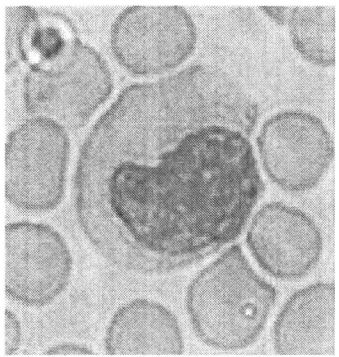

Figure 12.5. Monocyte

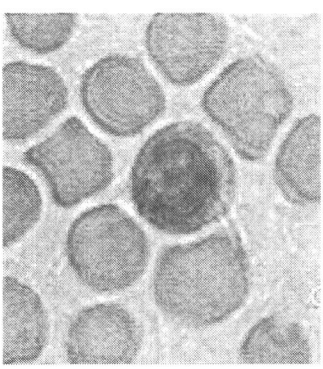

Figure 12.6. Lymphocyte

White blood cells can move between capillary walls and enter the tissue. They do so by a process known as diapedesis. In diapedesis an appendage of the cell first moves to an area. This is followed by the remainder of the cell.

White blood cells are attracted to an infected area by substances secreted by damaged cells. This is known as chemotaxis. The white blood cells then break up bacteria and form pus.

WBC Count

The typical WBC count is about 5,000 to 10,000 cells per cubic millimeter. A high count is called leukocytosis and can be caused by infection, exercise, loss of body fluids and emotional stress. A low count is called leucopenia and can be caused by viruses such as influenza, measles, mumps, chicken pox and toxins such as lead poisoning.

A test that breaks out the relative percentages of WBCs is called a differential.

Big Picture: Platelets

Platelets are fragments of cells that help to stop bleeding.

Platelets are cell fragments called thrombocytes. Hemocytoblasts (stem cells) differentiate into megakaryocytes that fragment into platelets. Platelets are about one half the size of red blood cells. There are about 130,000 to 360,000 platelets per cubic millimeter of blood.

Platelets help to stop bleeding by sticking together to form plugs and secreting the hormone serotonin which acts to vasoconstrict the vessels.

Blood Plasma

Plasma is the fluid portion of blood that carries the cells and platelets. Plasma is straw-colored and clear and contains water with a variety of substances. Plasma contains various proteins including fibrinogen, globulins and albumin. Plasma also contains dissolved gasses such as carbon dioxide and oxygen and nutrients such as carbohydrates, amino acids and lipids. Lipids are packaged in lipoproteins such as very low density lipoproteins (VLDL), low density lipoproteins (LDL), high density lipoproteins (HDL) and chylomicrons. Other constituents of plasma include electrolytes such as sodium, potassium, calcium, magnesium, chloride, sulfates, phosphates, and bicarbonate ions and nitrogenous substances such as uric acid, urea, creatine, and creatinine.

Big Picture: Hemostasis

There are three ways the blood system helps to stop bleeding.

1. *Damaged blood vessels constrict.*
2. *Platelets get sticky and form plugs.*
3. *Clotting.*

The blood system has some self-protection mechanisms built into it. These come into play during bleeding. The stopping of bleeding is called hemostasis. There are three basic mechanisms of hemostasis. These include blood vessel spasm, platelet plug formation and clotting.

Blood vessel spasm occurs in response to a damaged vessel. Blood vessels have a smooth muscle layer that constricts when the vessel is damaged. Platelets also release serotonin that facilitates constriction of the vessel. This action helps to stop the bleeding.

Platelets become sticky when they contact damaged blood vessels. They can stick together to form a plug. The platelet plugs help to plug small holes in vessels. This is called platelet aggregation.

Clotting is the third mechanism of hemostasis. There are two pathways consisting of a cascade of reactions involving molecules called clotting factors (designated by Roman numerals). The two pathways are called the intrinsic and extrinsic pathways. Both pathways converge at a common point to form a fibrin clot.

Hopefully you won't have to memorize the entire cascade but if you do we have some tips to help.

For one, think of the two pathways (intrinsic and extrinsic). The intrinsic pathway is characterized by blood remaining in the vessel. In other words a clot is formed inside the vessel without blood every leaving it. On the contrary the extrinsic pathway is activated when blood leaves the vessel.

Here is a diagram of the cascade (fig. 12.7).

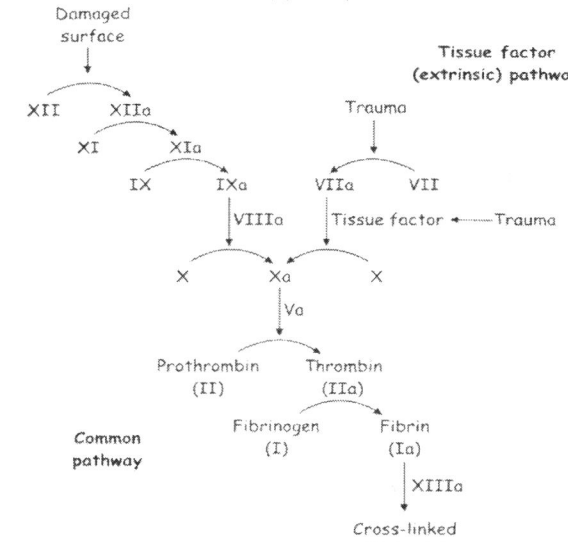

Figure 12.7. Clotting Cascade (detail)

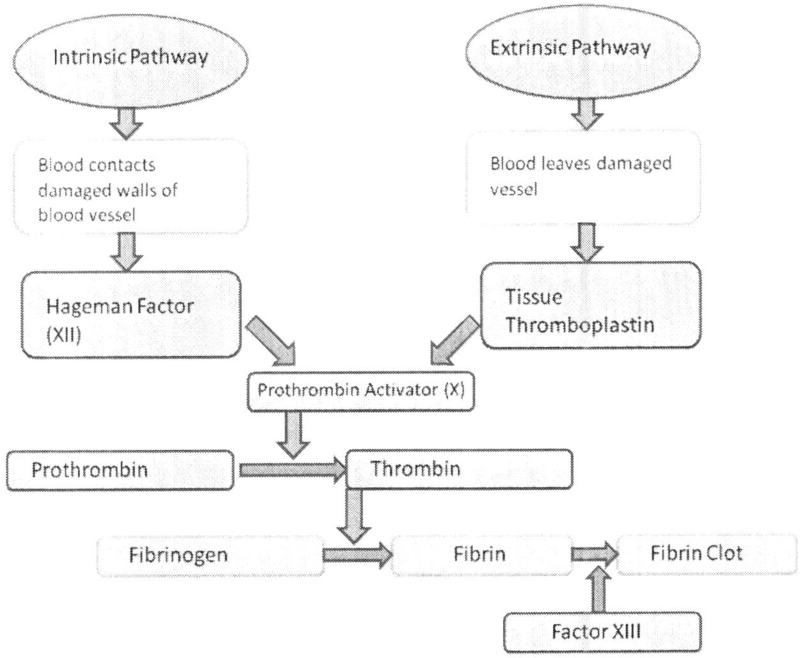

Fig. 12.8 Summary of Clotting Cascade

Notice the clotting factors in the intrinsic pathway run in a kind of sequence:

XII—XI—IX—X

Or: 12—11—9—10   (just like counting down from 12 and switching 9 and 10 at the end)

For the extrinsic pathway the factors run as follows:

7-10

There is a special factor here called tissue thromboplastin (factor VIII). Remember that in the extrinsic pathway blood has to leave the vessel and move into the tissue. This is where factor VIII is.

At this point both pathways converge at the common pathway:

Prothrombin---thrombin

Thrombin causes the conversion of fibrinogen to fibrin and we have a clot.

Clotting terms

Here are a few terms related to clotting.

- Hematoma—clot resulting from blood leakage
- Thrombus—clot forming in vessel
- Embolus—thrombus broken loose in bloodstream.
- Embolism—clot lodged in blood vessel cutting off circulation.
- Infarction—clot forming in vessel to organ (heart, lung, brain).
- Atherosclerosis—accumulation of fatty deposits in arterial linings. Causes thrombosis.

Blood Typing

Blood comes in different types. We can categorize blood according to the presence or absence of certain antigens on the surface of red blood cells.

There are only three antigens to remember: A, B, and Rh.

The Big Picture: Blood Typing

If your blood only has the type A antigen on the surface of your RBCs, then your blood type is A.

If your blood only has the type B antigen on the surface of your RBCs, then your blood type is B.

If you have both antigens, your blood type is AB.

If you have neither antigen, your blood type is O (like zero).

If you have a special antigen called the Rh antigen then you are Rh positive.

Now the other thing you need to know is that not only do you have antigens on the surface of your RBCs but you also have antibodies that can connect with the opposite antigen.

If you are type A you have the antibody that is against type B.

If you are type B you have the antibody that is against type A.

If you are type AB you have neither antibody.

If you are type O you have both antibodies.

The last bit of information you need to know is that if you are Rh positive you can receive Rh positive or Rh negative blood. However if you are Rh negative you can only receive Rh negative blood.

So now we can construct a compatibility chart.

First of all you are always compatible with your own blood type:

Blood Typing Compatibility

Type	A+	A-	B+	B-	AB+	AB-	O+	O-
A+	+							
A-		+						
B+			+					
B-				+				
AB+					+			
AB-						+		
O+							+	
O-								+

Now you can also receive type O blood in an emergency. Type O is known as the universal donor.

Blood Typing Compatibility

Type	A+	A-	B+	B-	AB+	AB-	O+	O-
A+	+						+	
A-		+						+
B+			+				+	
B-				+				+
AB+					+		+	
AB-						+		+
O+							+	
O-								+

And if you are type AB you can receive all blood types in an emergency. Type AB is known as the universal recipient.

Blood Typing Compatibility

Type	A+	A-	B+	B-	AB+	AB-	O+	O-
A+	+						+	
A-		+						+
B+			+				+	
B-				+				+
AB+	+		+		+		+	
AB-		+		+		+		+
O+							+	
O-								+

Notice we have not taken into account our last rule:

If you are Rh positive you can receive Rh positive or Rh negative blood. However if you are Rh negative you can only receive Rh negative blood.

Blood Typing Compatibility

Type	A+	A-	B+	B-	AB+	AB-	O+	O-
A+	+	+					+	+
A-		+						+
B+			+	+			+	+
B-				+				+
AB+	+	+	+	+	+	+	+	+
AB-		+		+		+		+
O+							+	+
O-								+

Now we have a complete compatibility chart.

Review Questions

1. All blood cells come from one cell called:

    a. Hemocytoblast
    b. Erythroblast
    c. Reticulocyte
    d. Osteoblast

2. Which blood cell is characterized by the biconcave disc shape:
    a. Neutrophil
    b. Red blood cell
    c. Platelet
    d. Monocyte

3. Which blood cell contains a segmented nucleus:
    a. Neutrophil
    b. Basophil
    c. Eosinophil
    d. Lymphocyte

4. Which blood cell would be prevalent in fighting off a parasitic infection:

    a. Basophil
    b. Eosinophil
    c. Neutrophil
    d. Monocyte

5. Which blood cell helps to moderate inflammation:

    a. Basophil
    b. Neutrophil
    c. Eosinophil
    d. Monocyte

6. The heme portion of hemoglobin is broken down into:

    a. Iron and biliverdin
    b. Iron and plasma proteins
    c. Potassium and bilirubin
    d. Biliverdin and bilirubin

7. Red blood cells live for about ____ days:

    a. 90
    b. 120
    c. 180
    d. 200

8. Red blood cells are recycled in:

    a. Liver and pancreas
    b. Spleen and adrenal gland
    c. Liver and spleen
    d. Ileum and thymus

9. A low RBC count is indicative of:

    a. Erythrocytosis
    b. Anemia
    c. Infection
    d. Parasites

10. Which of the following is not a method of hemostasis:
    a. Clotting
    b. Platelet plug formation
    c. Vasodilation
    d. Vasoconstriction

11. Which of the following blood types is not compatible with A+ blood:

    a. A-
    b. O+
    c. B-
    d. O-

12. Which blood type does not have either anti-A nor anti-B antibodies:
    a. A
    b. B
    c. AB
    d. O

Chapter 13

Emphatic Over the Lymphatic System

Chapter 13

Emphatic Over the Lymphatic System

Personally I don't get too excited about the lymphatic system. Yes, I know a good portion of the immune system resides there and immunity is very interesting but in and of itself I don't find the lymphatic system as sexy as, say, the cardiovascular system. But that's just my opinion. Many an A&P instructor will delightfully cover the minute details of the anatomy of a lymph node.

Don't get me wrong, the lymphatic system is important and vital to our health. Also, allied health practitioners will be assessing it by poking on lymph nodes and looking for swelling associated with lymph fluid (lymphedema). So let's get going and get emphatic about learning about lymphatics.

Big Picture: Lymphatic System

The lymph picks up fluid that has moved out of the circulatory system and carries it back into circulation while cleaning it up along the way.

Why do we have this system?

As I mentioned before a good portion of the immune system resides in the lymphatic system. A host of white blood cells stands ready to defend against nasty pathogens. We will cover immunity in another chapter so let's look at some of the other functions.

Circulatory system capillaries are leaky. It should be this way because substances should leak out of capillaries and into cells and tissues. Actually the function of capillaries is to bring good stuff like oxygen and nutrients to tissues so they can survive. The capillaries work much like a filter and we will see that filters work because of changes in fluid pressure. Well, when we calculate the pressures involved in pushing fluid through the capillaries we find that there is a net loss of fluid from the capillaries into the interstitium. This means that as blood flows through the capillaries some fluid is lost. Something's gotta pick up that fluid or we would swell up. That something is the lymphatic system.

Another important function is related to the digestive system. When you eat a Big Mac the carbs, fats and proteins get broken down by enzymes in the stomach and intestines. The broken down substances are then absorbed. Carbs and proteins move directly into the blood via digestive system capillaries. Fat however is packaged up in special packages called chylomicrons and gets dumped into, you guessed it, the lymphatic system. The lymphatic system then transports the chylomicrons to the blood.

How does it work?

Let's look at how the lymphatic system picks up interstitial fluid. Well it actually has its own set of capillaries called, you guessed it again, lymphatic capillaries. These babies intertwine with circulatory system capillaries and pick up that extra fluid. They are designed to allow fluid to flow one way, into them and off to the rest of the lymphatic system.

Part of that special design of lymphatic capillaries is that the layer of epithelium tissue forms a series of one way valves to only allow fluid in. The fluid flows from the capillaries to a series of vessels. The vessels have three layers of tissue and contain one way valves that only allow fluid to flow toward the circulatory system. The pressure to push the fluid through the vessels comes from contracting muscles. This acts like a pump to push fluid through the system. So if you want to move lymph fluid then you need to move some muscles.

Another way that lymph fluid moves is a bit trickier. As you inhale the pressure in the thoracic cavity becomes lower (actually it becomes a negative pressure). The lower pressure creates a suction that pulls lymph fluid into the thorax.

Lymph Nodes

If you've visited your doctor for the flu you would have probably been poked for swollen lymph nodes. Doc usually pokes around your neck and under your jaw for these. So you probably figured out that you have lymph nodes in your neck but you also have them in other locations as well.

These locations besides the neck (cervical) include:

Axilla (armpit)

Infraclavicular (under the clavicle)

Epitrochlear (elbow)

Inguinal (groin)

Popliteal (knee)

Pectoral (chest)

Lymph nodes are oval structures (fig. 13.1). They have a cortex (outer layer) and medulla (inner layer). Vessels carry fluid into the node (afferent vessels) and out (efferent vessels). There is a "dented" area called the hilus where the efferent vessel exits and blood vessels enter. There is also good ole reticular connective tissue forming a web-like structure throughout the inside of the node.

Lymph nodes contain white blood cells that help to "clean up" the fluid that flows through the node. These white blood cells include macrophages and lymphocytes. The macrophages "eat up" pathogens and the lymphocytes also attack pathogens by secreting chemicals that destroy them (more about this in the immunity chapter).

The lymphatic vessels eventually form larger structures known as lymphatic trunks. The lymphatic trunks drain specific portions of the body. The subclavian trunks drain the upper extremities. The jugular trunks drain the head and neck. The bronchomediastinal trunks drain the thoracic area. The intestinal trunks drain the abdomen. The lumbar trunks drain the lower extremities and pelvic area.

The lymphatic trunks connect with larger structures called lymphatic ducts which connect with the venous system at the subclavian veins. There are two ducts including the thoracic duct and right lymphatic duct. The jugular, subclavian and bronchomediastinal trunks connect to either the right internal jugular, right subclavian, or right brachiocephalic trunk. In some people the three trunks merge to form the right lymphatic duct.

The remaining trunks connect with the thoracic duct. The drainage of lymph fluid is therefore asymmetrical with respect to the arrangement of the right lymphatic and thoracic duct. In other words the right lymphatic duct drains the right side of the head, neck and trunk while the thoracic duct drains the left side of the head, neck and trunk as well as both lower extremities.

In some cases the intestinal and lumbar trunks merge to form a sac like structure called the cisterna chili.

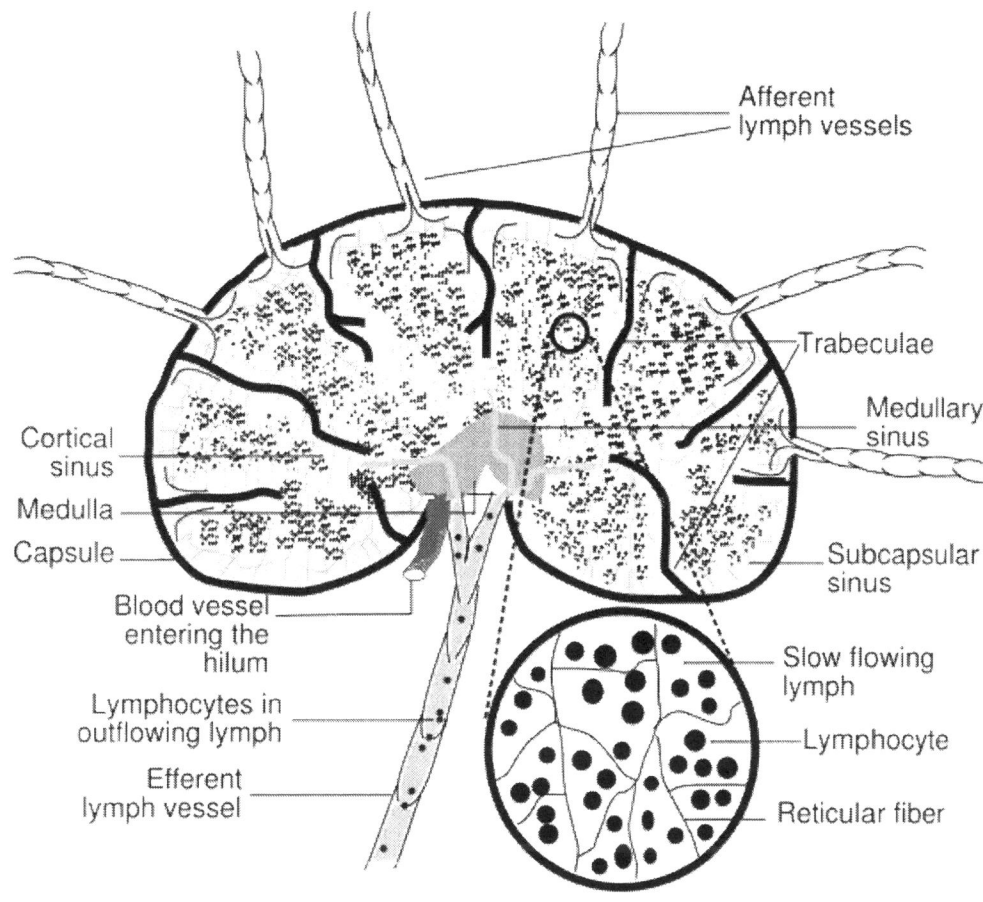

Figure 13.1. Lymph node

Lymphatic System Organs

There are two organs associated with the lymphatic system. These are the spleen and thymus. The organs contain lymphatic tissue consisting primarily of the white blood cells we mentioned earlier. As far as the lymphocytes are concerned there are two general types. These are the T and B lymphocytes. Both are produced in the bone marrow and carried to the lymphatic system. Activation of the immune system causes these cells to divide and attach pathogens.

Lymphatic tissue also contains reticular cells that produce reticular fibers. White blood cells connect with these fibers so that fluid moving through the tissue is exposed to the cells. The white blood cells can then destroy bacteria and debris.

Lymphatic tissue resides throughout the lymphatic system. When it is not located in a lymph node or organ such as in the mucous membranes of the digestive, urinary, respiratory and reproductive systems it has a fancy name known as Mucosa Associated Lymphoid Tissue (MALT). The tonsils are another example of MALT.

The spleen is located in the left upper quadrant of the abdominal area generally close to the diaphragm and is about as large as an adult fist (fig. 13.2). It consists of an outer connective tissue capsule. The inner portion has a trabeculated structure containing areas of red and white pulp. The spleen also contains venous sinuses.

White pulp consists of lymphatic tissue associated with arteries within lymphatic organs. Red pulp contains both white and red blood cells and is associated with veins.

The splenic artery and vein enter and exit the spleen at the hilum. Blood flows into the spleen and through the trabeculated network. The cells in the spleen work to destroy pathogens. Lymphocytes in the spleen can react to pathogens and trigger the immune system. The spleen also acts as a blood reservoir.

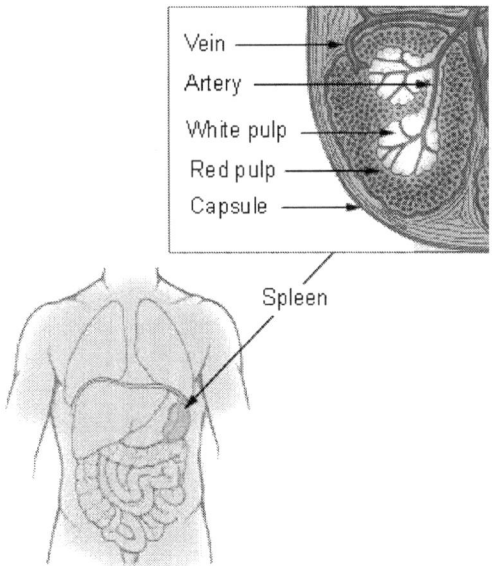

Figure 13.2. Spleen

The thymus is a gland located just deep to the sternum in the superior portion of the mediastinum (fig. 13.3). Early in life the thymus is larger and decreases in size with age although it continues to produce white blood cells.

The thymus has two lobes each surrounded by a connective tissue capsule. It contains an outer cortex and inner medulla. The internal region of the thymus is trabeculated and filled with lymphocytes. The thymus produces large numbers of T-lymphocytes that can travel to the blood.

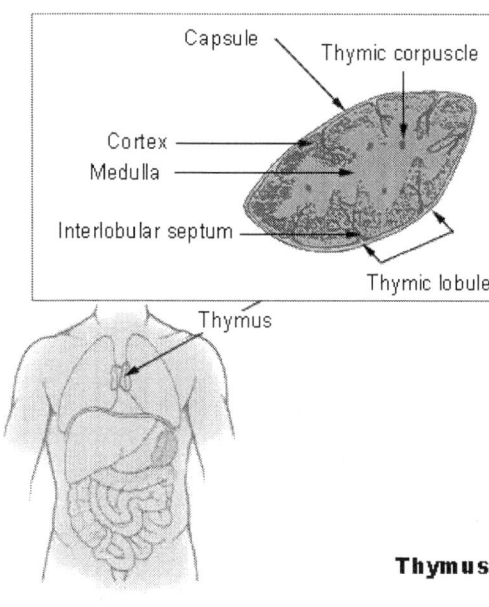

Figure 13.3. Thymus

Review Questions

1. Which of the following is not a function of the lymphatic system:
a. Transports dietary fats
b. Transports interstitial fluid
c. Contains white blood cells that function in immunity
d. Balances acids and bases

2. Lymphatic capillaries contain which type of tissue:

a. Cuboidal epithelium
b. Reticular connective
c. Simple squamous epithelium
d. Simple columnar epithelium

3. Which of the following is not a common location for lymph nodes:
a. Axilla
b. Cervical
c. Pectoral
d. Crural

4. Which lymphatic duct drains the right leg:
a. Thoracic
b. Right lymphatic
c. Left lymphatic
d. Cervical

5. The lymphatic system connects with the circulatory system at:
a. Carotid arteries
b. Jugular veins
c. Subclavian veins
d. Femoral arteries

6. Which best describes MALT lymphoid tissue:

a. Lymphatic tissue residing in mucous membranes of organs
b. Lymphatic tissue in lymph nodes
c. Located in the axillary region
d. Pre-cancerous tissue found throughout the lymphatic system

7. Which 2 organs are associated with the lymphatic system:

a. Stomach and spleen
b. Thymus and spleen
c. Liver and spleen
d. Pancreas and liver

Chapter 14

You Can't be Immune to the Immune System

Chapter 14

You Can't be Immune to the Immune System

We live in a germy world. For example the other day I was in a hurry and stopped at a local fast food establishment to get a breakfast sandwich. The attendant smiled at me and took my money with the latex gloves he was wearing. I don't think he took these off to handle the food after touching the register and who knows what else.

Thankfully we have an immune system to fight off our germy world. In fact our immune systems have two basic ways to defend our bodies from the onslaught of germs. These are known as non-specific and specific defense.

Big Picture: Non-Specific Defense

Non-specific defense is like having a zombie fence around your property (both zombies and humans are kept out).

Non-specific defense is the same kind of defense regardless of the type of pathogen. Let's say you wanted to isolate yourself from the world by living in a house with a large fence around it. The fence wouldn't care who wanted to come in, it would just keep everyone out. This is much like non-specific defense.

Examples of non-specific defense include mechanical barriers, chemical substances, inflammation and white blood cells.

Mechanical Barriers—include the skin and mucous membranes lining the various passages in the body.

Chemical Substances—include enzymes, cytokines and the complement system. Enzymes in your digestive system destroy pathogens. For example mucous is constantly moving from your respiratory system to your throat. When you swallow, this mucous moves into your stomach where the enzymes finish off the nasty pathogens. Cytokines are secreted by white blood cells and can cause fevers and interfere with viral reproduction. The complement system is a series of plasma proteins that become activated when exposed to pathogens. The end result is to destroy pathogens.

Inflammation—white blood cells release chemical substances like histamine and heparin that promote vasodilation and inhibit blood clotting. This brings more blood to the area and more pathogen-fighting blood cells.

White blood cells—known as macrophages wander through the tissue looking for pathogens to eat up or phagocytize.

The Big Picture: Specific Defense

Specific defense is like having a zombie fence with a guard at the gate (the guard specifically attacks zombies).

Specific defense differs from non-specific defense in that it recognizes pathogens and produces a specific response to them. An analogy would be the hiring of a guard at the gate of your zombie fence. The guard must recognize zombies and other human enemies and then produce a specific method of extermination without harming you.

There are two kinds of specific defense. In one type (cell mediated immunity) cells directly attack pathogens. In the other (antibody mediated immunity) cells make antibodies that do the dirty work of attacking pathogens.

The key to mounting a good defense against pathogens is to turn on or activate the immune system cells. The cells are white blood cells known as T and B lymphocytes. Before birth your body takes inventory of all of your cells and tissues. These are known as "self" tissues. Once the immune system knows all of the self cells it can easily recognize "non-self" cells or tissues. The non-self cells then activate the immune system.

The "T" in T-lymphocytes means they come from the thymus gland. These babies can become activated when directly exposed to pathogens. There are three basic types of T-cells as well. These include cytotoxic, helper and suppressor cells.

Cytotoxic T-cells—secrete pathogen-destroying cytokines when activated. These have a special protein called the CD8 protein.

Helper T-cells—secrete cytokines the help to activate B-cells and cytotoxic T-cells. These have a special protein called the CD4 protein.

Suppressor cells—actually suppress immune system activation to prevent overactivation of the immune system.

The "B" in B-lymphocytes stands for the Bursa of Fabricus in the chicken where the cells were first discovered (buck-buck). B-cells typically require costimulation through exposure to a pathogen and help from a helper T-cell. The helper T-cell secretes cytokines to help activate the B-cell.

Activated B-cells secrete antibodies that attack pathogens. This is the antibody mediated immunity we talked about earlier. The antibodies attach to pathogens and cause a variety of reactions including:

Agglutination—this is a clumping up of cells. These clumps attract other white blood cells for destroying pathogens.

Neutralization of pathogens—antibodies bind to pathogens and neutralize them.

Inflammation—antibodies can trigger inflammation.

Activation of complement system—antibodies can turn on the complement system.

The Complement System

The complement system is a series of about 20 plasma proteins (fig. 14.1). They include proteins that are named C1-C9 and factors B, D, P. They act much like the clotting cascade (see blood chapter) in that activation of the first complement protein causes the others to activate. There are two pathways in which to activate the complement system.

The alternative pathway is activated by the presentation of a pathogen to the body. The C3 protein is normally inactivated by the body's cells, however presentation of a non-self cell can cause it to remain active triggering a response.

Complement system responses include inflammation, phagocytosis from white blood cells attracted to the area, and attacking non-self cells. Activated complement system proteins can form membrane attack complexes that attack pathogens directly.

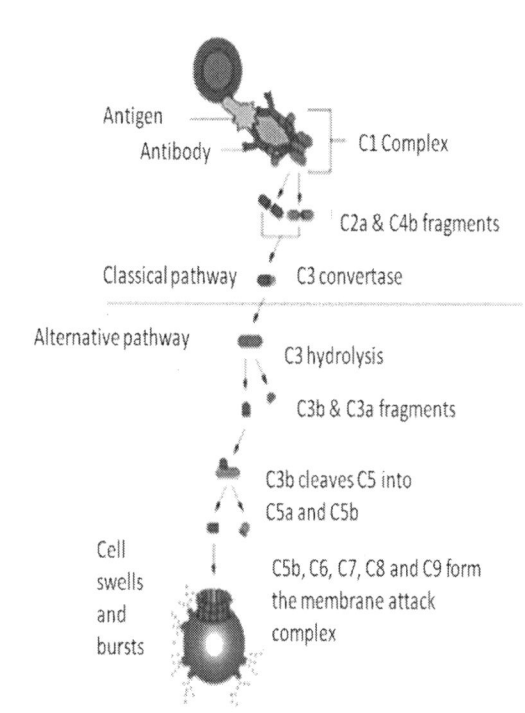

Figure 14.1. Complement System

Antibodies

There are lots of different antibodies in the body. They are generally organized into categories such as type A, B, and so on. Antibodies are also known as immunoglobulins (Ig). Some categories include:

IgA-- found in secretions such as tears, mucous and saliva and attacks pathogens.

IgE-- functions in allergic reactions.

IgG--largest category and accounts for 80% of all antibodies. IgG antibodies attack viruses and bacteria.

IgD antibodies bind to antigens on the surface of B-cells and help in activation of B-cells.

IgM--works with IgG antibodies to form immune complexes. IgM antibodies are also resident in plasma as anti-A and anti-B antibodies.

Attack of the Clones

Besides attacking pathogens and making antibodies, activated T and B-cells produce clones (sometimes called plasma cells). This army of clones jumps into action to help win the battle against the nasty pathogens. Some of these clones don't fully develop. These "memory cells" just hang out and wait until the next time that same nasty pathogen enters the body. When it does, the memory cells are ready to spring into action and destroy the pathogen lickity split.

Immune Responses

Let's say that I came down with a nasty flu. I was exposed to the flu which triggered my immune system and produced antibodies and memory cells. I was left with immunity to the same flu. This means that the next time I am exposed to the exact same flu my immune system will turn on and fight it off much faster than the first exposure. The problem is that flu viruses mutate so frequently that it is difficult to develop immunity to them.

Achooo—I'm Allergic

An allergy is an immune response to a seemingly inert substance—like pollen, peanuts, and wool. In allergies the body has a tendency to overproduce the IgE antibodies. The antibodies can produce a wide range of symptoms including inflammation, runny noses and itchy eyes. The worst of these reactions is the dreaded type I anaphylactic reaction.

The type I anaphylactic reaction can be life threatening. It is typically caused by insect stings but can be caused by just about anything. The first reaction is generally not severe because it takes time for the B-cells to produce the antibodies. However, the second reaction can be very severe and include symptoms such hives, constriction of bronchioles, and peripheral vasodilation that can cause shock.

Autoimmune Disorders

In some cases the immune system reacts to self cells and tissues and produces an immune response to them. B-cells produce antibodies known as autoantibodies. Examples of autoimmune disorders include rheumatoid arthritis, systemic Lupus erythematosis, insulin dependend diabetes, and thyroidosis.

Types of Immunity

Immunity is either innate or acquired by exposure to a pathogen. Active immunity is developed after exposure to a pathogen and production of antibodies. Passive immunity results from the presentation of antibodies from other sources.

Naturally acquired active immunity results from the exposure to a pathogen. The immune system is activated and produces antibodies and memory cells.

Artificially acquired active immunity results from exposure to pathogens given to the body in the form of vaccines. Vaccines contain inactive or attenuated pathogens that are just strong enough to produce an immune response.

Naturally acquired passive immunity occurs in utero with the passing of antibodies to the fetus from the mother. Antibodies are also passed to the infant through breast milk after birth.

Artificially acquired passive immunity occurs when antibodies are given to a person who has a damaged immune system. Antibodies must be injected periodically because of their short life span.

Review Questions

1. Which of the following is not an example of non-specific defense:

   a. Mucous membrane
   b. Digestive enzymes
   c. Antibodies
   d. Inflammation

2. Which of the following best describes a method of activation of T-cells:

   a. T-cells come into contact with antigen on pathogen
   b. T-cells activated by basophils
   c. T-cells activated by helper B-cells
   d. T-cells activated by enzymes

3. If a T-cell contains a CD8 protein it becomes a ____ when activated:

   a. Helper T
   b. Autogenic T-cell
   c. Cytotoxic T-Cell
   d. Antibody-secreting

4. Which of the following occurs with activation of both T and B cells:

   a. Antibodies are secreted
   b. Clones are produced
   c. Inflammation is produced
   d. Lymph nodes enlarge

5. Which of the following is characterized by the secondary immune response:

   a. Fast response
   b. Slow response
   c. Antigens produced
   d. Antibodies produced slowly

6. Which immune cell secretes antibodies when activated:

   a. Helper T-cells
   b. Cytotoxic T-cells
   c. B-cells
   d. Natural killer cells

7. Which antibody is prevalent in allergic reactions:
   a. IgA
   b. IgG
   c. IgD
   d. IgE

8. Which type of immunity is produced via a vaccine:
   a. Naturally acquired active immunity
   b. Artificially acquired active immunity
   c. Naturally acquired passive immunity
   d. Artificially acquired passive immunity

Chapter 15

Gettin the Blood to All the Right Places

Chapter 15

Gettin the Blood to All the Right Places

Our tissues need that good ole precious oxygen and it is the job of the heart and blood vessels to move it on through. This chapter will cover the exciting cardiovascular system in a way that I hope you will find interesting and learnable.

The Big Picture: Cardiovascular System

The heart pumps blood from the body to the lungs and the lungs to the body.

The cardiovascular system consists of the heart and blood vessels. The heart is located in the chest (thoracic cavity) just under the sternum. It angles toward the left and takes up a bit more space on the left. The lower "point" of the heart is known as the apex and the upper portion is called the base (fig. 15.1)

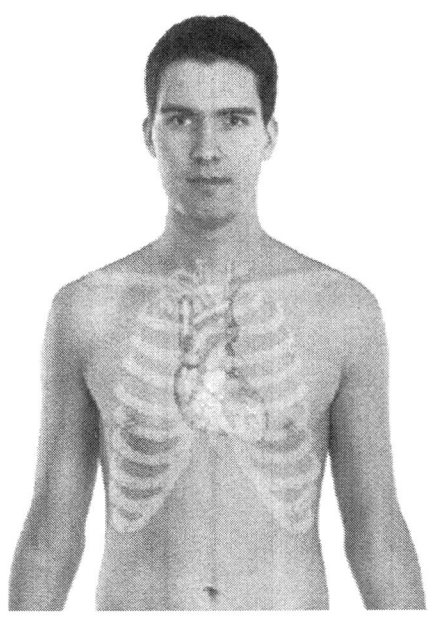

Fig 15.1. Location of the heart.

The Heart has 2 Jobs

We all know that the function of the heart is to pump blood, but did you ever think about where the heart pumps blood to? We can think of the heart as having two basic jobs.

Job 1—the heart pumps blood from the body to the lungs so the blood can get oxygen.

Job 2—the heart pumps blood from the lungs to the body to deliver oxygen.

That's it. If you understand this you can learn a lot about the structure of the heart. So, which job is harder? If you think about the distance the heart must push the blood, in job one the heart must just push the blood to the lungs. Since the lungs are right next to the heart it doesn't have to push the blood very far. Job two is a different story. In job two the heart has to push blood up to your head and out to your toes. So job two is harder than job one.

2 Jobs = 2 Sides

It turns out that each job is performed by one side of the heart. The heart has right and left sides that are very similar to each other. Job one (body to lungs) is performed by the right side of the heart while job two (lungs to body) is performed by the left side. Since job two is harder than job one the left side is a bit larger than the right.

Heart Chambers

All this pumping occurs inside hollow chambers in the heart. There are two receiving chambers (left and right) called atria. There are also two pumping chambers called ventricles. So blood flows into the atria to the ventricles and then out. The left and right atria are separated by a wall of tissue called the interatrial septum. Likewise the ventricles are separated by the interventricular septum (fig. 15.2).

Vessels

We need vessels to transport the blood to and from the heart. On the right side we have the superior and inferior vena cavae that deliver blood to the right atrium. On the left side we have the pulmonary veins that deliver blood to the left atrium.

We also have vessels that carry the blood from the ventricles toward their destinations. On the right side we have the pulmonary trunk and on the left side we have the aorta (fig. 15.2).

Valves

The heart has one way valves that only allow blood to flow in one direction (where it's supposed to go). There are valves between the atria and ventricles which include the tricuspid on the right and bicuspid on the left. There are also valves where the exiting vessels attach to the ventricles which include the pulmonary (right) and aortic (left)(fig. 15.2).

Layers and Membranes

The heart consists of three layer of tissue. These include the endocardium, myocardium, and epicardium. The endocardium is the internal layer consisting of simple squamous epithelium and connective tissue. This layer is consistent with the valves of the heart. The middle myocardium is a thick layer of cardiac muscle. The outer epicardium (aka) visceral pericardium consists of a thin serous membrane.

The heart is surrounded by a double layered sac consisting of two membranes. The outer membrane consists of fibrous connective tissue and is known as the fibrous or parietal pericardium. The inner membrane is thinner and consists of simple squamous epithelium. It is known as the visceral or serous pericardium. The visceral pericardium is consistent with the great vessels of the heart and the diaphragm.

Serous fluid known as pericardial fluid exists between the membranes. The fluid helps to reduce friction when the heart beats.

The visceral pericardium can become inflamed and produce extra fluid in a condition known as pericarditis. This can result from infection or diseases of the connective tissues. Pericarditis can also result from damage caused by radiation therapy. Pericarditis can cause severe sharp pains in the chest and back.

Blood Flow through the Heart

We now know enough to trace a drop of blood through the heart (a favorite question among anatomy professors). Here we go. Blood flows from…

Superior and inferior vena cava to…

Right atrium past the…

Tricuspid valve to…

Right ventrical past…

Pulmonary valve to…

Pulmonary trunk and ateries to…

The lungs (that as far as we will go)

AND blood flows from the lungs to…

Pulmonary veins to…

Left atrium past…

Bicuspid valve to…

Left ventricle past…

Aortic valve to…

Aorta to..

The Body!

Some Nitty Gritty Heart Structures

The bicuspid and tricuspid valves are connected to the walls of the ventricles via tendon-like structures aptly named chordae tendonae. These connect to papillary muscles on the ventricle walls.

Between the atria is a small dent-like structure which is a remnant of fetal circulation called the fossa ovalis. This used to be a hole with a flap called the foramen ovale but this closes at birth.

The coronary arteries supply the heart muscle with oxygenated blood. The right coronary artery branches off of the aorta and resides in the coronary sulcus (groove). It divides into the right marginal and posterior interventricular arteries. The right marginal supplies the right atrium and ventricle while the posterior interventricular supplies the posterior sides of both ventricles.

The left coronary also branches from the aorta and divides to form the anterior interventricular artery (aka left anterior descending artery), the left marginal artery and circumflex artery. The anterior interventricular artery supplies the anterior side of the ventricles. The left marginal artery supplies the lateral wall of the left ventricle and the circumflex artery supplies the posterior wall of the heart.

The left side of the heart muscle is drained by the great cardiac vein and the right side is drained by the small cardiac vein. Both veins empty into the coronary sinus which empties into the right atrium.

Lub-Dub the Cardiac Cycle

Big Picture: Cardiac Cycle

The cardiac cycle consists of 3 phases (Rest—Lub—Dub):

Rest: Heart rests and fills up with blood.

Lub: Atria contract pushing blood into ventricles.

Dub: Ventricles contract pushing blood out.

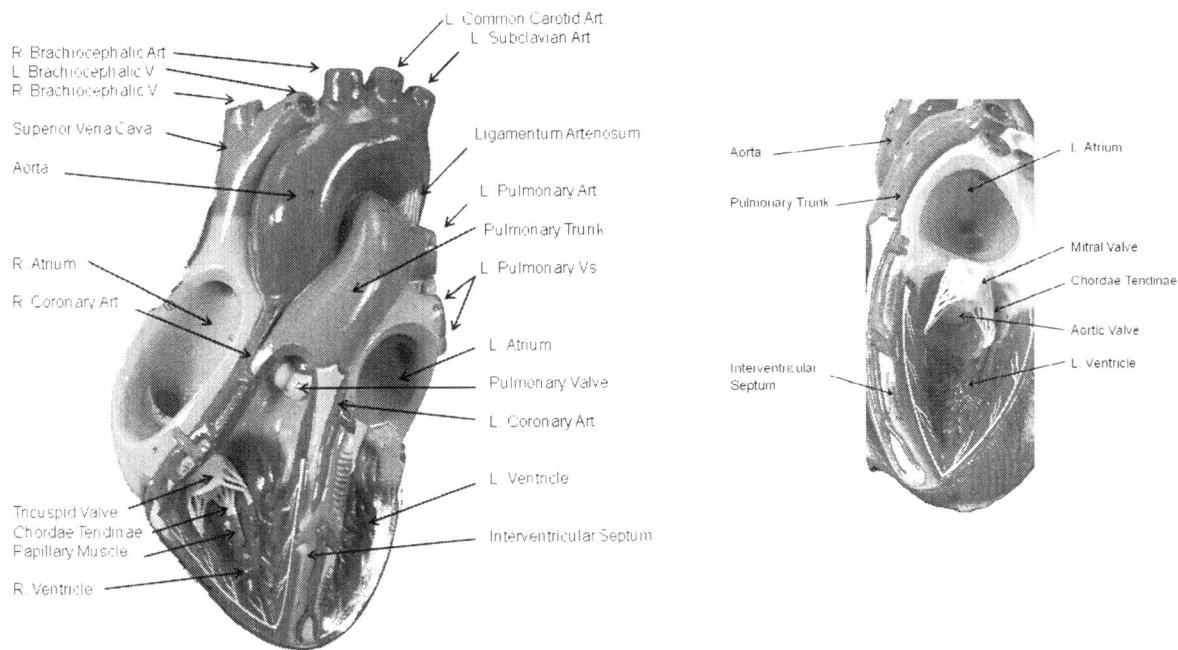

Figure 15.2. Heart

Just what is going on in the heart when it beats? There is a pattern to this lubbing and dubbing that is called the cardiac cycle. There are three phases to the cardiac cycle that repeat until the final beat (fig. 15.3). But before we describe the cycle we need to cover a couple of terms. These are "systole" and "diastole." Systole means to contract and diastole means to rest. So we can have atrial or ventricular systole and diastole. We also have to cover the concept that valves open and close by virtue of pressure changes. For example when you squeeze something (like in systole) the pressure inside it builds. Likewise when you stop squeezing, the pressure decreases. If you know the direction of blood flow then you can figure out which valves are open or closed during the different phases.

Here are the phases of the cardiac cycle in a nutshell:

Rest phase—atria and ventricles in diastole (resting, but not for long). The bicuspid and tricuspid valves are open and the aortic and pulmonary valves closed. It is interesting to note that about 70% of the blood flows into the heart during this phase.

Atrial Systole and ventricular diastole (phase 2)—atria are contracting so the pressure builds keeping the bicuspid and tricuspid valves open. The pressure in the ventricles is not high enough to cause opening of the aortic or pulmonary valves at this point so they remain closed.

Ventricular systole and atrial diastole (phase 3)—atria are relaxing and ventricles squeezing. Now the pressure builds up in the ventricles causing the bicuspid and tricuspid valves to close and the aortic and pulmonary valves to open.

When you listen to the heart (auscultation) what you hear is the closing of the valves. There are two points in the cardiac cycles where valves snap shut making an audible sound. The first heart sound known as S1 represents ventricular systole and is produced by the closing of the bicuspid and tricuspid valves. The second heart sound represents ventricular diastole and is produced by the closing of the aortic and pulmonary valves.

Arteries and Veins

The Big Picture: Arteries and Veins

Arteries always move blood away from the heart while veins move blood toward the heart.

Arteries are thicker than veins.

Veins have valves.

Blood is carried throughout the body by arteries and veins. In fact blood flows from arteries to smaller tubes called arterioles to the smallest capillaries. The capillaries have an arterial and venous side. Blood flows from the venous side of the capillaries to larger structures called venules to veins.

Arteries and veins have similarities and differences. One similarity is that they both have three layers. These layers, from inside to out are called the tunica interna, tunica media, and tunica externa. The difference is that arteries have a thick tunic media while veins have a much thinner middle layer. Another difference is in the direction of blood flow. All arteries carry blood away from the heart while veins carry blood to the heart. One mistake students make is to get used to seeing arteries colored red and veins colored blue. The problem is that there are some exceptions. For example the pulmonary arteries carry deoxygenated blood (blue) and the pulmonary veins carry oxygenated blood (red).

The third main difference is that veins have one way valves that only allow blood to flow to the heart. The reason for this is that there is not a lot of pressure generated internally in veins. Veins have to rely on external pressure generated by muscular contraction. Muscles contract and veins are squeezed increasing the pressure inside and moving the blood toward the heart via the one way valves.

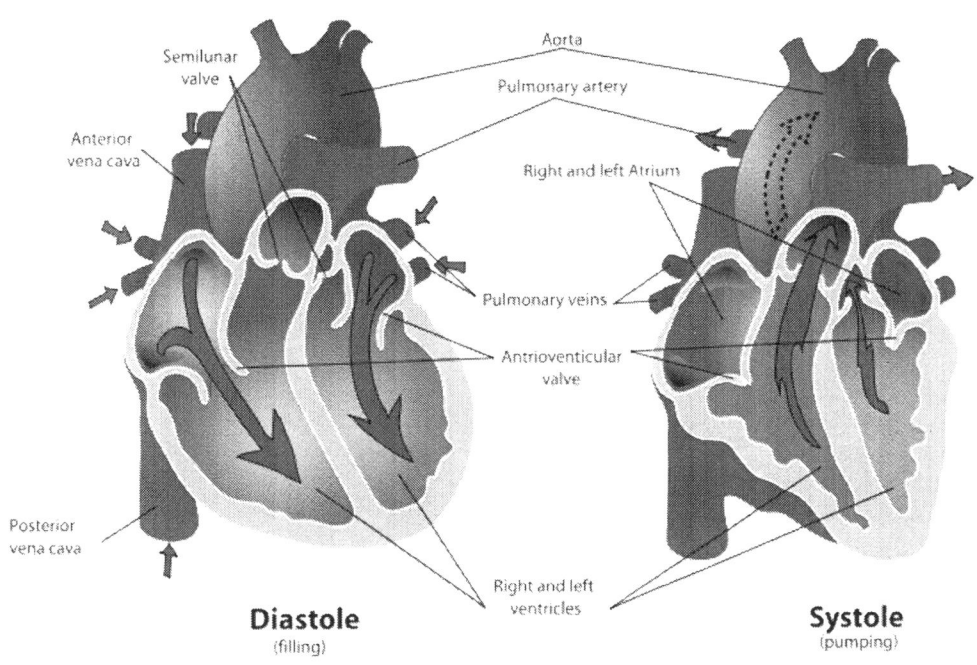

Fig. 15.3. Diastole and Systole.

Artery and Vein Mania

There are lots and lots and lots of arteries and veins. You will have to rely on the mercy of your professor to choose which of these you will need to learn. I like students to learn the main pathway of blood in the arterial system and a few of the more popular veins. This is what we will cover here. I would suggest learning it and then adding nitty gritty structures later.

The Main Flow

The circulatory system is a lot like the freeway system. There are major interstate highways (large arteries) that connect with smaller roads (arterial branches) that lead to, for example, your house (organ). You could even give directions starting from the heart and ending in a specific organ. Let's see, take the aortic arch and follow the thoracic aorta which will become the abdominal aorta. Don't turn off until you see the exit for the renal artery. Turn right and you will run right into the right kidney.

We will begin our arterial system tour with the first branches of the aorta (fig. 15.4). From right to left these are the brachiocephalic, left common carotid and left subclavian. These branches are a bit tricky because they are not symmetric. There is a common branch on the right called the brachiocephalic but two separate arteries on the left (common carotid and subclavian). One way to remember the brachiocephalic is to look at the words brachio (arm) and cephalic (head). One branch goes to the arm and the other to the head. On the right there are separate branches for the arm (subclavian) and head (common carotid).

The brachiocephalic then divides into the common carotid and subclavian arteries which continue to the head and arm.

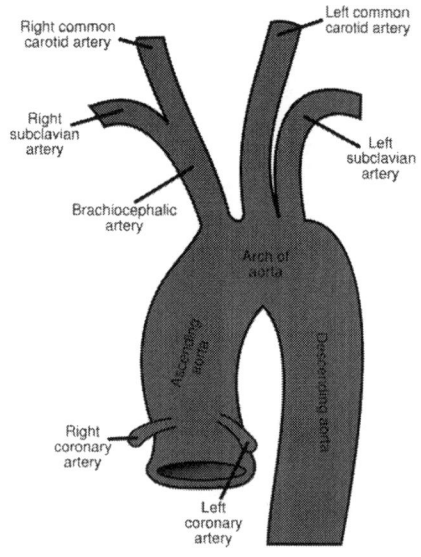

Figure 15.4. Aorta

Once you get these down you can move on to learning the major arteries for the head, upper extremity, thorax, abdomen, pelvis and lower extremity.

Here is the major flow to the left side of the head:

Aorta—brachiocephalic—left common carotid—internal/external carotids

Here is the major flow to the left arm: *Aorta—subclavian—axillary—brachial—ulnar/radial*

Here is the major flow to the trunk and leg:

Aorta—arch of aorta—thoracic aorta—abdominal aorta—common iliacs---external iliac—femoral—popliteal—anterior/posterior tibial.

What About Veins?

One nice thing about learning the arteries first is that many veins have the same name. There are deep veins and superficial veins (just look at your arm and you might see some superficial veins). Many of the deep veins have the same name as the arteries so you've killed two birds with one stone (so to say).

If you begin with the heart you might remember that the inferior and superior vena cavae bring blood to the right atrium. If we look at the superior vena cava we see that it divides into two brachiocephalic branches (symmetry at last!). Using our logic from before, we know that the brachiocephalic will divide into branches for the head and arm. These are the jugulars and subclavians (fig. 15.6).

If we follow the subclavian vein we see that it follows the same naming as the arteries:

Subclavian—axillary—brachial—radial/ulna

The difference is that we also have some superficial veins. There is one on the outside of the arm called the cephalic and one on the inside called the basilic. There is also a superficial vein in the elbow called the median cubital.

If we follow the inferior vena cava all the way down to the abdomen we see that it eventually divides into common iliacs just like the arteries. The naming is the same as the arteries. In fact these veins run right next to the arteries.

The difference is again the superficial veins. There is one important superficial vein that runs down the inside of the leg all the way to the ankle. This is the great saphenous vein.

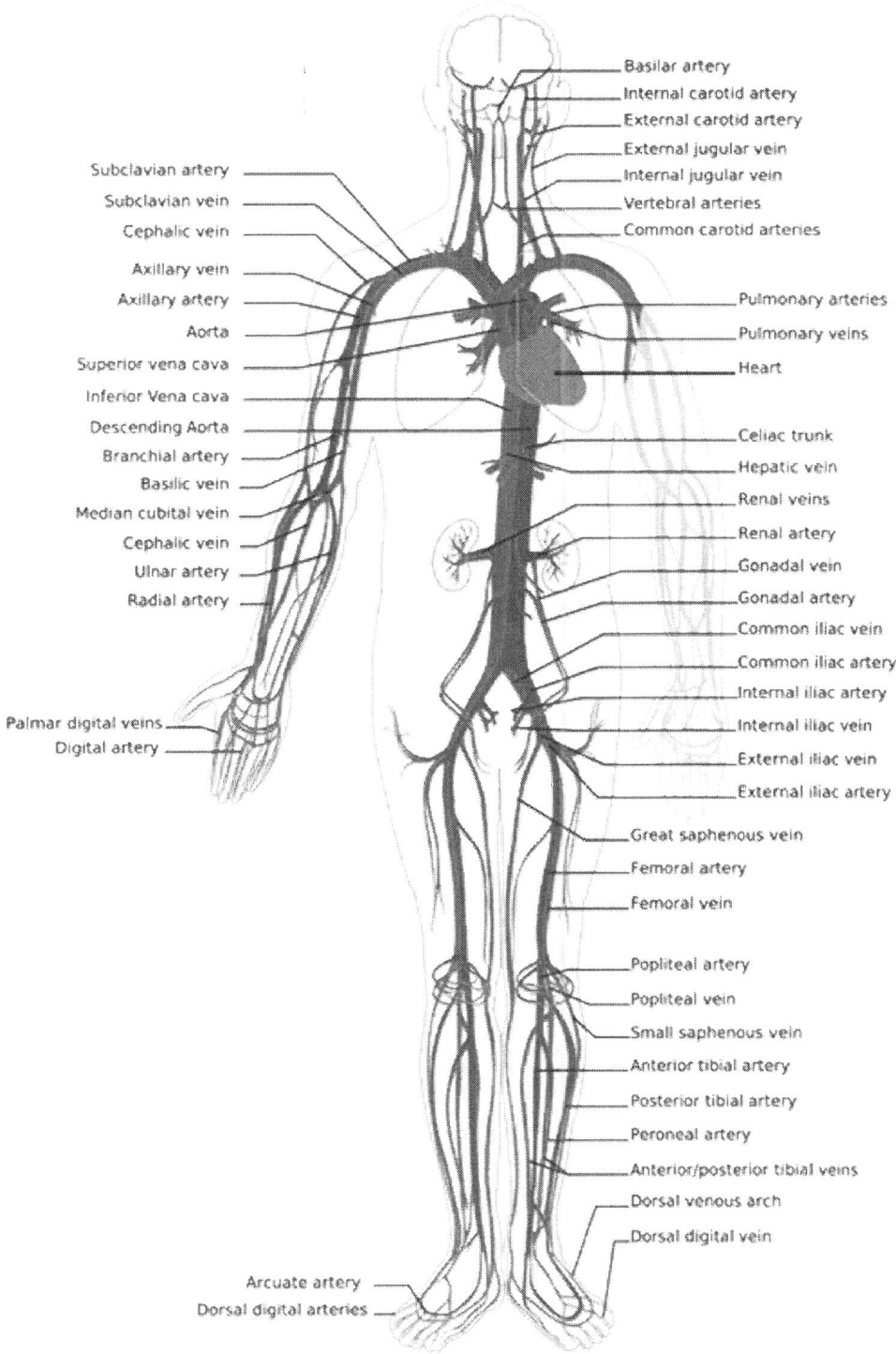

Fig. 15.5. Arteries and veins.

How does the cardiovascular system work?

Now that we've covered a good deal of cardiovascular anatomy it's time to get the big picture of the physiology. We will begin with heart muscle contraction.

One good thing about learning how cardiac muscle contracts is that it is very similar to skeletal muscle contraction (you might want to review this). There are some differences that we will cover here.

You might remember that cardiac muscle contains a special structure called an intercalated disc. These connect the cardiac muscle cells together and help to conduct electrical impulses across the heart. This provides for a more organized contraction of heart muscle (that's a good thing).

You might also wish to review the steps involved in producing an action potential in the nervous system chapter. Cardiac muscle contraction involves action potentials. Remember that nervous system cells exist at a resting membrane potential of -70 mV? Well cardiac muscle cells exist at a resting membrane potential of -90mV. Both will depolarize when sodium gates open and sodium rushes into the cell.

The difference between nervous system depolarization and cardiac muscle is that in nervous tissue the action potentials are generated fast. This is good because you don't want your brain to work slowly. This is not good for the heart because you don't want it to contract too soon before it can pump the blood out. So the depolarization needs to be delayed in heart muscle.

There is another gate that opens to help to maintain the depolarized state. This is the calcium gate. Calcium gates open and hold the cardiac muscle in a depolarized state. This keeps the heart contracting allowing it to empty.

Once the calcium channels close, good ole potassium channels work to repolarize the tissue. The process then repeats.

Cardiac Conduction

One very nice thing about cardiac tissue is that it contains areas that produce action potentials. In other words there are self-generating areas right in the heart. This means that theoretically you could take a heart and put it in a solution of electrolytes and it would beat totally on its own, without any signal from the nervous system.

The Big Picture: Cardiac Conduction

Electrical impulses flow across the heart from node to node.

The nodes are the SA, AV, and AV Bundle (fig. 15.6).

The self-generating areas of the heart are known as "nodes." The main node is located in the posterior wall of the atrium and is called the sinoatrial node (SA node). The SA node is often called the "pacemaker" node because it sets the pace of heart contraction. The SA node can generate impulses totally on its own at 60-100 beats per minute (bpm).

The SA node transmits impulses to the next node called the atrioventricular node (AV node). The AV node is located at the floor of the right atrium. If something goes wrong with the SA node the AV node can take over and generate impulses from 40-60 bpm.

The AV node transmits impulses to the atrioventricular bundle (Bundle of His) located between the atria and ventricles (hence the name atrioventricular). If something goes wrong with the AV node the AV bundle can generate impulses from 20-40 bpm.

The impulses then move down two branches called the right and left bundle branches (say that three times real fast). These carry the impulses to the ventricles via fast conducting fibers called Purkinjie fibers.

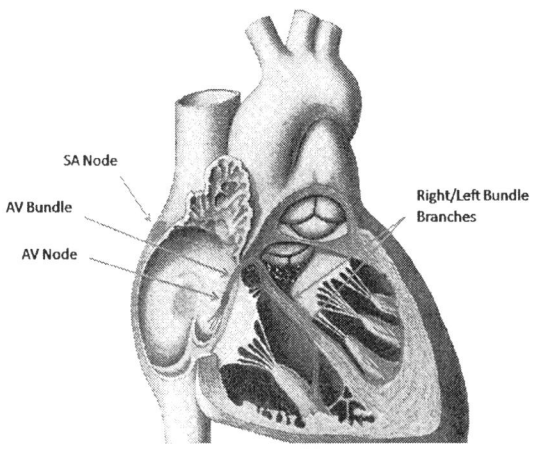

Fig. 15.6 Cardiac conduction system.

I guess we need a nervous system after all...

Even though the heart can contract all on its own it still needs to be regulated by the nervous system. So if I'm working out aerobically or sitting here on my behind writing my heart rate needs to be regulated to get just the right amount of oxygenated blood to my tissues. Hey that's the point of the cardiovascular system anyway, isn't it?

The heart is regulated by the nervous system via the sympathetic and parasympathetic divisions (remember those?). Way up there in the brain stem in the medulla oblongata is an area known as the cardiac control center. This baby sends impulses to the vagus nerve (cranial nerve X) and the cardiac accelerator nerves. The vagus nerve is a parasympathetic nerve that secretes acetylcholine to help slow the heart down while the cardiac accelerator nerves are sympathetic nerves that secrete norepinephrine that speed the heart up. Both work to regulate the heart rate depending on the body's needs.

In order to regulate heart rate something must be sensed so that the nervous system can respond. Fortunately there are pressure receptors called baroreceptors located in the aorta and carotid arteries.

Here's how they work. The alarm clock rings and I have to get my lazy behind out of bed. After a few snoozes I say to myself "one, two, three" and quickly sit up. I suddenly feel a bit dizzy for a few moments and have to sit on the side of my bed until the dizziness passes.

Here is what happened. When I quickly sat up the blood pressure in my head began to drop because my blood was obeying the law of gravity and moved out of my head. This drop in blood pressure was sensed by the baroreceptors in my carotid arteries (carotid sinus). This sent a message to my cardiac control center in my medulla oblongata. My medulla activated my sympathetic nervous system and sent a message via the cardiac accelerator nerves to the SA node of my heart. The nerves secreted adrenaline (norepinephrine) that told my heart to speed up. My heart responded bringing my blood pressure back to normal. Ah yes, I can think again!

Keepin that Blood Pressure Normal

Remember that the goal of the cardiovascular system is to get oxygenated blood to the tissues. In order to do this the system must create a force to push the blood to where it needs to go. That force is pressure. The system must create and maintain what is known as normal blood pressure in order to move the blood.

When you visit your doctor she takes your blood pressure by wrapping a cuff around your arm and pumping it up with air. She then lets the air out and listens for the familiar lub dub sound of the heart beat. The sound she first hears is known as the systolic blood pressure. This one is usually around 120 mm Hg (we measure pressure in millimeters of mercury). She then listens and notes when the sound disappears which should be around 80 mm Hg. This is the diastolic blood pressure.

The systolic pressure then represents the pressure in the system during contraction of the heart or systole. The diastolic pressure represents the pressure in the system when the heart is at rest or diastole.

There is also the mean arterial pressure. Mean arterial pressure is a measure of pressure in the arteries and is somewhere between the average systolic and diastolic pressures. Mean arterial pressure (MAP) can be determined by the following:

MAP = SV x HR X PR

SV = stroke volume (Stroke volume is the amount of blood ejected by a ventricle in one contraction.)
HR = heart rate
PR = peripheral resistance (resistance in vascular system)

This means that anything affecting stroke volume, peripheral resistance, or heart rate will also affect blood pressure. The mechanisms that control these variables will also work to control blood pressure.

We will shortly see that the heart has a lot of help in maintaining good blood pressure (to keep that blood moving). The main factors that affect blood pressure are:

Cardiac output

Blood volume

Peripheral Resistance

These are the big three when it comes to controlling blood pressure. Let's look at the big picture for each one.

Cardiac Output (Q)

Big Picture: Cardiac Output

Cardiac output is the amount of blood moved through the heart in 1 minute.

As the name implies, cardiac output represents how much blood is moved through the heart in one minute. We can describe cardiac output by using the following formula:

Q = SV x HR

Q = Cardiac output
SV = Stroke volume
HR = Heart rate

See, not too much math here, just a little multiplication. Generally SV is about 70-80ml. For example if HR is 70 bpm and SV is 80 ml then cardiac output is 5600 ml per minute or 5.6 L per minute.

But we don't need to be looking at numbers here. We can learn a lot about cardiac output by just looking at two fictitious cases.

Mary the Marathoner

Mary runs marathons and is in tip top shape. She is always training and just loves to run in all types of weather. Mary visits her doctor and while waiting is examined by the nurse. The nurse takes Mary's pulse and low and behold it is only 54 beats per minute.

Cedric the Congestive Heart Failure Patient

Cedric is suffering from congestive heart failure (CHF). He had a heart attack a few years ago and has a lot of health problems. Cedric visits his doctor and while waiting the nurse takes his pulse as well. Cedric's pulse is 102 beats per minute.

What is going on here?

It all has to do with good ole cardiac output. The key to understanding the difference between Mary and Cedric is in the stroke volume part. Mary has a strong heart with a huge stroke volume. Just like lifting weights, heart muscle will get stronger if exercised regularly. There is no better exercise for the heart than aerobic activities like running or bicycling. When Mary's heart beats it moves a lot of blood. So in order to maintain a given cardiac output with a large stroke volume the heart rate has to decrease. In other words if cardiac output remains the same then as stroke volume increases, heart rate decreases.

Likewise in poor Cedric's case his heart is weak and does not move much blood with every beat. So his heart has to beat much faster in order to maintain the same cardiac output as Mary's. In Cedric's case stroke volume has decreased causing his heart rate to increase.

The other thing we need to know about when it comes to cardiac output is that it directly affects blood pressure. In other words if cardiac output goes up, so does blood pressure. This means that whatever affects stroke volume and heart rate can also affect blood pressure.

Here is an example of how cardiac output affects blood pressure:

Jim is studying A&P and is under a lot of stress. His sympathetic nervous system is constantly turned on. Since we know all about the cardiac control center we know that Jim's sympathetic nervous system is continuously stimulating his heart to beat faster and stronger. Jim's cardiac output goes up and so does his blood pressure.

This is one way chronic stress leads to high blood pressure.

Blood Volume

Big Picture: Blood volume

Blood volume and blood pressure are related. If Blood volume goes up or down, so does blood pressure.

It should make sense that increasing the amount of blood in the vascular system can increase the pressure inside of it. So how does blood volume increase?

There is a close relationship between fluid volume and blood volume in the body. As fluid volume increases so does blood volume and consequently blood pressure. For instance, if you ate a lot of salt we know that by osmosis you will retain fluid (remember water follows salt). If your body can't compensate by getting rid of the extra salt your blood pressure will increase.

This means that the fluid volume regulating mechanism in the body will also work to regulate blood volume and blood pressure. The kidney ultimately works to control fluid volume so we will explain these systems in more detail when we get to the urinary system. However there is one fluid volume regulating system we will talk about here because it directly involves the heart.

As blood volume increases it stretches the atria of the heart. The walls of the atria respond by secreting a hormone known as atrial natriuretic hormone (sometimes called atrial natriuretic peptide). This hormone targets the kidneys telling them to remove sodium in the urine. When sodium is lost so is water (water follows salt) and when water is lost blood volume decreases.

Peripheral Resistance

Big Picture: Peripheral Resistance

Peripheral resistance is the resistance to blood flow in the cardiovascular system.

Peripheral resistance is related to blood pressure in that if peripheral resistance goes up or down, so does blood pressure.

Imagine the plumbing system in a house. The water supply line usually comes into the house in the basement so it is below ground. The water supply line then has to have enough pressure to push the water all the way to the second floor. The pressure required to push the water is just like peripheral resistance.

Peripheral resistance is the resistance in the vascular system that the heart must overcome to push the blood. Since we are talking about blood pressure here we can say that as peripheral resistance increases so does blood pressure.

Let's look at some examples of how peripheral resistance can affect blood pressure. We know that the sympathetic nervous system can increase heart rate but what we might not know is that the sympathetic nervous also causes constriction of the vessels (vasoconstriction). We know that when we squeeze a vessel the pressure inside increases. We can say that vasoconstriction increases blood pressure. Likewise vasodilation decreases blood pressure. Certain types of shock cause lots of vasodilation which cause a drop in blood pressure. Fainting (vasovagal syncope) also promotes vasodilation. The blood pressure drops and so does the patient.

Back to our house example. Let's say that we have two houses. House #1 is brand new with shiny new plumbing. House #2 is 100 years old with rusty, nearly clogged plumbing. Which one do you think will require a greater water pressure in order to push the water to the second floor? If you said house #2 you were right. But what does this have to do with the human body?

The "plumbing" in the human body can also become somewhat clogged. Increased levels of cholesterol and lipoproteins can cause deposits on the inner walls of arteries. These deposits work like the rusted, clogged plumbing in an old house. They increase peripheral resistance which in turn increases blood pressure.

Lastly, let's say that again we have two houses. House #1 is a small cozy bungalow with one bathroom. House #2 is a huge mansion with six bathrooms. Which one do you think needs more water pressure? Again if you said house two you were right. The bigger the house the more water pressure is needed to push the water through. We can say that increasing the area of the house increases the resistance. Likewise with regard to the human body gaining a few extra pounds will increase the body's surface area. As surface area increases so does peripheral resistance and likewise so does blood pressure. This is why many doctors tell their overweight patients with high blood pressure to lose a few pounds.

Pressure in the Capillaries

Big Picture: Capillaries

Capillaries are leaky. The arterial side loses fluid while the venous side gains fluid (fig. 15.7).

Since we are on the topic of blood pressure we can also cover how blood pressure helps the capillaries to function. Remember that the capillaries are the smallest blood vessels and their function is to exchange substances with tissues and cells. In order to achieve this goal, capillaries must operate via some kind of pressure system.

You can think of the capillaries as working like a filter. Think of how a water filter works. For example my home's water supply comes from a well. We have a pump that pushes the water though our house and the water moves through a filter to filter out sediment and some minerals. In order for the water to move through the filter the pressure on one side must be greater than the other side. Water always moves from higher to lower pressure in a system such as this.

The same kind of thing happens in capillaries. Blood must move from higher pressure inside the capillaries to lower pressure outside. The pressure inside the capillaries (blood pressure) is also known as hydrostatic pressure (this means fluid pressure). We can call this pressure capillary hydrostatic pressure (CHP). This pressure must be the highest in order to move fluid from inside to outside the capillary.

Since fluid moves out of the capillary and into the interstitium (tissue surrounding the capillaries) we have to consider any fluid pressure there. The interstitial fluid pressure is called the interstitial hydrostatic pressure (IHP). This pressure is negative (a sucking force) due to the sucking action of the lymphatic capillaries.

Now here's where it gets a bit complicated but stay with me (it's just a little math). In order to calculate the total fluid pressure pushing the fluid out of the capillaries, we have to combine the CHP and IHP. We do so by subtracting them to get what is called net hydrostatic pressure (NHP):

NHP = CHP - IFP

For example, CHP is usually around 30 mm Hg and IFP is usually around -3 mm Hg. We can calculate net hydrostatic pressure (NHP) as follows:

NHP = CHP - IFP

NHP = 30 mm Hg – (-3 mm Hg)

NHP = 33 mm Hg

This represents the pressure pushing fluid out of the capillaries.

This is not the whole story. There is also a mysterious sucking pressure that pulls fluid back into the capillaries. This pressure comes from osmosis (water follows salt). Floating around in the blood plasma are plasma proteins that act like a "salt" (our osmosis example). These plasma proteins create an osmotic force that pulls the fluid back into the capillary. Plasma proteins are also known as colloids so this pressure is called blood colloid osmotic pressure (BCOP).

There are also substances that act as "salts" in the interstitium. These will also create an osmotic sucking force that pulls fluid back into the interstitium. This pressure is called interstitial colloid osmotic pressure (ICOP).

We can combine both pressures to get the net osmotic pressure (NOP) by subtracting them. Here is an example:

The BCOP is usually around 28 mm Hg and the ICOP is usually around 8 mm Hg.

Net osmotic pressure (NOP) = Blood colloid osmotic pressure (BCOP) – interstitial colloid osmotic pressure (ICOP).

NOP = 28 mm Hg – 8 mm Hg

NOP = 20 mm Hg

Hang on, we're not done yet! We know the force pushing fluid out of the capillaries (NHP) and the sucking force pulling fluid back into the capillaries (NOP). So, you guessed it, we need to combine both pressures to determine the total pressure pushing fluid out of the capillary. This pressure is called the net filtration pressure (NFP).

Here's the next step:

NFP = NHP – NOP

NFP = 33 mm Hg – 20 mm Hg

NFP = 13 mm Hg

This represents the arterial end of the capillary. We see that there is a 13 mm Hg pushing force that pushes fluid out of the capillary. Thus there is a net loss of fluid from this end of the capillary bed (you know, I could have just told you that!).

The venous end of the capillary is a bit different. Fluid pressure decreases between the arterial and venous ends of capillaries. The capillary hydrostatic pressure at the venous end decreases to about 10 mm Hg.

The net hydrostatic pressure at the venous end is:

NHP = CHP − IFP

NHP = 10 mm Hg − (-3 mm Hg)

NHP = 13 mm Hg

The colloid osmotic pressures do not change because the movement of proteins from capillary to interstitium does not change.

To calculate net filtration pressure at the venous end:

NFP = NHP − NOP

NFP = 13 mm Hg − 20 mm Hg

NFP = -7 mm Hg

The negative pressure at the venous end of capillaries causes fluid to move into the capillaries. The various movements of fluid between the capillaries and interstitium maintain a balance. Disrupting this balance can result in edema.

So in a nutshell the arterial side of the capillary loses fluid while the venous side gains fluid.

The ECG

Now for something a bit less abstract. You've probably either have been to the doctor or have seen one of these on one of those medical drama shows on TV. No, I'm not talking about relationships or who broke up with whom. I am talking about the electrocardiogram or ECG (fig. 15.8). This is a cool test that consists of putting sticky electrodes on someone and recording the electrical signals from the heart (beep—beep—beep).

In order to learn the ECG you need to know at least part of the alphabet (P, Q, R, S, T). Each portion of the ECG represents something going on in the heart.

The P-wave is the first bump on the ECG and it represents depolarization of the atria. The atria depolarize shortly before they contract so the P-wave represents atrial systole.

The P wave is followed by the QRS complex. The QRS complex represents ventricular depolarization. Atrial repolarization is also occurring during this time but is overshadowed by the powerful ventricular signal. The T wave follows the QRS complex and results from ventricular repolarization.

Some common measurements include the P-R interval and the Q-T interval. The P-R interval extends from the beginning of the P wave to the beginning of the QRS complex. A prolonged P-R interval can indicate a conduction problem. The Q-T interval extends from the end of the P-R interval to the end of the T wave. The Q-T interval represents ventricular systole. A prolonged Q-T interval can indicate heart damage or electrolyte problems.

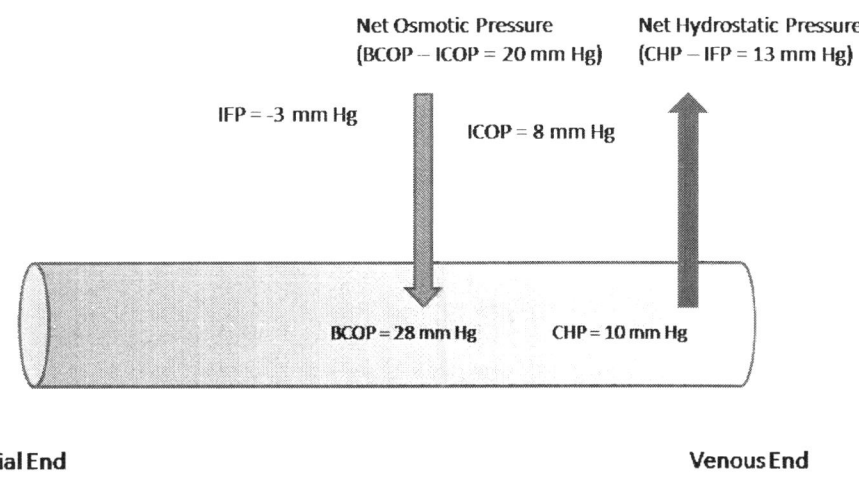

Figure 15.7. Fluid flows back into capillaries at the venous end. The net filtration pressure is -7 mm Hg.

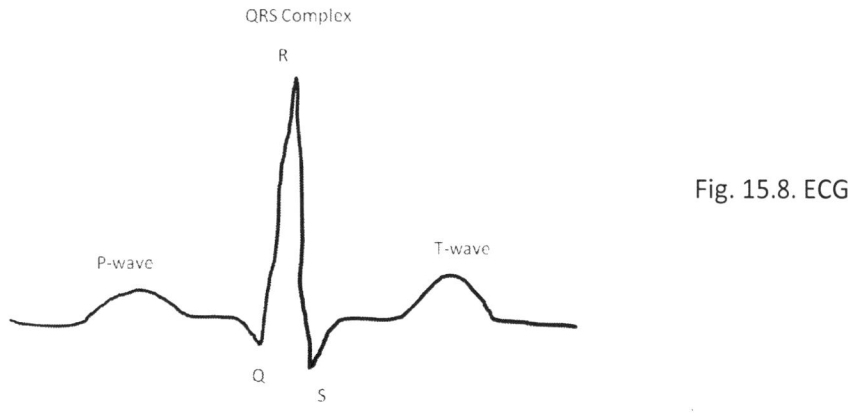

Fig. 15.8. ECG

Chapter 15 Review Questions

1. The heart contains ___ valves:

   a. 2
   b. 3
   c. 4
   d. 6

2. Blood flows from ___ to ___ through the heart:

   a. Right atrium, pulmonary trunk
   b. Left ventricle, superior vena cava
   c. Right ventricle, pulmonary trunk
   d. Left atrium, right atrium

3. Which of the following is a branch of the left coronary artery:
   a. Posterior interventricular
   b. Circumflex
   c. Brachiocephalic
   d. Common carotid

4. Which of the following is the thickest layer of the heart:

   a. Epicardium
   b. Myocardium
   c. Endocardium
   d. Pericardium

5. When the atria are in systole and ventricles in diastole the ___ valves are open and the ___ valves closed:

   a. Bicuspid/tricuspid, aortic/pulmonary
   b. Aortic/pulmonary, bicuspid/tricuspid
   c. Bicuspid/ aortic, tricuspid/pulmonary
   d. Tricuspid/aortic, bicuspid/pulmonary

6. How much blood passively flows into the heart during the rest phase of the cardiac cycle:
   a. 50%
   b. 60%
   c. 70%
   d. 80%

7. Which of the following contributes to the S2 heart sound:

   a. Bicuspid valve
   b. Aortic valve
   c. Tricuspid valve
   d. Rapid ventricular filling

8. Which of the following is not a difference between arteries and veins:

   a. Thick tunica media
   b. Valves
   c. Blood flow direction
   d. 3 layers

9. Which of the following is not a branch off of the aorta:

   a. Brachiocephalic
   b. Jugular
   c. Common carotid
   d. Subclavian

10. The common carotid artery divides into which arteries:

    a. Internal and external carotids
    b. Medial and lateral carotids
    c. Inferior and superior carotids
    d. External and internal jugulars

11. As the subclavian artery emerges from beneath the clavicle it becomes:

    a. Brachial
    b. Cephalic
    c. Axillary
    d. Basilic

12. Which of the following is a lateral superficial vein of the arm:

    a. Basilic
    b. Great saphenous
    c. Cephalic
    d. Brachial

13. Which branch becomes the femoral artery:
    a. Internal iliac
    b. External iliac
    c. Pudendal
    d. Common iliac

14. Which of the following is a superficial vein of the leg:

    a. Jugular
    b. Basilic
    c. Great saphenous
    d. Femoral

15. When the thoracic aorta descends below the diaphragm it becomes:

a. Inguinal
b. Common iliac
c. Abdominal
d. Femoral

16. Cardiac muscle has a resting membrane potential of:
a. -55mV
b. -70mV
c. -90mV
d. +30mV

17. Which of the following electrolytes is responsible for maintain depolarization in cardiac muscle:

a. Sodium
b. Potassium
c. Calcium
d. Chloride

18. The atrioventricular node is capable of producing an action potential at ___ bpm.

a. 60-100
b. 20-40
c. 40-60
d. 10-20

19. An area of abnormal tissue that generates its own impulses is known as:

a. Myocardial node
b. Ectopic pacemaker
c. Purkinjie fiber
d. Endocardial generator

20. In an ECG the T-wave represents:

a. Atrial depolarization
b. Ventricular depolarization
c. Ventricular repolarization
d. Atrial repolarization

21. Which of the following best describes cardiac output:

a. Amount of blood moving through the atria in one contraction
b. Amount of blood moving through the ventricles in one contraction
c. Amount of blood moving though the ventricles in one minute
d. Amount of blood moving though the atria in one minute

22. The cardiac control center is located here:

a. Cerebrum
b. Medulla oblongata
c. Pons
d. Midbrain

23. Parasympathetic impulses are carried to the heart by way of:

a. Vagus nerve
b. Cardiac accelerator nerves
c. Spinal accessory nerve
d. Phrenic nerve

24. The difference between systolic and diastolic blood pressures is known as:
a. Stroke volume
b. Pulse pressure
c. Mean arterial pressure
d. Peripheral resistance

25. Which of the following occurs with an increase in fluid volume:

a. ADH secretion
b. Activation of the renin-angiotensin system
c. ANH secretion
d. Retention of sodium

26. Which of the following does not increase peripheral resistance:

a. Vasodilation
b. Atherosclerosis
c. Weight gain
d. Increased sympathetic nervous system activity

27. Which type of shock results from fainting:

a. Cardiogenic
b. Hypovolemic
c. Neurogenic
d. Emotional

28. Which 2 pressures contribute to net hydrostatic pressure:
a. Blood colloid osmotic pressure and capillary hydrostatic pressure
b. Interstitial fluid pressure and blood colloid osmotic pressure
c. Capillary hydrostatic pressure and interstitial fluid pressure
d. Interstitial colloid osmotic pressure and interstitial fluid pressure

29. The net hydrostatic pressure at the arterial end of a capillary is:
   a. 10 mmHg
   b. 13 mmHg
   c. 15 mmHg
   d. 20 mmHg

30. Which of the following is a true statement about the net filtration pressure at the venous end of a capillary:
   a. It is negative
   b. It is positive
   c. It allows for stasis of blood
   d. It works to push blood back into the arterial side

31. Which of the following is a true statement about the net filtration pressure at the venous end of a capillary:
   e. It is negative
   f. It is positive
   g. It allows for stasis of blood
   h. It works to push blood back into the arterial side

Chapter 16

In With the Good Air, Out With the Bad

Chapter 16

In With the Good Air, Out With the Bad

The Big Picture: Respiratory System

Oxygen enters the lungs and moves from the lungs to the blood to the tissues.

Carbon dioxide is produced in the tissues and moves to the blood and back to the lungs.

Take in a deep breath and let that fresh country air fill your lungs. The oxygen molecules flow through your nose and into your lungs. They move from your lungs to your blood and are carried to your tissues that use the oxygen for chemical reactions that keep you alive. The reactions also produce carbon dioxide that moves from the tissues to the blood to the lungs. When you exhale you are releasing the carbon dioxide back into the atmosphere.

That's the respiratory system in a nutshell. However, your professor (or you) might want to cover a bit more detail in this fascinating system. So here we go…

What's nice about this system is that there isn't a heck of a lot of anatomy especially when compared to the skeletal or nervous systems. Also, many students are familiar with at least some of the structures (like the nose).

The respiratory system can be divided into the upper and lower respiratory systems. The upper respiratory system consists of the nose and nasal cavity, the sinuses and the pharynx. The lower respiratory system consists of the larynx, trachea, bronchi, bronchioles, lungs and alveoli (fig. 16.1).

Upper Respiratory System (fig. 16.1)

As air moves into the nose it enters the nasal cavity. The nasal cavity is lined with a mucous membrane that is highly vascularized (ever get a nose bleed?). There are bony bumps in the nasal cavity called the conchae (no, not sea shells). There are three on each side (superior, middle, inferior). The conchae create turbulent air flow that helps to clean and warm the air before it gets further down the respiratory tract.

Right down the middle of the nasal cavity is the nasal septum (sometimes it runs off to one side if deviated). Sometimes professors are pleased with students who know the bones that form the septum. These include the ethmoid bone (superior portion of the septum) and the vomer bone (inferior portion).

Once the air is nice and warm and clean it moves into the back of the nasal cavity on to the pharynx. The pharynx is a passageway that is common to the respiratory and digestive systems. Since it has to deal with both air and food traffic it needs some sort of management system. The pharynx's traffic management system consists of two structures. One is the thing that dangles from the roof of your mouth when you open it and say "ahhh". This is the uvula. The other is a cartilage flap that flips over the larynx called the epiglottis.

When you swallow the uvula flips up and closes off the nasopharynx while the epiglottis flips down and closes off the larynx. The system isn't perfect because if someone tells you an extremely funny joke while you are swallowing your laughter can cause whatever you are drinking to shoot out of your nose. The next time that happens you can say you learned about it in anatomy class.

The air can now leave the pharynx and enter the larynx (is it lair—inks or lair—nix?) (fig. 16.2). This is where the vocal cords live. There are some cartilages in the larynx as well. These include the Adam's apple (thyroid cartilage) and the smaller cricoid cartilage below it. Other cartilages include the arytenoids, corniculate and cuneiform cartilages. These cartilages are paired.

There are two parts to the vocal cords. These are the true and false vocal cords. When you speak the true vocal cords vibrate. The false vocal cords lie lateral to the true vocal cords. When the vocal cords relax they form a triangular space called the glottis.

The air exits the larynx and moves into the trachea (fig. 16.2). The trachea is a tube that contains cartilage rings that are open in the back. This allows for some flexibility while swallowing. The trachea ends with a structure called the carina. Coming off of the inferior end of the trachea are the right and left bronchi which connect with the lungs. The bronchi branch into secondary and tertiary branches that get smaller and smaller until they terminate at the alveoli. The alveoli are structures that exchange gas between the lungs and the blood. They are surrounded by capillaries.

The bronchi carry the air to the lungs. You probably know you have two lungs, a right and a left, but do you know the differences between them? The right lung is larger because it doesn't share the right side of the thoracic cavity with the heart. It actually has three lobes while the left lung has only two. Both lungs have angular fissures called oblique fissures. The right lung has an additional fissure called the horizontal fissure.

The left lung has an indentation for the heart called the cardiac impression. It also has an indentation for the aorta. Surrounding the lungs are two membranes. The visceral pleural membrane is closely adhered to the surface of the lungs while the parietal pleura lines the inside of the thoracic cavity. Both membranes secrete a slimy fluid to help reduce friction during breathing.

In with the good air…

Inhalation (breathing in) has to do with pressure changes inside and out of the lungs (fig. 16.3). You need to remember that air is a gas and gas moves according to pressure changes. Gas will always move from an area of higher to lower pressure.

If we measured the pressure outside of the lungs we might see that it is around 760 mm Hg (at sea level). This means that the pressure inside the lungs has to be less than 760 mm Hg in order for air to move from outside to inside the lungs.

If this is the case then there must be some way to manipulate pressure inside of the lungs. There is and it has to with what is known as Boyle's law. The key to understanding Boyle's law is to know that pressure and volume are inversely related. In other words when volume goes up, pressure goes down and vice versa.

Here's how it works. At the base of the thoracic cavity is a muscle called the diaphragm. The diaphragm moves downward when contracting. This action increases the volume of the thoracic cavity. When volume increases guess what happens to pressure? You're right, pressure decreases and air moves into the lungs.

That's fine and dandy for resting inhalation but what about taking in a deep breath like when you blow out your birthday candles? Well, using our logic you would need to expand the thoracic cavity even more. Fortunately we have some muscles (called accessory muscles of inspiration) that help make this happen. These include the following:

Sternocleidomastoids—helps to lift the ribcage and expand the thoracic cavity superiorly.

Pectoralis minors—lift and expand the ribcage superiorly and laterally.

External intercostals—muscles between the ribs that work to expand the ribcage.

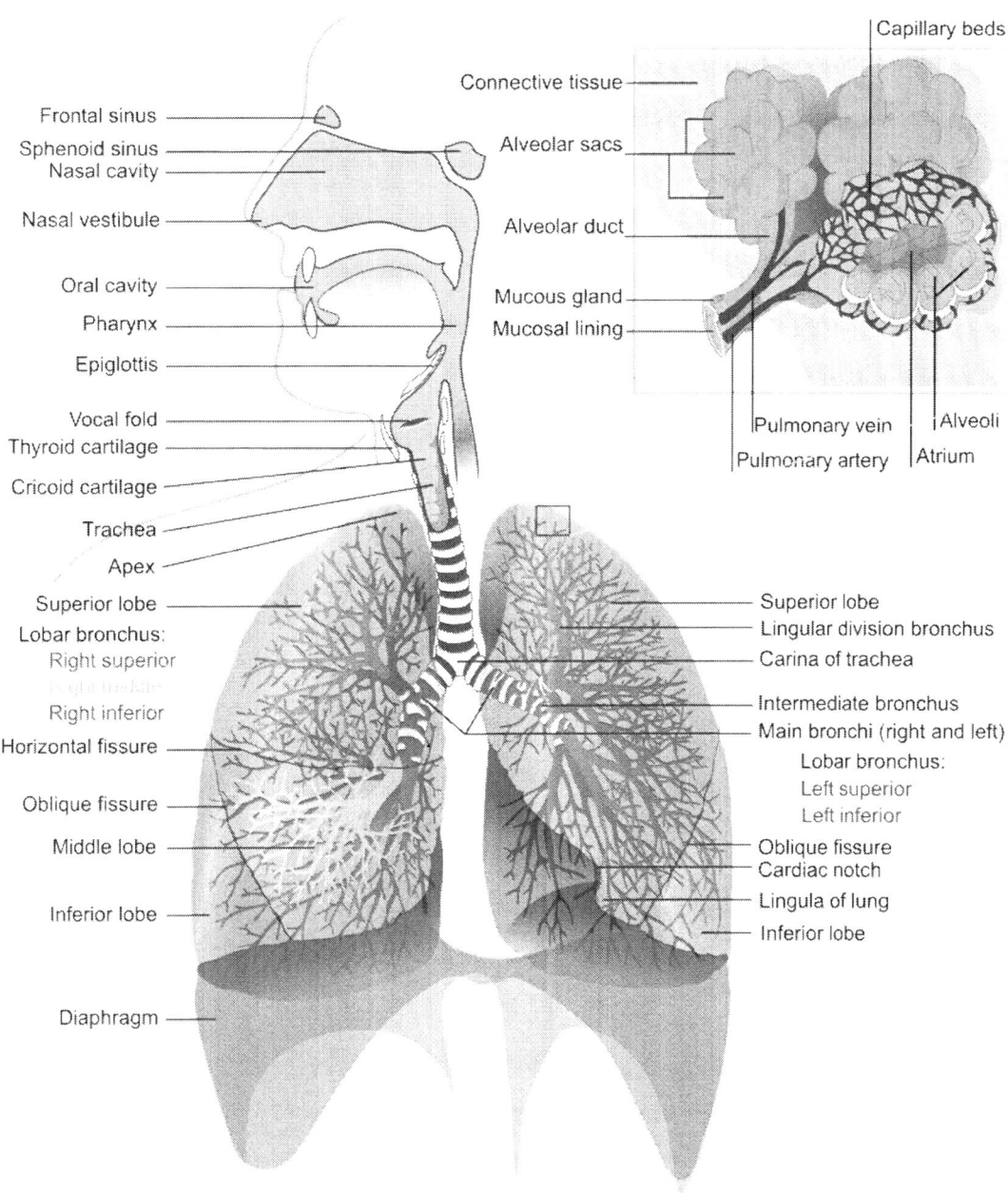

Figure 16.1. Respiratory System

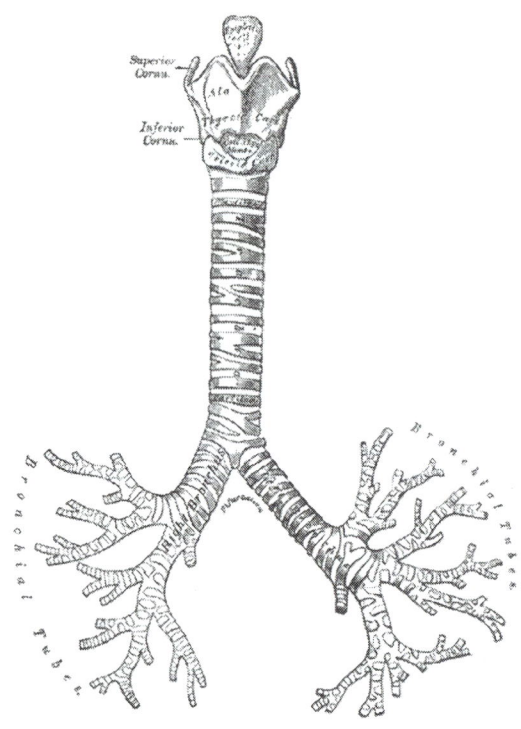

Figure 16.2. Trachea

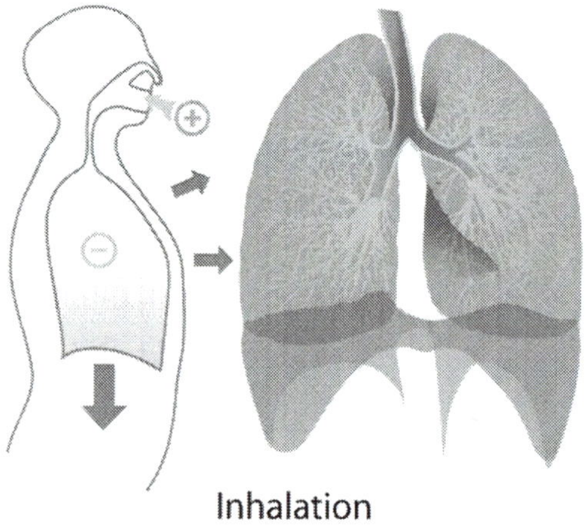

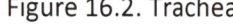

Figure 16.3. Inhalation

Out with the bad air...

To exhale the diaphragm relaxes and decreases the volume of the thoracic cavity (fig. 16.4). The lungs then spring back into their resting position and air is expelled from the lungs. The lungs have an elastic property called compliance. The elasticity of the lungs allows them to spring back into shape during exhalation. Since no energy is used exhalation is considered a passive process.

Just like in inhalation sometimes you need to exhale forcefully such as in blowing up a balloon. Again, there are some muscles to help out. The accessory muscles of exhalation include:

Internal intercostals—muscles between the ribs that work to squeeze the ribs together.

Abdominals—contract and help to decrease the volume of the thoracic cavity.

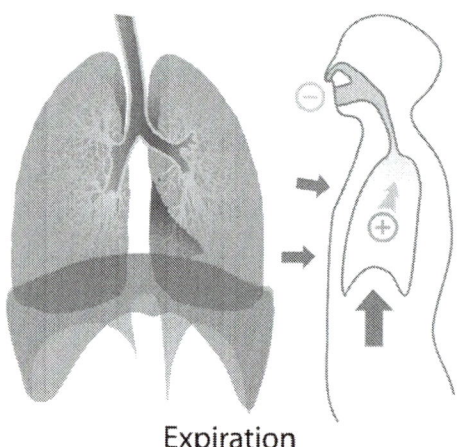

Figure 16.4. Exhalation

Respiratory Rates and Volumes

The typical adult breathes in and out about 12 to 18 times per minute. Kids are a bit more hyper and breathe in and out about 18 to 20 times per minute. Doctors (and physiologists) can measure the amount of air going in and coming out using a device called a spirometer. Back in my old school days these used to be large devices with revolving drums and pens that would record the volumes of air on a rotating graph. Also, back in those days a lot of my classmates smoked. This was great for physiology class as we could compare the smokers' lung volumes with us healthy geeks. Of course now-a-days we know better—eh? Also, thanks to the geeks modern spirometers can fit in the palm of your hand (go geeks!).

If I am just relaxing and, of course breathing, the amount of air going or coming out of my lungs is known as tidal volume (about half a liter or 500 ml).

If I take in a deep breath after a normal exhale, the amount of air going into my lungs in addition to tidal volume is known as inspiratory reserve volume (3.3 L in males and 1.9 L in females).

If I exhale as hard as I can after inhaling normally, the amount of air coming out of my lungs in addition to tidal volume is known as expiratory reserve volume (about 1 L).

No matter how hard I try I cannot blow out all of the air in my lungs (once in awhile students will ask me if they can exhale all of the air out of their lungs). There is always some air left in the lungs. This additional air is known as residual volume (1.2 L in males and 1.1 L in females).

What we have just covered are the basic respiratory volumes. We can add these together in various ways to get what are known as capacities.

Here are the basic respiratory capacities:

Vital capacity is the maximal amount of air that can move in and out of the lungs in a single breath. It is the sum of tidal volume, inspiratory reserve volume and expiratory reserve volume. It is about 4800 ml in males and 3400 ml in females.

Inspiratory capacity is the amount of air that can move into the lungs after resting inhalation and exhalation. Inspiratory capacity is the sum of tidal volume and inspiratory reserve volume.

Functional residual capacity is the air remaining in the lungs after a resting inhalation and exhalation. Functional residual capacity is the sum of expiratory reserve volume and residual volume.

Total lung capacity is the total volume of air in the lungs. It is the sum of vital capacity and residual volume. It is about 6000 ml in males and 4500 ml in females.

Partial Pressure of a Gas

We tend to think of air as containing oxygen. Yup, it's in there but there are also a few other gasses in air. Actually air is mostly nitrogen (78.6%). Oxygen accounts for about 20.9% and carbon dioxide about 0.4%.

Remember our 760 mm Hg pressure of air at sea level. Well not all of that pressure is due to oxygen. In fact since air is only about 21% oxygen then oxygen only accounts for 21% of the 760 mm Hg. This works out to about 159 mm Hg. This is known as the partial pressure of oxygen or PO_2.

Why do we need to know about partial pressures? Well, it is useful to know about partial pressures when we study how oxygen and carbon dioxide are exchanged between the lungs, blood and tissues. This is our next topic.

Big Picture: Respiratory Physiology

Everything goes where it needs to go...

Yes, I know it sounds a bit stupid and simplistic but bear with me. Let's elaborate:

Oxygen travels from the lungs to the blood to the tissues.

Carbon dioxide travels from the tissues to the blood to the lungs.

So how do oxygen and carbon dioxide move from lungs to tissues to blood to lungs, etc? They follow partial pressure gradients. We can think of partial pressures as being analogous to concentration. Substances always move from areas of higher to lower concentration. We can apply this thinking to partial pressures and say that oxygen and carbon dioxide move from areas of higher partial pressure to lower partial pressure. Or, to summarize:

Oxygen moves from areas of higher PO2 to areas of lower PO2.

Carbon dioxide moves from areas of higher PCO2 to areas of lower PCO2.

So we just need to follow oxygen and carbon dioxide through the system. Let's start with oxygen. The partial pressure of oxygen is greatest at the source of oxygen which is the lungs. Alveolar PO2 is the greatest PO2 in the system at about 104 mm Hg. As deoxygenated blood moves into the lungs (from the pulmonary arteries to the capillaries surrounding the alveoli) it has a PO2 of about 40 mm Hg. Oxygen can then diffuse from the higher PO2 (104 mm Hg) to the lower PO2 (40 mm Hg)(fig. 16.5).

Oxygen diffuses into the blood causing the PO2 to rise to 104 mm Hg. The blood mixes with some blood from the bronchial veins which causes the PO2 to decrease to about 95 mm Hg. This is the PO2 that leaves the lungs and moves to the rest of the body.

When blood reaches the tissues it needs to diffuse from the capillaries to the tissues so the tissue PO2 must be less than that of the blood. Tissue PO2 is about 40 mm Hg so oxygen can easily diffuse from blood to the tissues. Since oxygen moves out of the blood the PO2 decreases back to 40 mm Hg.

The tissues need oxygen for chemical reactions and they produce carbon dioxide which must be removed. So let's look at how carbon dioxide moves through the system.

Let's start at the source of carbon dioxide where the PCO2 is the greatest. The source of carbon dioxide is the tissues with a PCO2 of about 45 mm Hg. The PCO2 of oxygenated blood is about 40 mm Hg. When oxygenated blood meets the tissues, carbon dioxide diffuses from the tissues to the blood. This brings the PCO2 of deoxygenated blood leaving the tissues up to 45 mm Hg.

The deoxygenated blood moves to the lungs where the alveolar PCO2 is at about 40 mm Hg. When the deoxygenated blood with a PCO2 of 45 mm Hg meets the alveoli (at PCO2 of 40 mm Hg) carbon dioxide diffuses into the lungs and is exhaled out of the body (fig. 16.6).

The process then starts all over again.

One way to think of this is my dump truck analogy. I thought of this a few years ago while playing with my daughter. We have two dump trucks. One truck is a bright red oxygen truck while the other is a pretty blue carbon dioxide truck. The dispatcher calls the oxygen truck and asks it to fill up with oxygen at the lungs. The driver says that he can because he only has a partial load of 40 units. He drives to the lungs (following the veins and pulmonary arteries) to the lungs where he "fills up" with oxygen to 104 units. He then drives to the tissues (following the pulmonary veins and systemic arteries) to "dump" the oxygen. Along the way the road is bumpy so he loses a bit and he arrives at the tissues with only 95 units. He only delivers a partial load and still has 40 units when he is finished. He then travels back to the lungs and gets another load and so on.

The pretty blue carbon dioxide truck has a similar job. He has to go to the tissues to "fill up" with carbon dioxide. His truck is partially full as well and he begins with only 40 units of carbon dioxide. He follows the systemic arteries to the tissues where he "fills up" partially to 45 units. He then heads to the lungs following the systemic veins and pulmonary arteries where he partially "dumps" his load to 40 units. He then heads back to the tissues and starts over (vroom vroom).

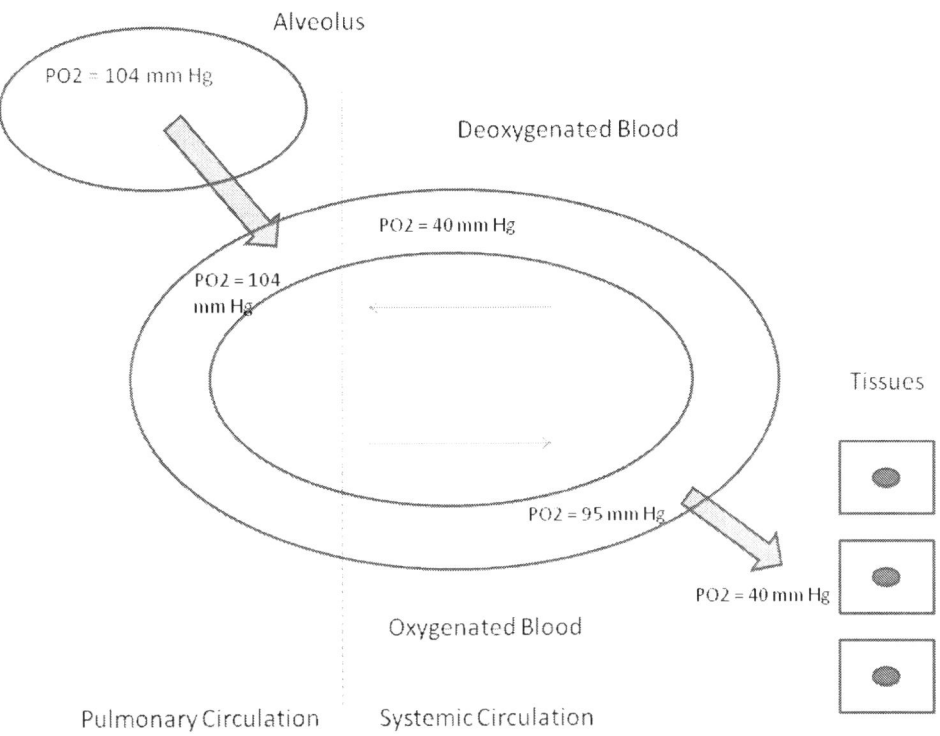

Fig. 16.5 Oxygen exchange in the respiratory system.

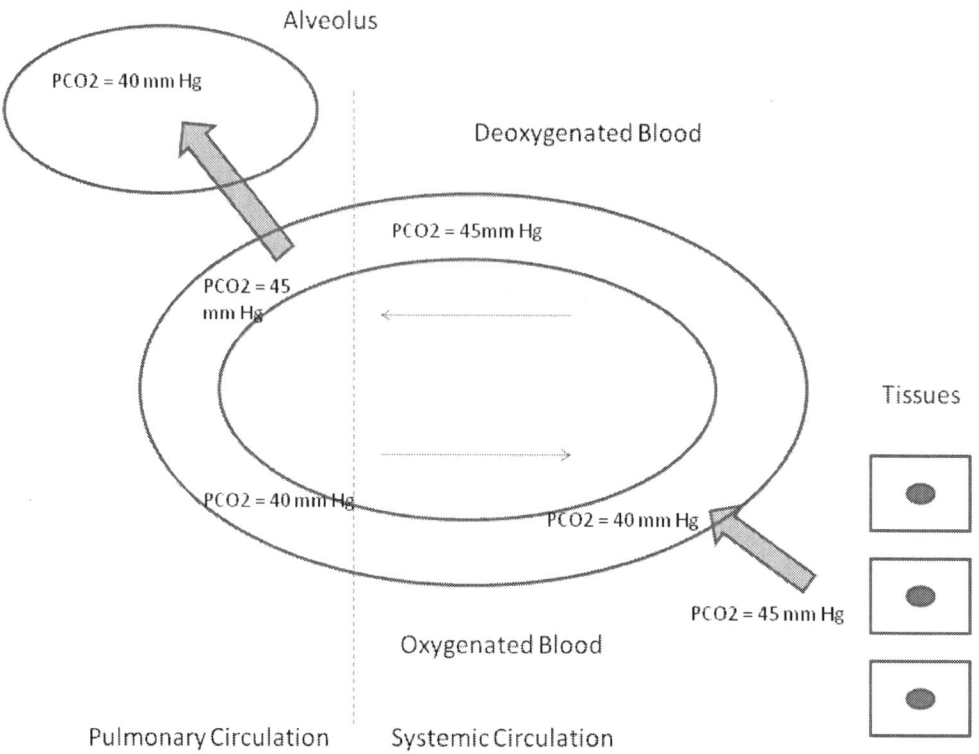

Fig. 16.6 Carbon dioxide exchange in the respiratory system.

Big Picture: Carbon dioxide storage

Carbon dioxide is stored three ways in the blood:

1. *Dissolved in plasma*
2. *Attached to hemoglobin*
3. *Hidden in the bicarbonate ion*

Now that we've seen how carbon dioxide moves into the blood, what happens to it once it gets there? Carbon dioxide is transported in the blood three different ways. Two of these are relatively simple and the other is a bit trickier.

One way carbon dioxide is transported is that some of it is dissolved in blood plasma. Only about 7% of the carbon dioxide carried in blood is dissolved in plasma. Another way that carbon dioxide is transported is by way of a hemoglobin compound called carbaminohemoglobin (use this word to impress your friends or win at scrabble). Red blood cells contain hemoglobin which is famous for transporting oxygen. Hemoglobin can also transport some carbon dioxide (about 23%).

The final way is the tricky one, and the most important of all. Most of the carbon dioxide is transported in blood by way of the bicarbonate ion (HCO_3^-).

Here is the way it works. Carbon dioxide moves into the blood and comes in contact with red blood cells where it diffuses into the cells. Red blood cells contain an enzyme called carbonic anhydrase. When carbon dioxide meets carbonic anhydrase and water it form a molecule of carbonic acid. The carbonic acid is ionically bonded so it can dissociate (break apart) in water to form the bicarbonate ion and hydrogen ions (fig. 16.7, 16.8).

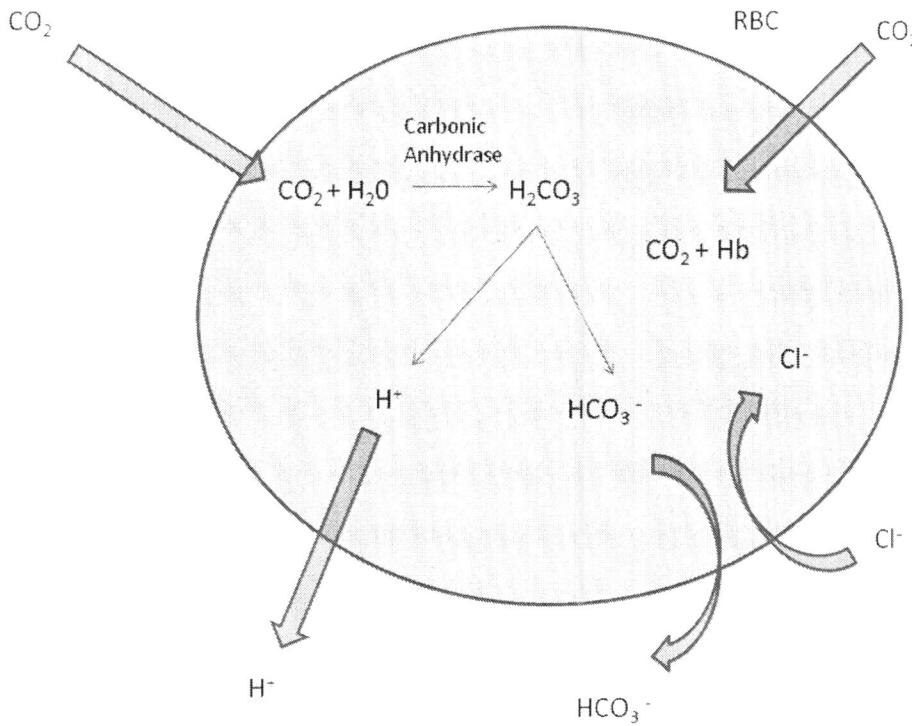

Figure 16.7. Storage of carbon dioxide in bicarbonate ions. Carbon dioxide combines with water and carbonic anhydrase to form carbonic acid that dissociates into bicarbonate and hydrogen ions. Both move out of the red blood cell with chloride ions moving in to maintain ionic stability. Carbon dioxide also enters the red blood cell and combines with hemoglobin. This process occurs in areas of high PCO2 where carbon dioxide needs to be transported.

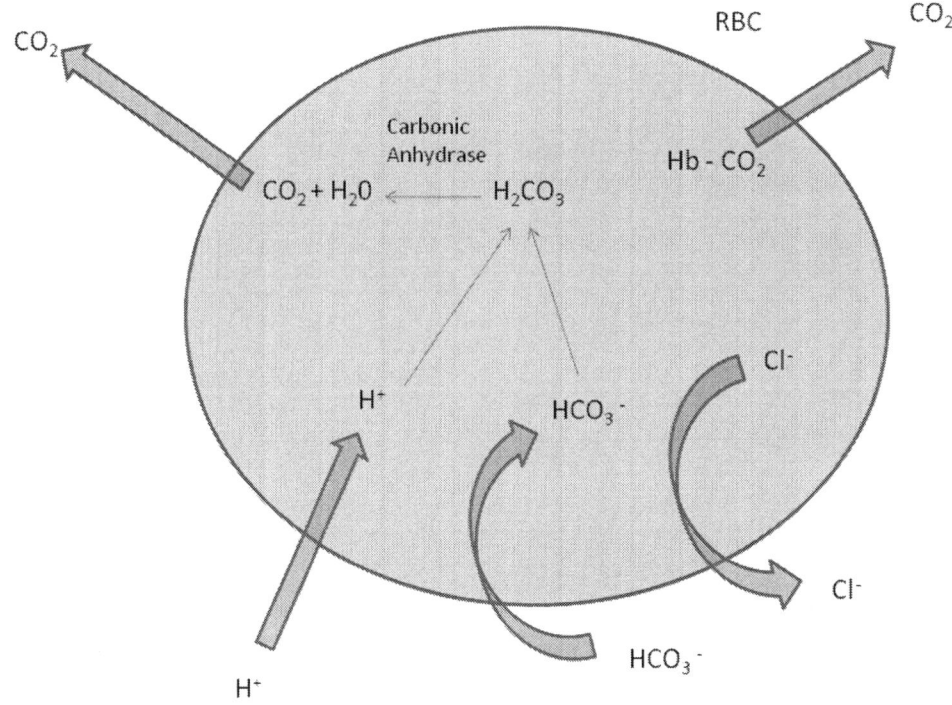

Figure 16.8. In areas of low PCO2 the process reverses. Bicarbonate and hydrogen ions enter the red blood cell and are converted to carbonic acid which converts to water and carbon dioxide. Hemoglobin also releases carbon dioxide. Both mechanisms work to release carbon dioxide for diffusion into the lungs so that it can be expelled by exhalation.

This brings up an interesting point. The more carbon dioxide builds up in the blood the more it is stored in the bicarbonate ion and the more hydrogen ions are produced. Remembering your chemistry you know that a high concentration of hydrogen ions makes things acidic. So when carbon dioxide builds up in the blood the blood also becomes acidic. We then have a condition called respiratory acidosis.

Likewise the reaction can go the other way. Let's say that for some reason the lungs are getting rid of lots of carbon dioxide.

The hydrogen ion concentration will decrease causing the blood to become alkaline. We then have a condition called respiratory alkalosis.

You can induce a state of respiratory acidosis or alkalosis right now if you'd like. To produce respiratory acidosis, simply hold your breath. Your tissues still produce carbon dioxide (hey, you're still alive) but it can't get out of your lungs so it builds up in the blood. Fortunately your nervous system will sense the buildup of carbon dioxide and hydrogen ions and cause an overwhelming urge to inhale.

Likewise you can induce a state of respiratory alkalosis by hyperventilating. Your lungs will get rid of lots of carbon dioxide which decreases the hydrogen ion concentration making your blood more alkaline.

Okay, we've talked a lot about carbon dioxide but what about oxygen? A little bit of oxygen dissolves in blood plasma but by far the majority of oxygen is transported via hemoglobin. Up to four oxygen molecules can bind with one hemoglobin molecule and there is something like 300 million hemoglobin molecules in one red blood cell.

One way to understand how hemoglobin works is to look at the oxygen-hemoglobin saturation curve (fig. 16.9).

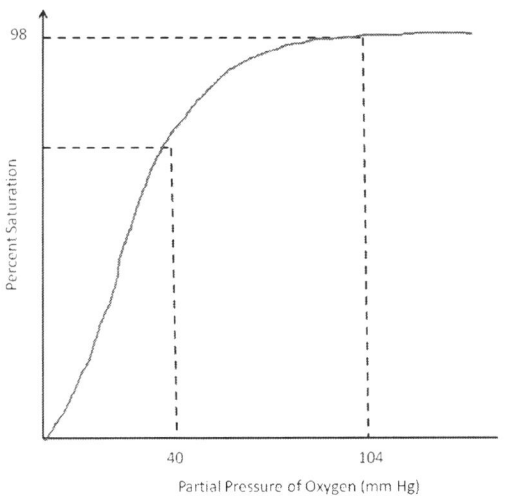

Figure 16.9. Oxygen-hemoglobin saturation curve.

Notice that PO2 is on the horizontal axis and percent saturation of hemoglobin in on the vertical axis. In order to further understand what's going on here I will invoke one of my universal physiology rules one more time:

Everything goes where it needs to go...

Let's apply the rule to the curve. First of all in areas of low oxygen what do you think hemoglobin should do? It can either hang on to the oxygen or let go of it so it moves into the tissues. Our tissues need oxygen so you should say that hemoglobin lets go of oxygen so it can move into the tissues. Now let's use fancy physiology terminology to describe this:

Hemoglobin decreases its affinity for oxygen binding in areas of low PO2.

Likewise what should hemoglobin do in areas of high levels of oxygen such as in the lungs? Should hemoglobin want to bind with oxygen or let go of it? Well we know that oxygen needs to go from the lungs to the blood to the tissues so if you said hemoglobin wants to bind with oxygen you were right. Again using fancy terminology we can say:

Hemoglobin increases its affinity for oxygen binding in areas of high PO2.

In fact hemoglobin is almost completely saturated (98%) in the lungs.

Now hemoglobin is a bit tricky but that's a good thing. Let's say that I am doing my aerobic workout. Logically my tissues would need more oxygen, right? Well, my hemoglobin can help me out with this. It can actually change how it works based on the body's need for oxygen. In the case of exercise hemoglobin actually releases more oxygen in areas of low PO2 (my oxygen-wanting tissues) while still becoming almost completely saturated in the lungs. We can say that the saturation curve shifted to the right. The change in hemoglobin's affinity for oxygen binding is called the Bohr effect.

At the same time my tissues need more oxygen they are also making more carbon dioxide. Hemoglobin helps out here too. Remember that some carbon dioxide binds with hemoglobin. Well, during periods of increased demand for oxygen such as in exercise hemoglobin also wants to bind with more carbon dioxide in areas of higher carbon dioxide (PCO2). Hemoglobin helps us out by helping to transport more carbon dioxide from the tissues to the blood so that it can be removed by the lungs. This is known as the Haldane effect.

Exercise isn't the only thing that triggers the change in hemoglobin. Increases in temperature can also cause hemoglobin to change how it functions.

How Respiration is controlled by the Nervous System

We need a control system in order to adjust our breathing to maintain just the right amount of oxygen and carbon dioxide in our blood. The control system is located in the nervous system, specifically in the brainstem. There are four main respiratory centers located in the pons and medulla oblongata.

The medulla contains the medullary respiratory center. The medullary respiratory center consists of two groups of neurons called the dorsal and ventral respiratory groups.

The dorsal respiratory group consists of two groups of neurons located in the posterior area of the medulla oblongata. This group is primarily responsible for contraction of the diaphragm for regulation of breathing rate. The neurons receive input from other parts of the brain and receptors that sense changes in concentrations of gases and pH.

The ventral respiratory group stimulates the external and internal intercostals and abdominal muscles. This group works to regulate breathing rhythm.

The pons contains the pneumotaxic center (now called the pontine respiratory group). This center works with the centers in the medulla and helps to fine tune breathing rate and rhythm. The pneumotaxic center also receives input from other centers in the brain.

The apneustic center also resides in the pons. The pneumotaxic center inhibits the apneustic center to help control exhalation. However if damage to the brainstem occurs the person can exhibit what is known as apneustic breathing.

This consists of a very slow respiration rate with a deep inhalation held for ten to twenty seconds followed by shallow and brief exhalations that provide little pulmonary ventilation.

Breathing is not entirely unconscious. We can decide to take in a deep breath or hold our breath. The cerebral cortex provides connections to the brainstem centers for breathing. The limbic system also affects breathing. For example strong emotions elicited in the limbic system can speed up breathing.

All of the above respiratory centers innervate the phrenic and intercostals nerves.

A Protective Mechanism

Have you ever wondered whether you could burst your lungs by taking in a deep breath? I remember taking a fitness test in my younger years and I really thought my lungs were going to burst from my breathing so hard. Fortunately we have a protective mechanism that does not allow our lungs to burst.

The Hering-Breuer reflex is a protective mechanism and prevents overinflation of the lungs. Stretch receptors on the walls of the bronchi and bronchioles send impulses to the vagus nerve to the medulla oblongata. The impulses inhibit the respiratory centers and produce exhalation.

The Role of Surfactant

Big Picture: Surfactant

Surfactant is secreted by alveolar cells in order to reduce surface tension.

Before we get into what surfactant does, we need to cover the concept of surface tension. Surface tension is a force that exists on the surface of a fluid (like water). It is produced by chemical bonds that occur between water molecules. You might remember from chemistry that water molecules are polar, meaning they have a slight negative charge on one end and a slight positive charge on the other. (Fig. 16.10)

The positive and negative charges cause the water molecules to stick together by forming polar covalent bonds. These bonds produce a force called surface tension. If you were to spill a drop of water on a table, you would see that the drop is in the shape of a dome. This is caused by the surface tension pulling the water molecules toward the center.

So what the heck does this have to do with the lungs? Well it turns out that there is a law discovered by a scientist by the name of Pierre-Simon Laplace (1749-1827) called, you guessed it, Laplace's law. Laplace's law relates the shape of the alveolus to the pressure caused by surface tension. The actual law (Simon-Laplace Law) is a non-linear differential equation (a bit much for this book). The main idea is that assuming the alveolus is a spherical structure there is an inverse relationship between the small size of the alveolus and the surface tension. In other words, surface tension works to cause collapse of the alveolus.

So why are we not walking around with collapsed lungs? Well fortunately Mother Nature finds a way to prevent this. Our alveoli have special cells (Type II cells) that secrete a substance that reduces surface tension. This substance is known as surfactant.

So why do we need to know this? Well, it turns out that premature infants risk lung collapse because their type II cells are not developed enough to secrete surfactant. They run the risk of what is known as respiratory distress syndrome (RDS).

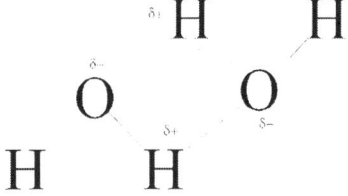

Fig. 16.10 Water molecules have a partial positive and partial negative charge. The dotted lines represent the polar covalent bonds between water molecules.

Review Questions

1. Which of the following structures is not part of the upper respiratory system:
   a. Nasal cavity
   b. Larynx
   c. Pharynx
   d. Sinuses

2. Which 2 bones make up the floor of the nasal cavity:
   a. Palatine and maxilla
   b. Ethmoid and maxilla
   c. Sphenoid and palatine
   d. Maxilla and sphenoid

3. The posterior portion of the soft palate is known as:
   a. Epiglottis
   b. Larynx
   c. Pharynx
   d. Uvula

4. The Adam's apple is known as:
   a. Cricoid cartilage
   b. Thyroid cartilage
   c. Arytenoids cartilage
   d. Epiglottis

5. When the vocal cords relax they form a triangular space called the:
   a. Glottis
   b. Epiglottis
   c. Uvula
   d. Pharyngeal triangle

6. Which type of epithelium lines the trachea:
   a. Simple squamous
   b. Stratified squamous
   c. Pseudostratified columnar
   d. Simple cuboidal

7. Cells lining the alveoli secrete a soapy substance known as:
   a. Mucous
   b. Surfactant
   c. Emulsifier
   d. Cytosol

8. Which of the following is not a structure of the left lung:
   a. Superior lobe
   b. Oblique fissure
   c. Inferior lobe
   d. Horizontal fissure

9. During resting inhalation:
   a. Volume increases and pressure decreases
   b. Volume and pressure increase
   c. Volume and pressure decrease
   d. Volume decreases and pressure increases

10. Which of the following is not an accessory muscle of inspiration:
    a. External intercostals
    b. Pectoralis minor
    c. Sternocleidomastoid
    d. Pectoralis major

11. At the base of the trachea is a structure known as:
    a. Secondary bronchi
    b. Thyroid cartilage
    c. Carina
    d. Arytenoid cartilage

12. The typical volume of air moved in and out of the lungs in one minute in an adult is about:
    a. 10 liters
    b. 7.5 liters
    c. 5.5 liters
    d. 4 liters

13. A person is breathing 15 times per minute with a tidal volume of 400 ml. What is their alveolar ventilation:
    a. 1 liter
    b. 800 ml
    c. 750 ml
    d. 500 ml

14. Inspiratory capacity is represented by:
    a. Inspiratory reserve volume and tidal volume
    b. Vital capacity and tidal volume
    c. Residual volume and tidal volume
    d. Vital capacity and inspiratory reserve volume

15. Oxygen produces how much of the total pressure of air:
    a. 50.2%
    b. 30.4%
    c. 20.9%
    d. 12.7%

16. The PCO2 of tissues is about:
    a. 104 mm Hg
    b. 45 mm Hg
    c. 40 mm Hg
    d. 95 mm Hg

17. Oxygen moves out of the alveoli and into the blood by way of:
a. Osmosis
b. Diffusion
c. Active transport
d. Filtration

18. Which of the following is not a transport mechanism for CO2:
a. Carbaminohemoglobin
b. Dissolved in plasma
c. Bonds with plasma proteins
d. Bicarbonate ion

19. Carbon dioxide combines with water and carbonic anhydrase to form:
a. Carbonic acid
b. Hemoglobin
c. Hydrogen ions
d. Tricarbonate

20. Holding your breath would cause a state of:
a. Respiratory alkalosis because of the buildup of hydrogen ions
b. Respiratory alkalosis because of the elimination of hydrogen ions
c. Respiratory acidosis because of the elimination of hydrogen ions
d. Respiratory acidosis because of the buildup of hydrogen ions

21. The oxygen-hemoglobin saturation curve represents:
a. Hemoglobin's affinity for oxygen binding in various partial pressures of oxygen
b. Hemoglobin affinity for oxygen binding over various times
c. Partial pressures of oxygen during various times
d. Hemoglobin's release of oxygen and carbon dioxide over time

22. The change in affinity for carbon dioxide binding in hemoglobin is known as:
a. Carbon dioxide release effect
b. Bohr effect
c. PCO2 difference effect
d. Haldane effect

23. Which respiratory center causes slow, deep breathing when stimulated:
a. Ventral respiratory group
b. Dorsal respiratory group
c. Apneustic center
d. Pneumotaxic center

Chapter 17

There's A Lot to Making Pee Pee

Chapter 17

There's A Lot to Making Pee Pee

The urinary system in a nutshell—not a lot of anatomy but lots of physiology. Those little ole kidneys may look simple but they are packed with complex systems for making urine. This system can be the downfall of many a student, at least if they don't get the big picture first.

Anatomy of the Urinary System (fig. 17.0)

Hey, how hard can it be? Kidneys, ureters, urinary bladder, urethra and that's it? Well, there's a bit more detail to go over. Let's start with the kidneys (fig. 17.1).

The kidneys are bean shaped (you know, kidney beans) organs located behind the abdominal cavity. We say they are retroperitoneal. The peritoneum is the membrane that lines the abdominal cavity so they are *behind* the abdominal cavity (watch out for those kidney punches).

They are located on the sides of the low back around the twelfth rib and extend to about the third lumbar vertebra. The kidneys are surrounded by fat (perirenal fat) and a tough fibrous membrane (renal capsule).

If we look inside we will see two major divisions. The outer portion is called the renal cortex and the inner portion is called the medulla (kinda like the brain). Some parts of the cortex extend into the medulla (renal columns). The medulla contains triangular structures called renal pyramids (where's the pharoh?). At the tip of each pyramid is a structure called the renal papilla. Urine drains from the renal papilla to the minor and major calyces to the renal pelvis and finally to the ureter.

Blood enters the kidney at a dent called the renal hilus. The renal artery brings blood in and the renal vein brings it out. The renal artery branches upon entering the kidney. The branches include:

Renal artery—segmental—interlobar—arcuate—interlobular

If you need to memorize these branches use the following mnemonic:

Read Several Interesting Articles Indeed

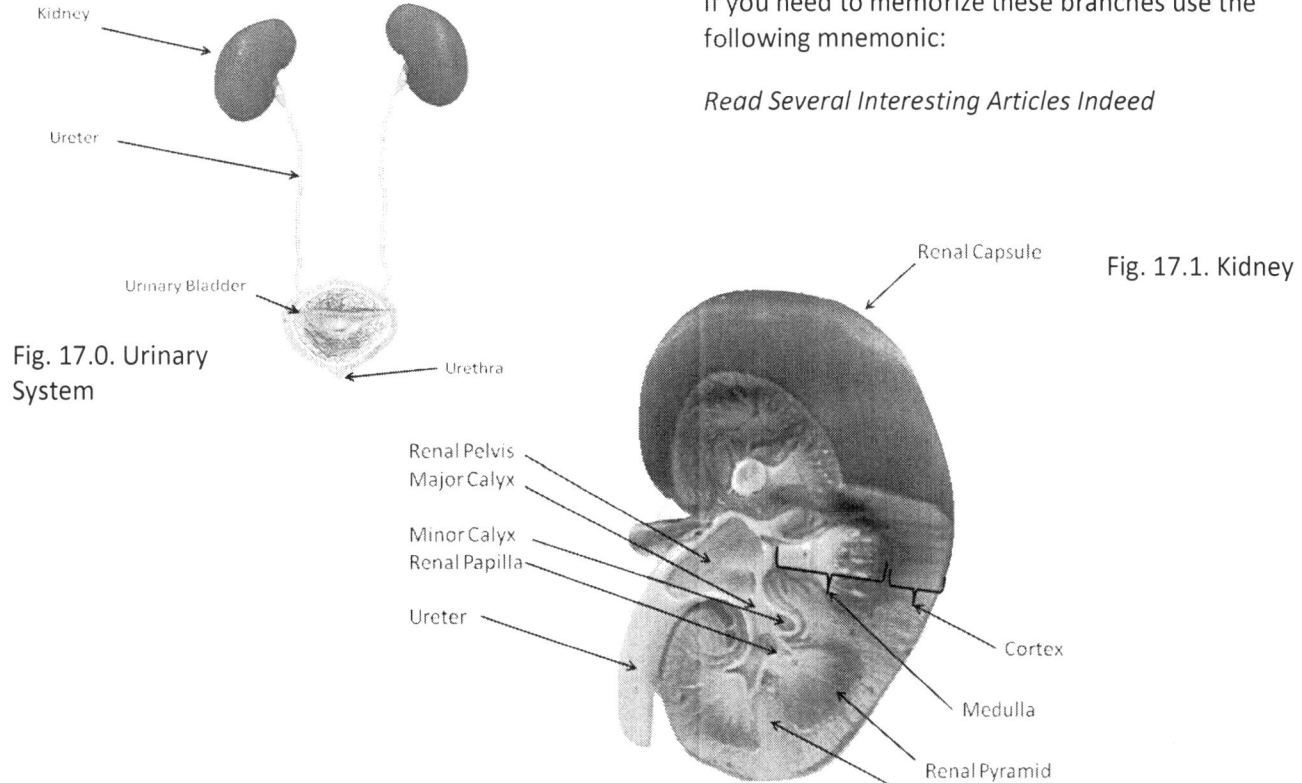

Fig. 17.0. Urinary System

Fig. 17.1. Kidney

Getting down to the microscopic structures we see one all-important structure in the kidney. This is the nephron which is the structure that makes urine (fig. 17.2). There are around one million nephrons in one kidney and they do a great job making urine. Some nephrons lie near the medulla and are called juxtamedullary nephrons. These nephrons extend deep into the medulla. Other nephrons reside in the cortex and only minimally extend into the medulla. These are known as cortical nephrons.

The afferent arteriole brings blood to the nephron and the efferent arteriole brings blood out. Between the afferent and efferent arterioles is a tuft of capillaries called the glomerulus. The glomerulus is surrounded by a fibrous capsule called the glomerular capsule (Bowman's capsule). The glomerular capsule connects with the proximal convoluted tubule which connects with the nephron loop (loop of Henle). The nephron loop connects with the distal convoluted tubule which in turn connects with the collecting duct which transports urine to the renal papilla.

The ureters carry the urine from the kidney to the bladder. The ureters have a smooth muscle layer that is capable of producing peristaltic contractions that occur once every two to three minutes. The parasympathetic nervous system increases these contractions and the sympathetic nervous system inhibits them.

The urinary bladder is a hollow organ that resides in the pelvic cavity (fig. 17.3). The area on the inside of the bladder between the two ureter connections and the urethra is called the trigone.

The urinary bladder and ureters are internally lined with transitional epithelium. The bladder also has a thick smooth muscle layer sometimes called the detrusor muscle. Contraction of the detrusor muscle increases the internal pressure of the bladder and causes urine to be expelled.

Male bladders contain an area of smooth muscle and elastic tissue called the internal urinary sphincter. This area is not present in females. The function of this structure is to keep semen from entering the urinary bladder during intercourse. Both males and females have an external urinary sphincter located in the urethra that controls the flow of urine.

The male urethra consists of three parts. The prostatic urethra exits the bladder and extends to the inferior prostate gland. It then becomes the membranous urethra until it enters the penis where it becomes the penile urethra.

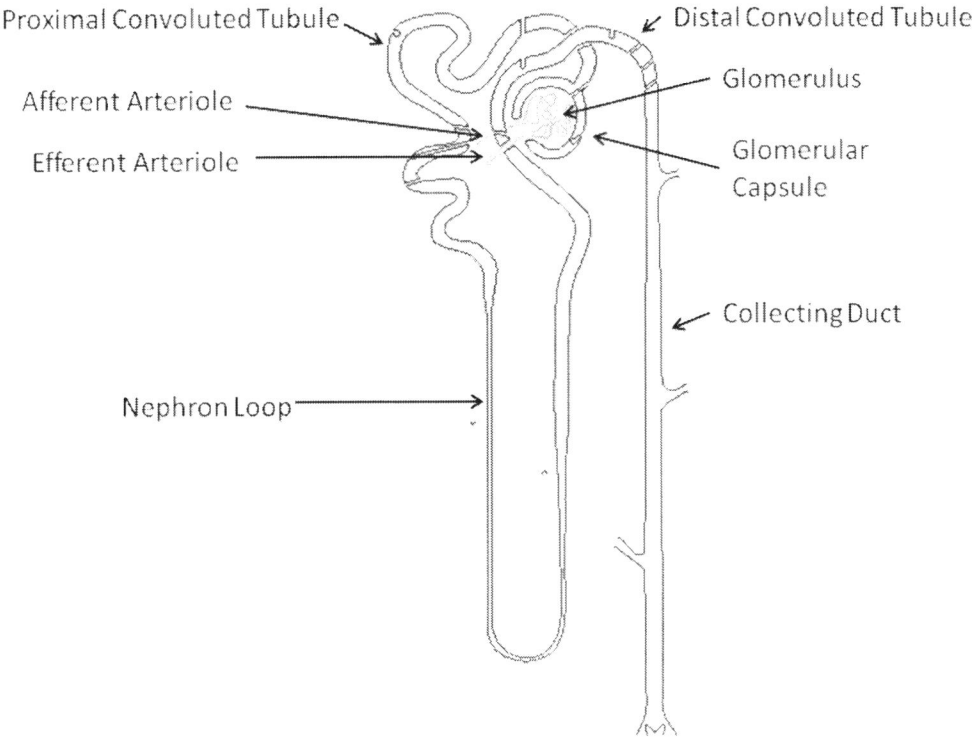

Figure 17.2. Nephron

Urinary Bladder

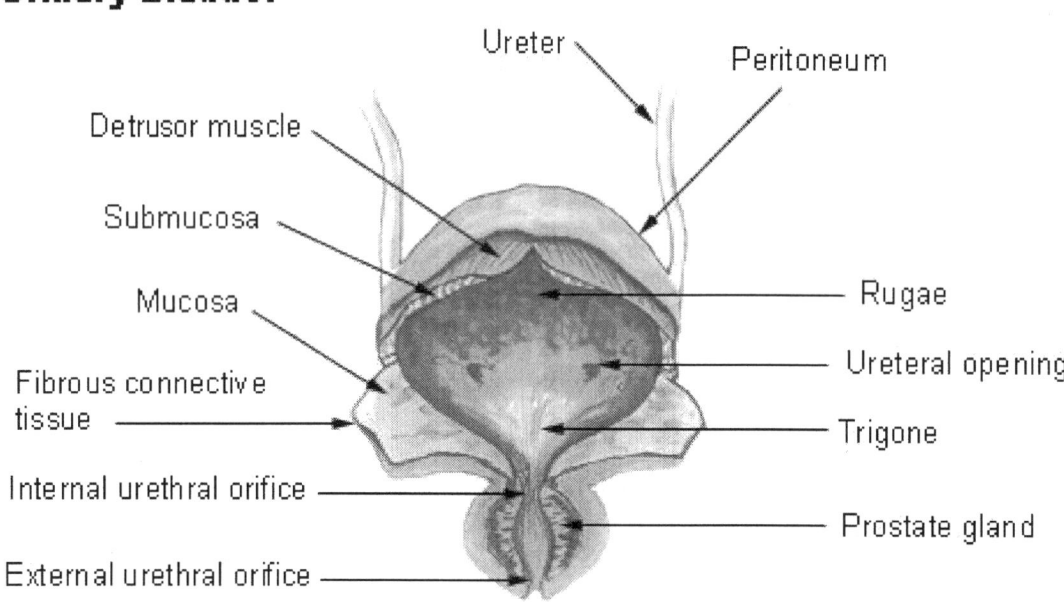

Figure 17.3. Urinary Bladder

The Big Picture: Urinary Physiology

Makin Pee Pee

There are 3 primary processes of urine formation:

1. *Filtration*
2. *Tubular reabsorption (move stuff from kidney to blood)*
3. *Tubular secretion (move stuff from blood to kidney)*

So we have a good idea of how the system is put together so now it's time to get into some of that physiology I told you about earlier.

We can gain a good deal of insight into how the kidneys work by examining the inputs and outputs. Blood flows into the kidney and urine and blood flow out. So the kidneys must somehow make urine from blood. The blood enters via the renal artery and exits via the renal vein. The urine exits by way of the ureters and flows to the bladder, urethra, and out of the body.

One of the simplest ways to make urine from blood is to filter the blood. This is actually the first stage of urine formation (filtration). Filters work by the movement of substances from areas of higher to lower pressure across a filtration membrane. The filtration membrane sorts substances based on size. You could think of it as being filled with holes. Smaller substances pass through the holes while larger substances do not. Smaller substances that are filtered include water, electrolytes and glucose.

There would be a problem if filtration were the only mechanism of urine formation. Our bodies need many of the filtered substances and they would be lost in the urine. So there must be some other mechanisms that help to maintain the balance. Fortunately there are and these include tubular reabsorption and secretion.

Tubular reabsorption and secretion work together to reclaim substances like water, glucose and electrolytes after they have been filtered. Tubular reabsorption employs a number of mechanisms in order to move filtered substances back into the blood. Tubular secretion also uses a number of mechanisms to move substances from the blood to the urine. Besides reclaiming filtered substances, both of these processes fine tune electrolyte, water and pH balance.

Besides maintaining fluid, electrolyte and pH balance the kidneys also monitor blood oxygen levels. They secrete the hormone erythropoietin in response to low oxygen levels. The hormone travels to the bone marrow to stimulate the production of red blood cells. The kidneys also work to control vitamin D synthesis.

Urine Formation Process #1 Filtration

The kidneys make a heck of a lot of filtrate each day. They typically produce about 123 ml of filtrate per minute which adds up to about 180 Liters per day. If all of this filtrate ended up as urine you would spend your days in the bathroom and drinking water! Fortunately most of that filtrate is reabsorbed leaving around one to two liters of urine per day (which allows you some time away from the bathroom).

In order to move substances through the filter there must be a pressure gradient (oh no, here we go again). Substances must move from an area of higher pressure to lower pressure. The pressure gradient is called filtration pressure or net filtration pressure. Net filtration pressure is directly proportional to the glomerular filtration rate. So if for some reason net filtration increases or decreases, so does glomerular filtration rate, and so does the amount of filtrate produced.

Net filtration pressure is the combination of a series of pressures that exist in the renal corpuscle. These include glomerular capillary hydrostatic pressure, glomerular capsular hydrostatic pressure and colloid osmotic pressure.

Glomerular capillary hydrostatic pressure is the blood pressure inside the capillaries. It is usually about 50 mm Hg and must be greater that the pressure inside the glomerular capsule known as glomerular capsular hydrostatic pressure. This one is the main pressure and must be greater than the total of the other pressures in order for the filter to work.

The glomerular capillary hydrostatic pressure is controlled in part by the diameter of the afferent and efferent arterioles. The efferent arterioles have a smaller diameter than the afferent arterioles. The smaller diameter works to decrease blood flow through the efferent arterioles increasing the pressure inside the glomerular capillaries. Changing the diameter of the afferent and efferent arterioles changes the glomerular capillary hydrostatic pressure. For example, increasing the diameter of the afferent arteriole or decreasing the diameter of the efferent arteriole increases the capillary pressure.

The pressure inside the glomerular capsule is called the glomerular capsular hydrostatic pressure. This pressure is created by fluid inside the capsule as well as downstream in the tubules. It is usually about 10 mm Hg. The glomerular capsular hydrostatic pressure works against filtration.

Colloid osmotic pressure is produced by the presence of plasma proteins in the blood called colloids. The colloids produce a pulling force causing water to move back into the glomerular capillaries. This pressure is usually about 30 mm Hg (fig. 17.4).

We can calculate the net filtration pressure by the following:

Net Filtration Pressure = Glomerular capillary hydrostatic pressure – Glomerular capsular hydrostatic pressure – Colloid osmotic pressure

If we plug in the normal values:

NFP = 50 mm Hg – 10 mm Hg – 30 mm Hg

NFP = 10 mm Hg

So the net filtration pressure is about 10 mm Hg.

The kidneys are constantly working to keep the amount of filtrate relatively constant despite changes in mean arterial pressure. For example as systemic blood pressure increases the afferent arterioles vasoconstrict keeping the glomerular capillary hydrostatic pressure constant. Likewise when blood pressure decreases the afferent arteriole dilates.

The sympathetic nervous system affects the afferent arteriole by causing it to vasoconstrict under intense sympathetic activity such as when exercising strenuously or when in shock.

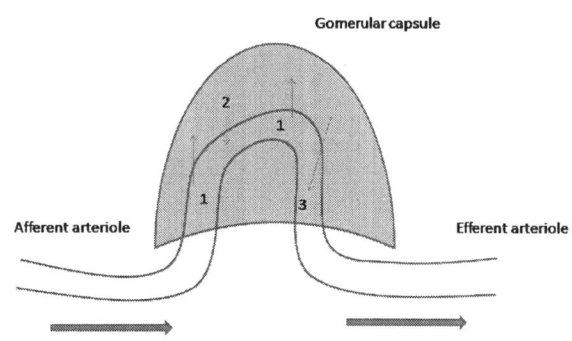

Figure 17.4. Glomerular filtration.

1. Glomerular capillary hydrostatic pressure.

2. Glomerular capsular hydrostatic pressure.

3. Colloid osmotic pressure

Urine Formation Process #2 Tubular Reabsorption

During filtration substances were just sorted based on size. This means that lots of stuff moves through the filter that our bodies need, like glucose and water for example. There must be some other process that works to reclaim these important substances. There is and it's called tubular reabsorption (figs.17.5-17.11).

One important thing to remember is that in tubular reabsorption substances move from the tubule to the blood.

Substances move by various ways in tubular reabsorption and we will cover a few.

Symporters

Big Picture: Symporters

Symporters are like mooching friends.

Some cells lining the kidney tubules contain special transport proteins called symporters. To describe the action of symporters I like to use my analogy of the mooching friend. Let's say my friend and I went to the movies. When it came time to pay my friend stated that he forgot his wallet and asked if I could "spot" him some money to get in. So I provided the energy (money) to get into the movie. My friend (who needs friends like him) just went along for the ride.

This is how the sodium glucose symporter works. Sodium provides the energy in the form of a gradient. There's lots of sodium coming out of the filter and moving into the tubules compared to inside the cells lining the tubules. So sodium moves through the transport protein and glucose (the mooching friend) goes along for the ride (fig. 17.5).

So, what would happen if lots and lots of glucose was produced by the filter and moved into the tubules? Well, the amount of glucose would exceed the number of symporters. This would cause glucose to flow into the urine and out of the body in a condition called glucosuria. This happens in diabetes.

Amino acids also move from tubule to blood via symporters.

Passive Movement of Substances

Other substances move passively from the tubules to the blood. Sodium is again one of these as well as calcium, magnesium, potassium. Since so much sodium moves into the interstitium surrounding the tubules water also follows sodium via good ole osmosis. This is how much of the water is reabsorbed (fig. 17.7).

Urine Formation Process #3 Tubular Secretion

Tubular secretion involves moving substances from the blood to the tubules. Unlike tubular reabsorption that moves substances in order to maintain fluid and electrolyte balance, tubular secretion primarily works to eliminate toxic substances or byproducts of metabolism.

Big Picture: Antiporters

Antiporters are like revolving doors.

Tubular secretion can involve active or passive transport. An example of passive transport is the sodium hydrogen antiporter. This transport protein uses the sodium gradient to move sodium from the tubule to inside the cell while at the same time moving excess hydrogen ions out of the cell and into the tubule.

In describing the antiporter I like to use the analogy of the revolving door. Let's say that I am staying in one of those fancy hotels with a big revolving door. I enter the door and push on the glass to move the door. At the same time someone is exiting the hotel and moves through the door going in the opposite direction and not pushing at all. I am providing the energy while the other person moves in the opposite direction getting a free ride. In the case of the anitporter, sodium provides the energy in the form of a gradient while hydrogen goes along for the ride (fig. 17.8).

Other examples of secreted substances include ammonia, potassium, penicillin, and para-aminohippuric acid. These substances are not normally produced in the body.

Action of Aldosterone

Big Picture: Aldosterone

Aldosterone tells the kidneys to 'hang on' to sodium.

Certain tubule cells are "leaky" to sodium and potassium. In other words when aldosterone attaches to receptors on these cells it increases their permeability to sodium and potassium. So, aldosterone causes more sodium to be reabsorbed, while at the same time causing potassium to be secreted (fig. 17.11).

Atrial Natriuretic Hormone

Big Picture: Atrial Natriuretic Hormone (ANH)

ANH tells the kidneys to 'get rid' of sodium.

Atrial Natriuretic Hormone (ANH) is secreted by the wall of the right atium in the heart in response to atrial stretch. ANH has the opposite action of aldosterone and inhibits sodium and water reabsorption in the kidney tubules.

Bicarbonate Reabsorption

Big Picture: Bicarbonate Reabsorption

Bicarbonate is reabsorbed

Yes, it sounds pretty simple. Bicarbonate ions do move from the kidney tubules to the blood. However, there are a few steps along the way that make it seem complicated.

So, here we go with good ole bicarbonate reabsorption. Remember that hydrogen is secreted into the tubule by way of the sodium-hydrogen antiporter. Hydrogen runs into bicarbonate ions to form carboxylic acid. The carboxylic acid dissociates into carbon dioxide and water (remember this from the respiratory system?). The carbon dioxide then diffuses into cells lining the lumen of the tubule and combines with water in the presence of carbonic anhydrase to form carboxylic acid (again). The carboxylic acid once again dissociates into hydrogen and bicarbonate ions. The hydrogen gets antiported back out to keep the process looping while the bicarbonate moves out of the cell and into the blood and viola! Reabsorption! (figs. 17.9-17.10)

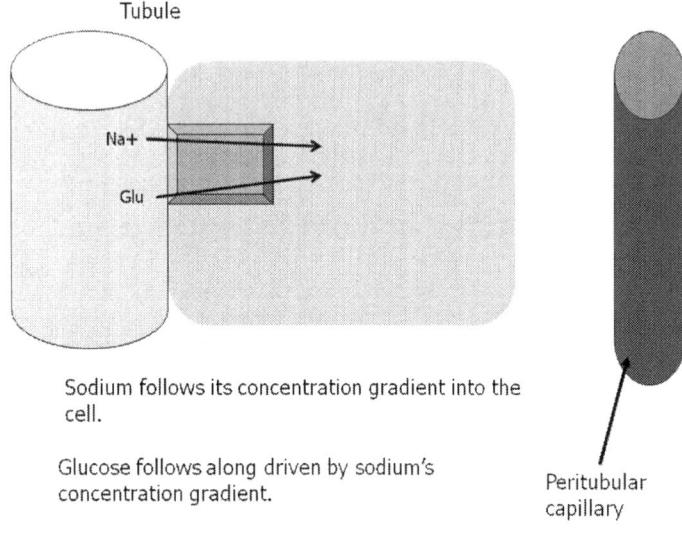

Figure 17.5. Sodium-Glucose Symporter

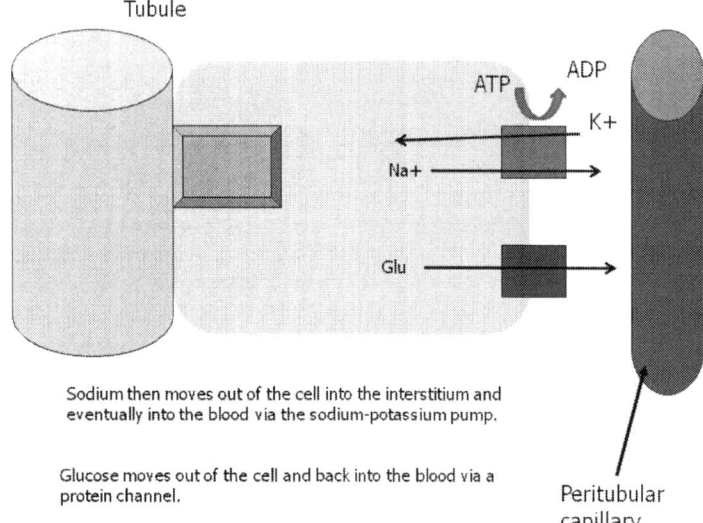

Figure 17.6. The sodium-potassium pump has a role in reabsorption.

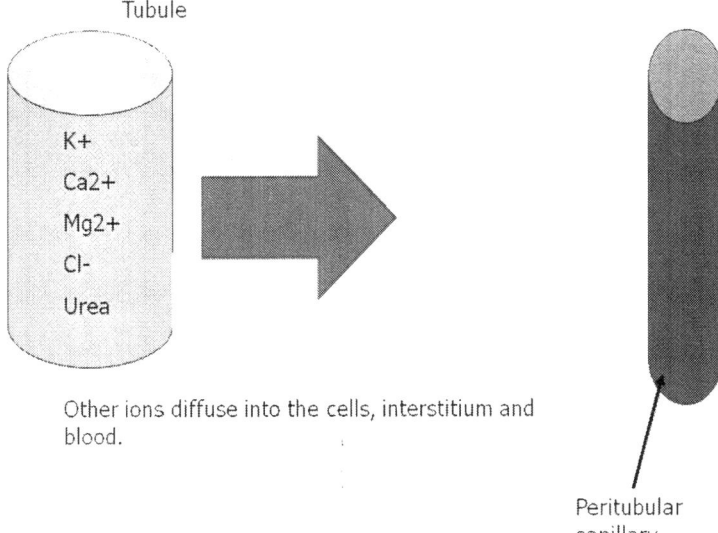

Figure 17.7. Some electrolytes are reabsorbed via diffusion.

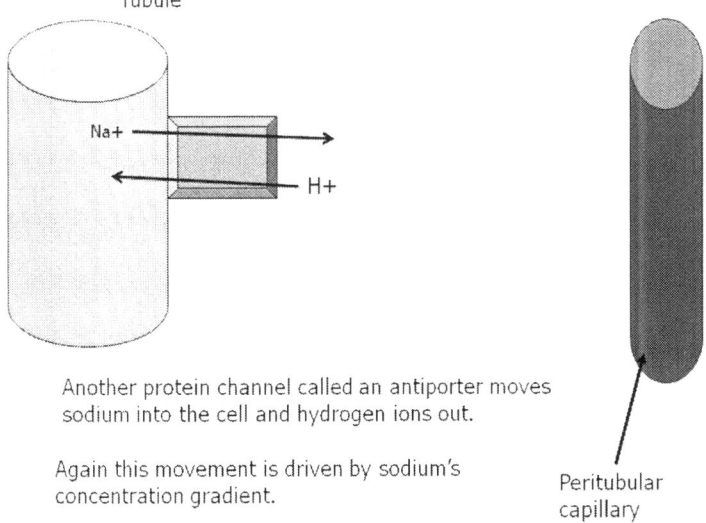

Figure 17.8. Sodium-hydrogen antiporter.

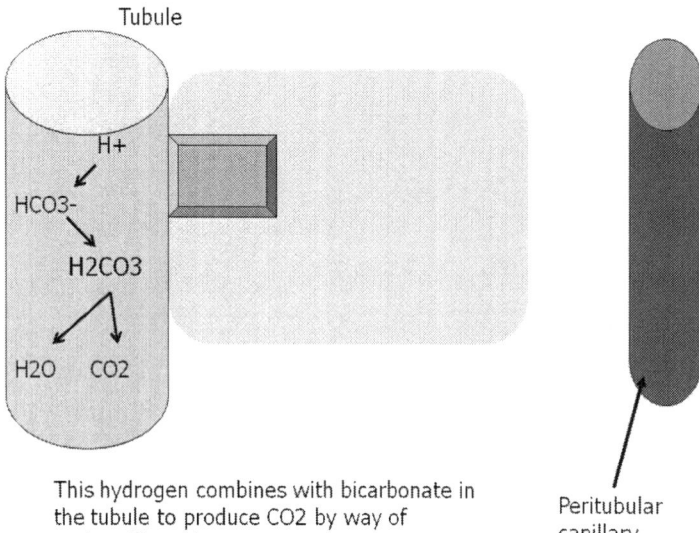

Figure 17.9. Hydrogen in the tubule.

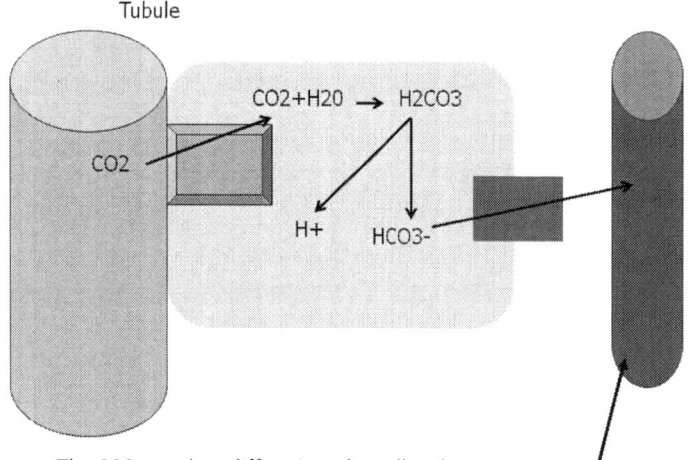

Figure 17.10. Bicarbonate reabsorption.

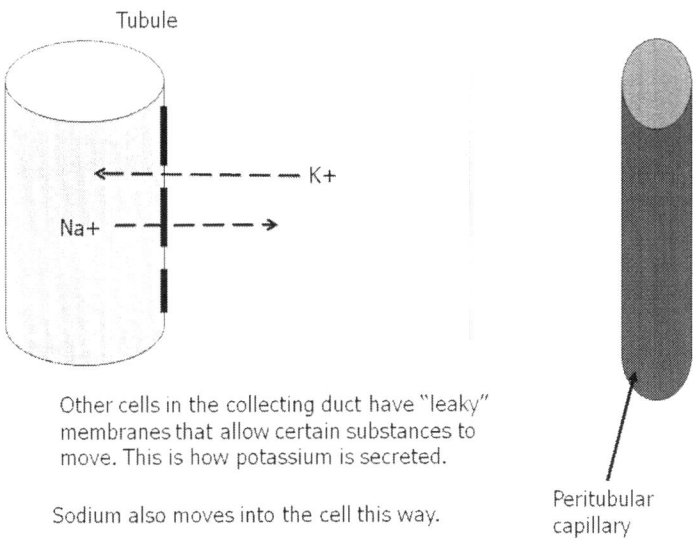

Figure 17.11. Reabsorption of sodium and secretion of potassium.

The Juxtaglomerular Apparatus (what the heck?)

Big Picture: Juxtaglomerular Apparatus (JG apparatus)

The JG apparatus helps to control blood pressure and the amount of urine produced.

The juxtaglomerular apparatus is a group of cells residing at the junction of the afferent arteriole and distal ascending limb of the nephron loop (fig. 17.12). The juxtaglomerular apparatus consists of two different types of cells. Juxtaglomerular cells are located on the afferent arteriole side

So in a nutshell, the juxtaglomerular cells monitor blood pressure. When blood pressure decreases the cells secrete something that will increase it. That something is renin. Renin then triggers the renin-angiotensin system. The end result is the formation of a substance called angiotensin II. Angiotensin II promotes

vasoconstriction (which raises blood pressure) and secretion of the adrenal cortex hormone aldosterone. Aldosterone promotes sodium reabsorption and since water follows salt (osmosis again) we get fluid retention which also raises blood pressure.

The other cells in the juxtaglomerular apparatus are the macula densa cells. These babies monitor the filtrate in the kidney tubule. Usually they keep the afferent arteriole open by secreting nitric oxide (a vasodilator), but if the filtrate gets too dilute (oops, we're making too much urine) the secretion of nitric oxide is inhibited. This causes vasoconstriction of the afferent arteriole and subsequent decrease in the production of filtrate.

This one works like one of those lawn sprinklers. The water source is the hose (afferent arteriole). If there is too much water flowing out of the sprinkler we could step on the hose (vasoconstrict) to decrease the flow.

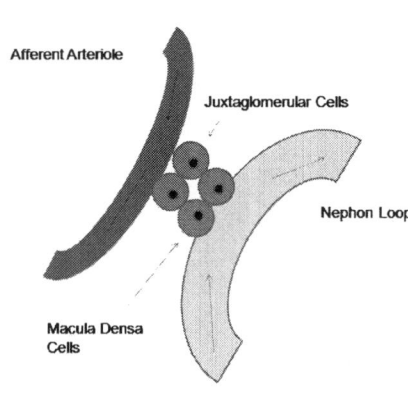

Figure 17.12. Juxtaglomerular apparatus. The juxtaglomerular cells secrete renin while the macula densa cells secrete nitric oxide.

The Nephron Loop

Big Picture: Nephron Loop

The nephron loop contains two parts. These are the descending and ascending limbs.

In the descending limb water moves out but salt stays in (concentration increases).

In the ascending limb salt is pumped out and water stays in (concentration decreases).

The nephon loop consists of two segments including a descending and ascending limb each with different characteristics (fig. 17.13).

The descending limb contains a thin layer of epithelium that is more permeable to water than the thick portion of the ascending limb. An isotonic fluid (about 300 mOsm) enters the descending limb. As it progresses down the limb, water diffuses into the interstitium causing the concentration to dramatically increase. The fluid concentration can increase to as high as 1200 mOsm (very hypertonic).

The thick segment of the ascending limb inhibits the passage of water by diffusion and contains a series of active transport proteins that selectively move substances. Sodium and chloride are moved out of the ascending limb and into the intestitium by way of these active transport proteins. As fluid moves up the ascending limb the concentration decreases. A 100 mOsm hypotonic solution exits the ascending limb and enters the distal convoluted tubule.

The countercurrent consists of the "current" of water moving in one direction and the "current" of sodium chloride moving in the opposite direction. The high "salt" gradient is maintained by the active transport of sodium and chloride in the ascending limb. Urea also diffuses into the descending limb adding to the increased concentration.

Antidiuretic Hormone (ADH)

Big Picture: Antidiuretic hormone (ADH)

ADH causes water retention by increasing permeability in the distal convoluted tubule.

ADH is secreted by the posterior pituitary gland in response to an increase in blood solute concentration as senses by osmoreceptors in the hypothalamus (remember the endocrine chapter?). ADH targets the kidney, particularly the distal convoluted tubule.

ADH affects the distal convoluted tubule by making it more permeable to water. Remember that the fluid exiting the nephron loop is hypotonic. The hypotonic fluid enters the distal convoluted tubule that is surrounded by the interstitium and peritubular capillaries which are isotonic. If the tubule is impermeable to water then dilute urine is produced. If the tubule is made more permeable to water, then water moves to the more highly concentrated interstitium and blood. ADH then plays an important role in maintaining fluid balance.

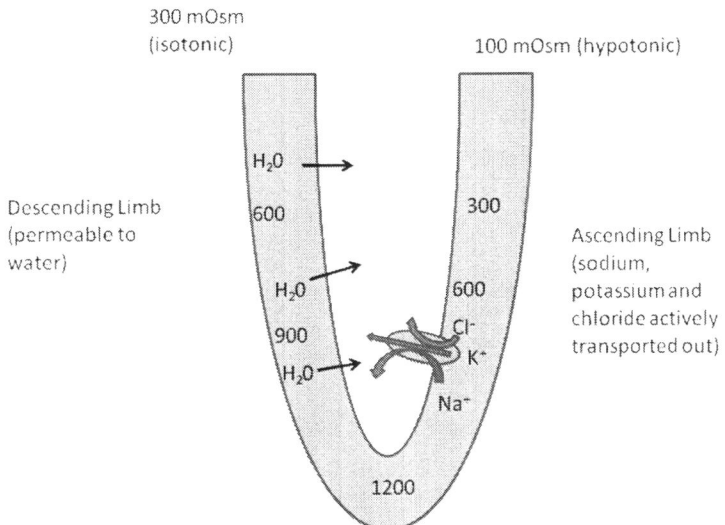

Figure 17.13. Nephron Loop.

Urine Composition

Adults produce about one to two liters of urine daily. Pathologies such as diabetes or some medications can produce a larger urine output known as polyuria. A urine output of less than 500 ml/day is known as oliguria and an output less than 100 ml is known as anuria.

Urine is the final product of the kidney. It is mostly water (95%) with a few other solutes including nitrogenous wastes, electrolytes, pigments, and toxins. It can also contain abnormal substances including glucose, albumin, bile and acetone.

Urine is usually clear or straw colored. An abnormal color may indicate presence of blood, bile, bacteria, drugs, food pigments, or high-solute concentration. Urine will become cloudy after standing due to a buildup of bacteria. Pus from problems such as kidney infections will also make urine cloudy.

Urine has a slight odor. It will develop an ammonia odor after standing due the breakdown of urea. An acetone odor may indicate diabetes. The pH of urine varies between 4.6 and 8.0. The specific gravity is between 1.001 and 1.035.

Micturation (Peeing)

Urine continuously flows from the kidney to the bladder. The bladder acts as a storage reservoir for urine and can store up to one liter. At about 300 ml the urge to urinate becomes evident. Once the wall is stretched the micturation reflex is stimulated. Stretch of the bladder sends impulses to sensory neurons in the pelvic nerves to the sacral segments of the spinal cord. Micturation is under parasympathetic control and parasympathetic impulses cause the bladder to contract. The motor impulses for micturation originate in a micturation center in the pons. The center also receives input from the cerebral cortex (so one can decide whether or not to micturate). Contraction of the bladder increases the internal pressure pushing urine into the urethra.

The micturation reflex is an involuntary reflex in infants. Voluntary control of the reflex does not occur until around age 2-3 years.

Renal Clearance

Big Picture: Renal Clearance

Renal clearance represents how much blood has to pass though the kidney to completely remove a substance.

Renal clearance is used to determine kidney function. Renal clearance is the volume of blood plasma from which a substance is completely removed in one minute. Renal clearance reflects the three processes of urine formation which include glomerular filtration, tubular reabsorption and tubular secretion. We can use an indirect method to determine renal clearance that includes the rate of urine output and the concentration of the substance in blood plasma and urine.

For example let's say we are determining the renal clearance for a substance 'X'. We know the concentration of X in the urine is 5.0 mg/ml and the concentration of X is .3 mg/ml in the plasma. We also know the rate of urine output equals 2 ml/min. The renal clearance can be determined by the following:

Renal Clearance (C) = UV/P

U = concentration of substance in urine

V = rate of urine output

P = concentration of substance in plasma

For our example:

C = (5.0)(2)/.3

C = 33.33 ml/minute

This can be interpreted as 33.33 ml of blood plasma is cleared of substance X every minute. Got it!

Review Questions

1. Which of the following is not a function of the urinary system:
a. Maintain electrolyte balance
b. Control blood pressure
c. Remove wastes
d. Provide nutrition to the body

2. At the tip of each renal pyramid is a structure known as:
a. Major calyx
b. Renal column
c. Renal papilla
d. Distal convoluted tubule

3. Urine flows from the proximal convoluted tubule to this structure in the nephron:
a. Collecting duct
b. Glomerulus
c. Afferent arteriole
d. Nephron loop

4. Which of the following is not a mechanism of urine formation:
a. Enzymatic action
b. Filtration
c. Tubular reabsorption
d. Tubular secretion

5. Which best describes the location of the kidneys:
a. Flank area in the peritoneal cavity
b. Inguinal area outside of the pelvic cavity
c. Retroperitoneal flank area
d. Just below 12th rib in the midline

6. Which type of tissue lines the inside of the bladder:
a. Transitional epithelium
b. Stratified squamous epithelium
c. Skeletal muscle
d. Dense connective tissue

7. Micturation is controlled by:
a. Sympathetic nervous system
b. Diencephalon
c. Parasympathetic nervous system
d. Pons

8. Blood leaves the nephron via:
a. Efferent arteriole
b. Renal artery
c. Acuate arteries
d. Afferent arteriole

9. Which of the following is not a part of the male urethra:
a. Penile
b. Prostatic
c. Membraneous
d. Corporus

10. The female urethra is about how long:
a. 1 cm
b. 5 cm
c. 4 cm
d. 3 cm

11. The renal output is:
a. 100 ml/min
b. 10 ml/min
c. 2 ml/min
d. 1.5 ml/min

12. How much filtrate is produced by the kidneys each day:
a. 2 liters
b. 50 liters
c. 120 liters
d. 180 liters

13. Glomerular capillary hydrostatic pressure is normally about:
a. 10 mm Hg
b. 20 mm Hg
c. 50 mm Hg
d. 60 mm Hg

14. A typical net filtration pressure is about:
a. 10 mm Hg
b. 20 mm Hg
c. 50 mm Hg
d. 60 mm Hg

15. An increase in colloid osmotic pressure would have which effect on urine production:
a. It would increase
b. It would decrease
c. No effect
d. It would decrease glomerular capillary hydrostatic pressure

16. During exercise urine production _____ because _____:
a. Increases, sympathetic effects
b. Decreases, sympathetic effects
c. Increases, parasympathetic effects
d. Decreases, parasympathetic effects

17. Renin is secreted by the juxtaglomerular apparatus when this occurs:
a. Systemic blood pressure drops below 80 mm Hg
b. Systemic blood pressure increases beyond 100 mm Hg
c. Increase in solute concentration of filtrate
d. Decrease in solute concentration of filtrate

18. Macula densa cells secrete ____ which promotes ____:
a. Norepinephrine, vasoconstriction
b. Nitrous oxide, vasoconstriction
c. Nitric oxide, vasodilation
d. Acetylcholine, vasodilation

19. Which of the following substances is secreted:
a. Glucose
b. Sodium
c. Water
d. Potassium

20. Which of the following best describes the action of aldosterone:
a. Increases reabsorption of hydrogen ions
b. Increases reabsorption of sodium
c. Decreases reabsorption of potassium
d. Decreases reabsorption of hydrogen ions

21. The ascending limb of the nephron loop is:
a. Permeable to water but not solute
b. Permeable to water and solute
c. Permeable to solute but not water
d. Impermeable to water and solute

22. What type of fluid exits the nephron loop:
a. Isotonic
b. Hypertonic
c. Hyperosmotic
d. Hypotonic

23. Which best describes the action of ADH:
a. Decreases membrane permeability to water in the distal convoluted tubule
b. Creates hypotonic solution in distal convoluted tubule
c. Increases membrane permeability to water in the distal convoluted tubule
d. Creates isotonic solution in the distal convoluted tubule

Chapter 18

You Already Know About Fluids and Electrolytes, How About Some Acid-Base Balance

Chapter 18

You Already Know About Fluids and Electrolytes, How About Some Acid-Base Balance

You've come a long way so far. In fact you already know a great deal about fluids and electrolytes. This chapter will review some concepts we've already covered and fill in the gaps with some new ideas about our body's regulation systems.

The Big Picture: Fluids and Electrolytes

Fluids carry electrolytes so conditions that affect fluid volume can also affect electrolytes.

Fluids and electrolytes go together. This is because of concentration. When we describe a solution in terms of concentration we are defining the amount of solute in the solution. In many cases the electrolytes represent the solute. Fluids also move across membranes by way of osmosis. For example, if we were losing water on one side of a membrane we would say that this area is becoming hypertonic. Water would then move across the membrane toward that area by osmosis.

There are many electrolytes in the body. The most important of these however are sodium, potassium and calcium so you really want to know how these are regulated.

Acid base balance is also important as the blood is kept at a narrow range of pH. If it becomes too acidic a condition called acidosis develops. Likewise if the blood becomes too basic a condition called alkalosis develops. Both acidosis and alkalosis can cause severe problems.

Fluids

Let's begin with an overview of fluids. Your body contains a lot of fluid. In fact, the human body is more than 70% water. Fluids are in tissues, membranes and many other structures. They also play an important role in the body's chemistry. Your fluid levels must be maintained in order for your body to function properly. The same can be said for electrolytes (remember sodium, potassium, etc). Since electrolytes are carried by fluids, the balance of both fluids and electrolytes is of extreme importance in keeping your body alive.

Physiologists like to think of the body as divided up into two fluid compartments. There is a compartment inside the cells called the intracellular compartment (ICF) and a compartment outside the cells or extracellular compartment (ECF). If we were to generalize fluid and electrolyte problems we could say that substances lost from the ECF are compensated for up to a point. In other words your body can handle some loss of fluid or electrolytes. But if the gain or loss exceeds your body's ability to compensate then the gain or loss affects the cells causing larger problems.

The total fluid in a normal adult body is about 40 liters. So where do you think there is more fluid, in the cells or not in the cells? If we were to measure the fluid in the ICF and ECF we would see that the ECF contains about 37% of the total fluid by volume. The ICF contains about 63%. So there is more fluid in the cells particularly in the cytoplasm. The ECF consists of a number of areas including interstitial fluid, plasma, lymphatic fluid, and transcelluar fluids. These include the cerebral spinal fluid, aqueous and vitreous humors, synovial fluid and glandular secretions.

The mechanisms allowing fluid movements include filtration and osmosis (remember these). The amount of fluid intake must equal the amount of fluid lost in order to maintain fluid balance. Normally about 1.5 liters of fluid is gained and lost per day. Where does the fluid come from? Most comes from drinking (1500 ml) and eating (750 ml). But some also comes from cellular metabolism (250 ml). Water is a byproduct of many metabolic processes of the body including aerobic metabolism.

How does your body lose fluid? You lose fluid by urinating (1500 ml), sweating (150 ml), feces (150 ml) and through respiration (700 ml).

Fluid Regulation

What happens when your body loses fluid? Well one of the first things that happens is that you get thirsty. Your body has a thirst regulating mechanism. Osmoreceptors in the hypothalamus sense an increase in solute concentration and stimulate the thirst center. A loss of as little as 1-2% of body fluid can stimulate thirst. Once you drink some fluid your stomach stretches. This feeds back to the hypothalamus and inhibits the thirst center.

Hormonal Regulation

Certain hormones have a powerful influence on regulating fluid volume. We've seen all of these before. These include antidiuretic hormone, aldosterone and atrial natriuretic peptide.

In case you've forgotten here's a review of these three hormones. Antidiuretic hormone (ADH) was covered in the urinary chapter. ADH is secreted by the posterior pituitary gland in response to increased solute concentration as sensed by osmoreceptors in the hypothalamus. ADH then travels through the blood to the kidneys where it causes an increase in tubular permeability particularly in the distal convoluted tubule of the nephron. When this occurs water moves from the tubule to the interstitium and eventually into the blood. The primary effect is to conserve fluid volume. ADH also stimulates the thirst center in the hypothalamus.

Aldosterone is a hormone secreted by the adrenal cortex in response to activation of the renin-angiotensin system (RA system) and adrenocorticotropin hormone (ACTH). The RA system is triggered by the secretion of renin from the juxtaglomerular cells of the juxtaglomerular apparatus in the nephron of the kidney. Aldosterone targets cells in the kidney tubules causing them to increase their permeability to sodium. When this occurs sodium is reabsorbed and water follows by osmosis. Water moves into the interstitium and blood.

Atrial natriuretic peptide (ANP) is secreted by the right atrium of the heart in response to an increase in atrial stretch. The atria stretch when there is an increase in blood volume. This increase in plasma volume relates to an increase in fluid volume. ANP also inhibits the thirst mechanism.

Fluid Regulation Problems

Dehydration (Oh no, I'm drying up!)

When you lose too much water you can get dehydrated. From a physiology standpoint we can say that water is first lost from the ECF causing the osmotic pressure to rise due to a rise in solute concentration. If your body can't compensate then the increase in osmotic pressure in the ECF causes a subsequent movement of water out of the ICF. In other words your body will try to compensate for some fluid loss but if you lose too much fluid the cells become dehydrated which causes larger problems.

Signs of dehydration include thirst, dizziness, weakness, mental confusion, delirium and coma. You can get dehydrated by not drinking fluids, vomiting, diarrhea, severe sweating known as profuse diaphoresis, or profuse urination from diseases such as diabetes.

Remember how fluid volume is related to blood volume and blood pressure? Well, dehydration can cause general hypovolemia which can result in low cardiac output, electrolyte imbalances and acid-base abnormalities.

To fix the problem you will need to restore your intake of fluids and electrolytes either orally or through an intravenous route.

Water Intoxication

Believe it or not alcohol is not the only thing that causes intoxication. Water intoxication or hyperhydration occurs when fluid intake exceeds water loss. It is a rare occurrence in adults and is more likely to occur in newborns given dilute formula or water. Newborns do not have a fully developed system for decreasing fluid volume. Water is gained first in the ECF which causes the compartment to become hypotonic. This causes subsequent loss of fluid from the ICF. Water intoxication can be severe and result in muscle cramping, convulsions, confusion, coma, and brain edema.

Electrolytes

Big Picture: Electrolytes

There is always more sodium outside the cells than in and more potassium inside the cells than out.

Sodium Balance (Fig. 18.1)

Sodium balance depends on the intake versus excretion of sodium. A normal adult human has a sodium intake of about 1.1 to 3.4 grams of sodium per day. You get sodium primarily from foods and you get rid of excess sodium via urination and sweating. The kidneys are the main regulators of sodium in the body.

Sodium is regulated by aldosterone (remember this one?). As discussed in the urinary system chapter, aldosterone increases the reabsorption of sodium in the kidney tubules. In other words aldosterone tells the body to "hang on" to sodium.

Sodium and water are often transported together as water moves by osmosis. This helps to keep the sodium concentration constant. For example, ingesting a large amount of sodium causes a subsequent increase in water absorption via osmosis. The additional water ends up in plasma and increases blood volume. The increase in blood volume results in an increase in blood pressure. This is why people with hypertension are told to limit their sodium intake.

Hyponatremia

Too little sodium in the body is called hyponatremia. This happens when sodium concentration is reduced to below 130 mEq/L (milliequivalents per liter). Hyponatremia can result from prolonged and severe sweating, vomiting, diarrhea, renal disease, and a condition of the adrenal gland called Addison's disease. In hyponatremia the ECF becomes hypotonic causing swelling of the ICF. Symptoms include muscle spasms, postural blood pressure changes, nausea, vomiting, convulsions, confusion, and coma.

Treatment for hyponatremia ranges from water restriction in mild cases to the administration of sodium orally or intravenously in more severe cases.

Hypernatremia

Too much sodium in the body is known as hypernatremia. This happens when the sodium concentration exceeds 145 mEq/L. Hypernatremia results from severe uncorrected diabetes insipidus or mellitus, severely high sodium intake, lack of fluid intake, diarrhea, heart disease or renal failure.

The signs and symptoms of hypernatremia include thirst, disorientation, lethargy, and central nervous system problems. Hypernatremia is treated with a hypotonic solution which lowers sodium concentration.

Potassium Balance

We can't forget about potassium. Most of the potassium in the body is located in the ICF (98%). You get potassium by eating foods (and some beverages) and your body excretes potassium in the urine. Potassium is mainly regulated by aldosterone (didn't we just see this one?). Remember that aldosterone regulates both sodium and potassium.

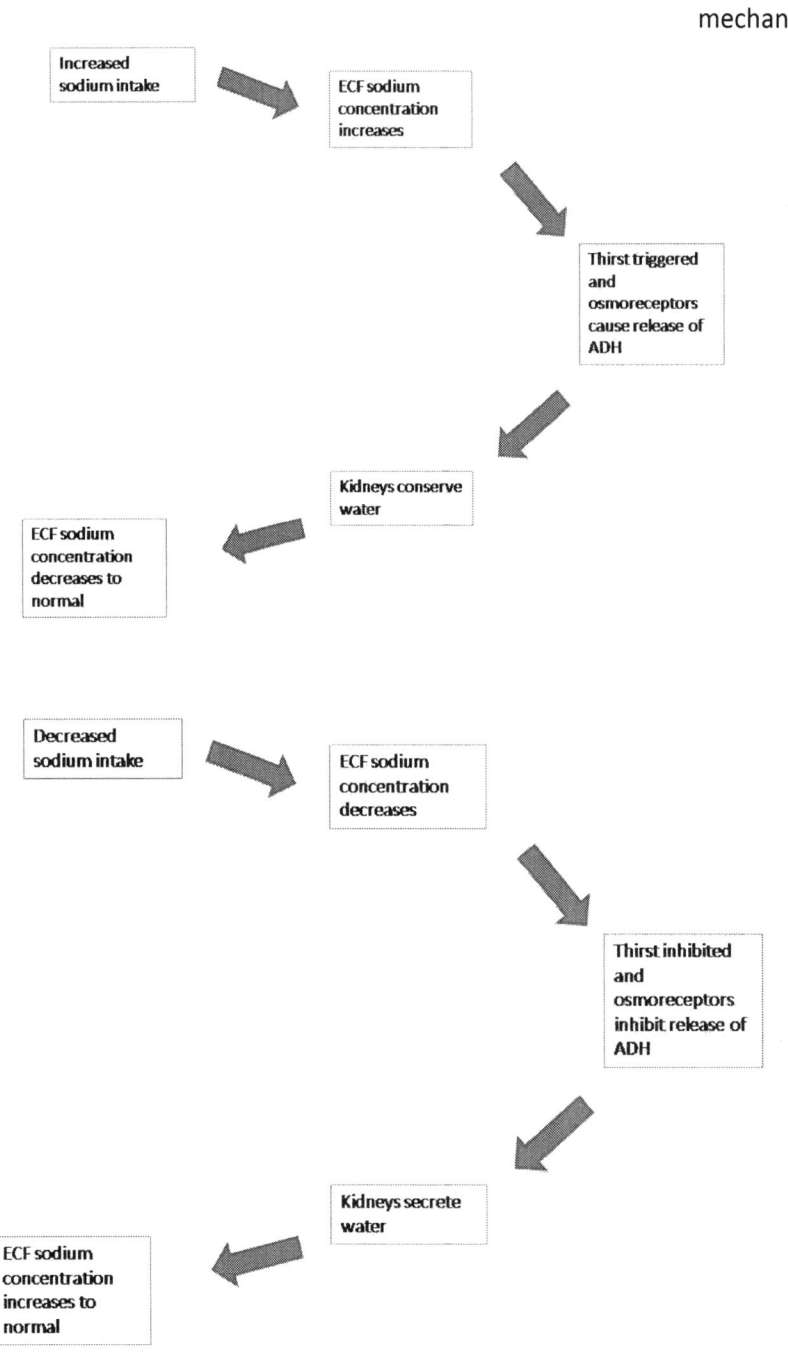

Fig. 18.1 Sodium balance mechanism.

Aldosterone causes secretion of potassium as well as reabsorption of sodium. High concentration of potassium in blood plasma stimulates the release of aldosterone.

Too little potassium is known as hypokalemia. This happens if potassium levels drop below 3.5 mEq/L. The causes of hypokalemia include Cushing's disease (affecting the adrenal cortex causing increased aldosterone levels resulting in potassium loss), potassium wasting diuretics, increased urine output, gastric suctioning (without potassium replacement) and vomiting.

Signs and symptoms of hypokalemia include muscle weakness, paralysis, atrial or ventricular arrhythmias, and respiratory problems. Potassium disorders can be very dangerous and result in life-threatening conditions.

Hypokalemia can cause peaked T-waves on ECG, ventricular dysrhythmias , cardiac arrest, muscle weakness, failure of respiratory muscles, intermittent diarrhea, and intestinal colic.

Hyperkalemia exists if potassium levels exceed 5.5 mEq/L. This condition is very dangerous and can also be life threatening. Causes of hyperkalemia include kidney disease, vomiting-diarrhea, potassium sparing diuretics, extensive tissue damage, severe infections, and Cushing's syndrome.

The treatment for hyperkalemia ranges from dietary restriction of potassium in mild cases to intravenous administration of calcium gluconate to correct cardiac problems along with dialysis to remove excess potassium. The underlying problem must also be corrected.

Calcium Balance

A typical adult human has about 1-2kg of calcium. Calcium plays an important role in skeletal and cardiac muscle contraction as well as in the transmission of nervous system impulses. There is a dynamic interplay between calcium in blood and bone. Calcium can be deposited or removed from bone when necessary.

Calcium enters the body through the digestive tract and is excreted in the kidneys with a small portion excreted in bile. A typical adult needs about .8 to 1.2 gm/day of calcium.

Calcium is regulated by the hormones calcitonin and parathyroid hormone (remember these from the endocrine chapter?). Calcitonin lowers calcium levels in blood by stimulating certain cells called osteoblasts that remove calcium from blood and deposit it into bone. Calcitonin also inhibits other cells called osteoclasts. Osteoclasts remove calcium from bone so that it is available in the blood.

Parathyroid hormone (PTH) has the opposite effect of calcitonin. PTH increases blood calcium by increasing osteoclastic activity and decreasing osteoblastic activity. Both calcitonin and PTH work together to maintain calcium balance.

When calcium levels decrease to lower than 4 mEq/L a state of hypocalcemia exists. Hypocalcemia results from hypoparathyroidism which produces a low level of parathyroid hormone, vitamin D deficiency, and renal failure. Signs and symptoms include muscle spasms, convulsions, and cardiac arrhythmias.

When calcium levels exceed 11 mEq/L a state of hypercalcemia exists. This can result from hyperparathyroidism and some cancers. Signs and symptoms include confusion, fatigue, arrhythmias and calcification of the soft tissues of the body.

Magnesium Balance

Magnesium is need for a number of metabolic reactions including phosphorylation of glucose and in muscle contraction. Magnesium is reabsorbed in the proximal convoluted tubule. Excess magnesium is excreted in the urine. The typical adult needs about .3-.4 g/day of magnesium.

Phosphate Balance

Phosphates are stored in the skeleton and used for phosphorylation of ADP. Phosphates are reabsorbed in kidney tubules and excreted in the urine. The typical adult needs about .8-1.2 g/day of phosphate.

Chloride Balance

Chloride ions are the most numerous negative electrolytes in the body. Chloride ions are absorbed in the digestive tract and cotransported with sodium ions. Chloride ions are reabsorbed in the kidney tubules. The typical adult requires about 1.7-5.1 g/day of chloride.

Acid Base Balance

Big Picture: Acid-Base Balance

The body maintains a narrow pH by using buffer systems.

When we talk about acids and bases we need to use the pH scale. You might remember from chemistry that pH is a measure of acidity or alkalinity in the body. To review, the pH scale ranges from 0 to 14 with 7 being neutral (like water). If you start at 7 and move toward 0 things become more acidic. Likewise if you begin at 7 and move toward 14 things become more alkaline.

The body maintains a narrow range of pH of the blood that is between 7.35 and 7.45. It must maintain this pH despite the constant release of acidic substances from metabolic processes and minute changes in pH associated with the respiratory system.

Here is an important concept about acids and bases. Acids are substances that release hydrogen ions. Bases are substances that combine with hydrogen ions in order to neutralize them. Many bases release hydroxide ions that combine with hydrogen ions to form water.

There are strong acids that completely dissociate (break apart) in solution. For example hydrochloric acid is considered a strong acid. This one will release lots of hydrogen ions:

$$HCl \rightarrow H^+ + Cl^-$$

Weak acids do not completely dissociate in solution. For example carbonic acid is considered a weak acid. This one will release some hydrogen ions:

$$H_2CO_3 \leftrightarrow H^+ + HCO_3^-$$

Notice the double arrow. This means that the reaction reaches equilibrium in solution and that means that only a portion of the carbonic acid molecules will dissociate.

Acid-base balance is maintained by the respiratory and urinary systems as well as buffer systems in the blood. We have seen that the kidneys secrete hydrogen ions and reabsorb bicarbonate ions (see urinary chapter). These actions help to regulate pH. We have also seen that the respiratory system works to adjust pH by carbon dioxide storage (see respiratory physiology chapter). Although the kidneys have a large effect on pH, they tend to work slowly over a period of hours or days. Buffer systems work instantly to adjust pH.

Most metabolic reactions in the body tend to release more hydrogen ions than combine with them. Hydrogen ions are released in the aerobic and anaerobic respiration of glucose, the incomplete oxidation of fatty acids, oxidation of amino acids containing sulfur and the hydrolysis of phosphoproteins and nucleic acids.

One important big picture concept is that buffer systems essentially all work the same way. If there are too many hydrogen ions buffers combine with them. Likewise if there are too few hydrogen ions buffers release them.

Buffer systems are bidirectional chemical reactions that either release or combine with hydrogen ions in order to control pH. There are several important buffer systems in the body. These include the carbonic acid system, proteins, phosphates and ammonium compounds.

Let's take a look at the carbonic acid system.

$H_2CO_3 <-> H^+ + HCO_3^-$

When there are too many hydrogen ions the reaction moves to the left. In other words the hydrogen ions combine with bicarbonate ions to form carbonic acid. This helps to raise pH. Likewise when there are too few hydrogen ions the reaction moves to the right. Carbonic acid tends to dissociate into hydrogen ions and bicarbonate.

The carbonic acid buffer system reacts quickly to the addition of substances such as lactic acid and carbon dioxide during periods of intense activity.

Proteins also play a role in acting as buffer systems. These include cellular proteins such as histones and plasma proteins such as hemoglobin. The ability of proteins to act as buffers has to do with the presence of the carboxyl and amine groups. The carboxyl group acts as a weak acid while the amine group acts as a weak base.

When conditions become acidic hydrogen ions bind to the amine group. Likewise when conditions become basic hydrogen ions are released from the carboxyl group.

Respiratory System Regulation of Acid-Base Balance

We saw that carbon dioxide was stored in bicarbonate by the following reaction:

$CO_2 + H_2O <-> H_2CO_3 <-> H^+ + HCO_3^-$

In areas of higher concentrations of carbon dioxide the reaction moves to the right. Carbon dioxide combines with water to form the intermediate carbonic acid which, in turn dissociates into hydrogen ions and bicarbonate. Likewise when the concentration of carbon dioxide decreases, the reaction moves to the left. Typically the reaction moves to the right in the tissues as they produce higher levels of carbon dioxide. The reaction moves to the left in the alveoli where carbon dioxide levels are lower causing the release of carbon dioxide so that it can be removed by the lungs.

As the blood becomes more acidic respiratory centers in the brainstem respond by increasing the rate and depth of breathing. This causes the reaction to move more readily to the left. Hydrogen ions are removed by recombining with bicarbonate to form carbonic acid and subsequent carbon dioxide.

As blood becomes more alkaline respiratory centers are inhibited causing an increased concentration of carbon dioxide. This causes the reaction to move to the right allowing the release of hydrogen ions. Thus respiratory rate and depth of breathing act to maintain pH.

In a nutshell we can say that if my blood becomes too acidic I will breathe faster and heavier. Likewise if my blood becomes too alkaline I will breathe slower.

Acid-Base Imbalances

Respiratory Acidosis

We learned about respiratory acidosis in the respiratory physiology chapter but it's always good to review it so here we go. Respiratory acidosis generally occurs from an inability of the lungs to get rid of carbon dioxide. Carbon dioxide then builds up in the blood and is converted to bicarbonate and hydrogen ions. The increased concentration of hydrogen ions causes the acidosis.

Respiratory acidosis is caused by injuries to the respiratory centers in the brainstem, obstructions in air passages, and decreases in gas exchange such as with emphysema and pneumonia. The symptoms of respiratory acidosis include drowsiness, disorientation, stupor, coma and even death in severe cases.

Respiratory acidosis is treated with an intravenous infusion of sodium lactate. The lactate ions are converted to bicarbonate ions in the liver and the bicarbonate ions help to buffer the hydrogen ions.

Respiratory Alkalosis

Respiratory alkalosis can develop from fever, hyperventilation and salicylate (aspirin) poisoning. An increased amount of carbon dioxide is removed from the lungs decreasing the hydrogen ion concentration in the blood.

Respiratory alkalosis results from aspirin poisoning because aspirin stimulates the respiratory centers in the medulla causing an increased respiratory rate. This causes a subsequent increase in removal of carbon dioxide from the lungs.

Symptoms of respiratory alkalosis include lightheadedness, dizziness, tingling sensations in the hands and feet and tetany of muscles in severe cases.

Metabolic Acidosis

Not all acidosis is from the respiratory system. Metabolic acidosis can occur from an accumulation of acids or loss of bases from the body. Examples of conditions that can cause metabolic acidosis include kidney disease (kidneys fail to secrete acids), prolonged vomiting (losing alkaline substances from the GI tract), prolonged diarrhea and diabetes mellitus (production of ketone bodies that lower pH). The symptoms are the same as respiratory acidosis and so is the treatment. The underlying cause of the problem must be treated as well as the symptoms.

Metabolic Alkalosis

Metabolic alkalosis occurs from a loss of hydrogen ions or gain of bases in the body. Examples of conditions that can cause alkalosis include gastric lavage, prolonged vomiting, diuretics, and taking too much antacid. The symptoms are the same as respiratory alkalosis and so is the treatment.

Review Questions

1. Which of the following is not part of the extracellular compartment:
   a. Interstitium
   b. Lymph
   c. Plasma
   d. Cytosol

2. The typical adult body contains about ___ liters of fluid:
   a. 20
   b. 30
   c. 40
   d. 50

3. The thirst mechanism resides here:
   a. Pons
   b. Medulla oblongata
   c. Parietal lobe
   d. Hypothalamus

4. Which best describes the mechanism by which aldosterone affects fluid volume:
   a. Causes sodium retention and water is retained by osmosis
   b. Causes potassium secretion and water is removed by osmosis
   c. Causes active transport of water to kidney capillaries
   d. Increases filtration of water in kidney tubules

5. Which of the following is not a result of dehydration:
   a. Dizziness
   b. Mental confusion
   c. Hypovolemia
   d. Vasodilation

6. A person with a sodium level of 160 mg/dl has:
   a. Nothing, this is normal
   b. Hyponatremia
   c. Hypernatremia
   d. Hypokalemia

7. Which is the most severe problem:
   a. Hyponatremia
   b. Hypernatremia
   c. Hypokalemia
   d. Hyperkalemia

8. Which of the following is treated with calcium gluconate:
   a. Hyponatremia
   b. Hypernatremia
   c. Hypokalemia
   d. Hyperkalemia

9. A decrease in blood calcium levels will facilitate the secretion of which substance:
   a. Parathyroid hormone
   b. Calcitonin
   c. Aldosterone
   d. Atrial natriuretic hormone

10. A decrease in hydrogen ion concentration can produce:
    a. Hyperkalemia
    b. Acidosis
    c. Hyponatremia
    d. Alkalosis

11. Prolonged vomiting producing an increase of hydrogen ions is most likely:
    a. Respiratory acidosis
    b. Metabolic acidosis
    c. Respiratory alkalosis
    d. Metabolic alkalosis

12. A low PCO_2 is typically associated with:
    a. Respiratory acidosis
    b. Metabolic acidosis
    c. Respiratory alkalosis
    d. Metabolic alkalosis

Chapter 19

Taking the Indigestion out of the Digestive System

Chapter 19

Taking the Indigestion out of the Digestive System

Next time you sit down to enjoy a burger and fries (or tofu and salad) think of the miracle of digestion. The food is chewed up, swallowed and embarks on a journey through a long and winding tube to become part of your body. You really are what you eat.

The digestive system can be thought of as a long tube beginning at the mouth and ending at the anus. The length of the tube ranges from 15-23 feet long. The tube is called the alimentary canal. The canal has the same number of layers but different tissue characteristics throughout (fig. 19.1).

Here are the basic processes of digestion:

1. Breakdown of food chemically and mechanically.
2. Absorption of microscopic constituents of food.

Different parts of the tube are designed for different functions. It is a good idea to keep this in mind when learning the digestive system.

Big Picture: Digestive system

Mouth—designed for breaking down food and beginning digestion of carbohydrates.

Stomach—designed for chemical digestion of proteins and some fats and very little absorption.

Duodenum (small intestine)—designed for lots of digestion and some absorption.

Jejunum and Ileum (small intestine)—designed for finishing up digestion and lots of absorption.

Colon (large intestine)—designed for almost no digestion or absorption. Mostly removing water and moving what's left through.

Those are the differences so here is one similarity. The "tube" has four layers throughout. From inside to out these are:

Mucosa—inner mucous membrane.

Submucosa—deep to the mucosa. Contains blood and lymphatic vessels.

Smooth Muscle (two layers)—one layer runs lengthwise while the other encircles the tube.

Serosa—outer serous membrane that secretes a slimy, serous fluid.

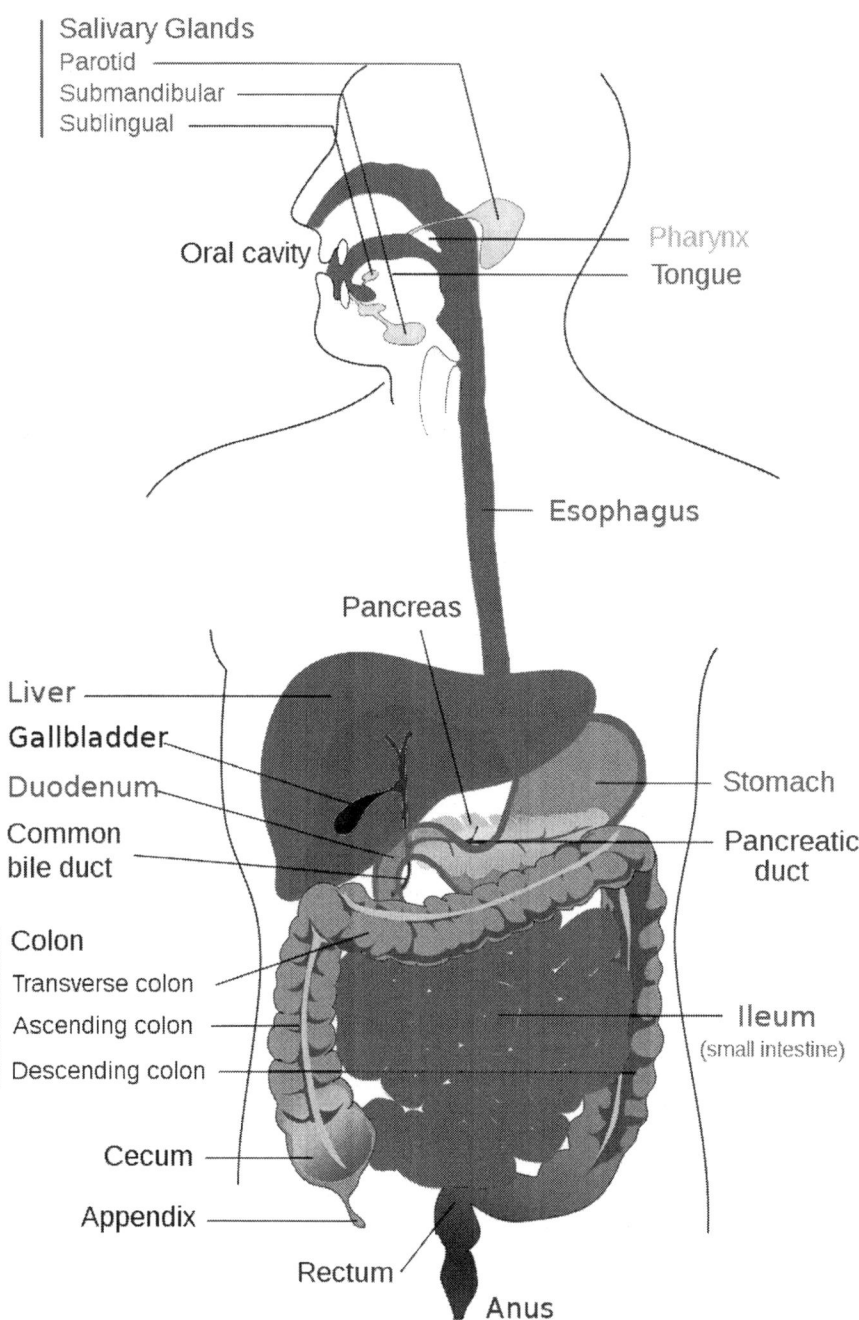

Figure 19.1. The gastrointestinal system.

Journey through the "Tube"

Let's begin our journey through the digestive tube by starting with the mouth. The mouth contains the oral cavity which of course contains the teeth and tongue. There are three sets of salivary glands on each side of the head that secrete saliva into the mouth. Saliva helps to get the food ready for digestion and contains an important enzyme that begins carbohydrate digestion. This enzyme is aptly named salivary amylase. This is why you can dissolve lollipops in your mouth. Lollipops are sugar (carbs) and salivary amylase breaks down carbs.

Here are the three salivary glands (fig. 19.2):

Parotids—located just in front of and below your ear. They secrete mucous and salivary amylase and connect to the mouth via ducts called Stensen ducts.

Sublinguals—these are under your tongue and secrete mucous.

Submandibulars—these are under your jaw and secrete enzymes and mucous.

The mouth also contains the teeth which are the hardest structures in the body (fig. 19.3). You begin cutting your teeth as a baby and finish as a young adult with the wisdom teeth. There are 20 primary (baby) teeth and 32 secondary teeth (fig. 19.4).

The mouth connects with the pharynx (fig. 19.5). The pharynx is a shared passageway for the respiratory and digestive systems. Every time you put food in your mouth your brain and mouth work together to chew the food and roll it up into what is known as a bolus. The bolus is then pushed to the back of the mouth where it enters the pharynx for swallowing.

The Esophagus

The pharynx connects with the esophagus (fig. 19.6). The esophagus is a muscular tube extending from the pharynx to the stomach. It lies behind the trachea.

The esophagus has two circular sphincter muscles. The upper esophageal sphincter keeps the esophagus closed during breathing to keep air from moving into the digestive tract. The lower esophageal sphincter (cardiac sphincter) is located at the inferior end of the esophagus where it pierces the diaphragm at the esophageal hiatus. The lower esophageal sphincter remains closed until swallowing occurs. In some cases the diaphragm is weakened near the hiatus and the sphincter enlarges. This is known as a hiatal hernia and can allow contents of the stomach to enter the esophagus and cause gastric reflux.

The esophagus empties into the stomach which is a curved pouch-like organ (fig. 19.7). There are four major divisions of the stomach. These include the cardiac region, body, fundus and pylorus. The cardiac region is the superior portion just after the esophagus. The fundus is an upward bulge that is located on the left side. The body is the central portion and the pylorus the inferior portion.

The outer portion of the stomach contains a concave and convex curve. The concave curve is called the lesser curvature while the convex curve is the greater curvature.

The stomach contains two sphincters at each opening. The lower esophageal (cardiac) sphincter allows substances to enter while the pyloric sphincter allows substances to exit.

Inside of the stomach are tube-like openings called gastric pits. Secretions from deeper gastric glands flow through the pits to the inside of the stomach. There are also a good deal of mucous secreting cells that secrete an alkaline mucous to help to protect the lining. If you were to look inside of the stomach you would see that it was lined with folds. These folds are called rugae. The rugae increase the surface area and help in mixing the contents of the stomach.

There are a number of different kinds of cells on the inside of the stomach. Here is an overview:

Parietal cells --secrete hydrochloric acid and intrinsic factor.

Chief cells-- secrete a precursor enzyme called pepsinogen. Pepsinogen combines with hydrochloric acid to become an active form known as pepsin. Pepsin digests proteins.

Mucous secreting cells—secrete, you guessed it, mucous.

Endocrine cells--secrete hormones that help to control digestion. An example is intrinsic factor that combines with vitamin B12 to help in absorption.

The combination of all of the stomach secretions is known as gastric juice. Food enters the stomach and combines with gastric juice to form a pasty substance called chyme. Chyme then leaves the stomach by way of the pyloric sphincter and enters the duodenum.

So, food goes in and chyme comes out.

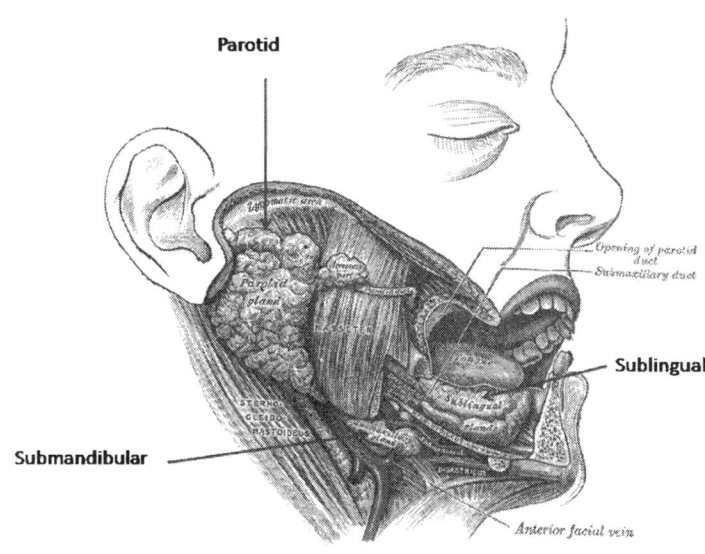

Figure 19.2. Salivary glands

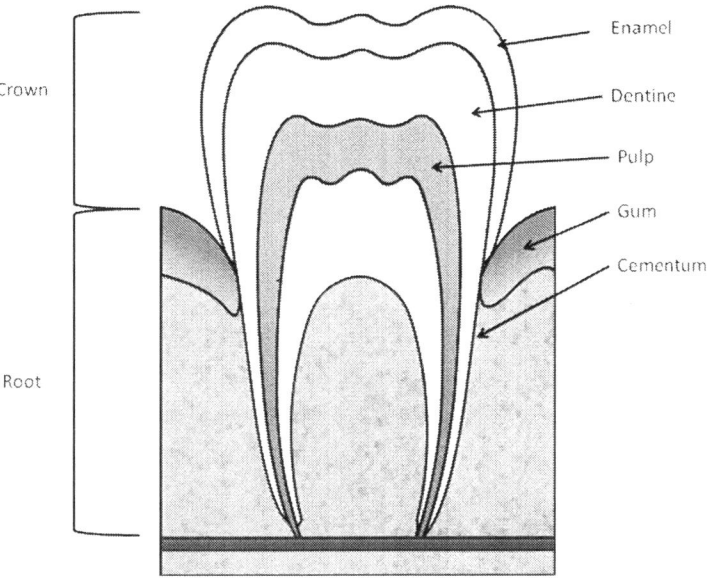

Figure 19.3. Tooth

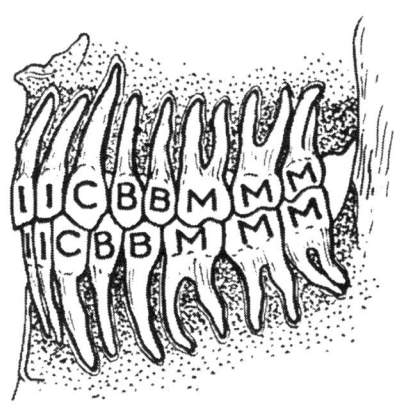

Figure 19.4. Teeth

I = incisor

C = canine

B = bicuspid

M = molar

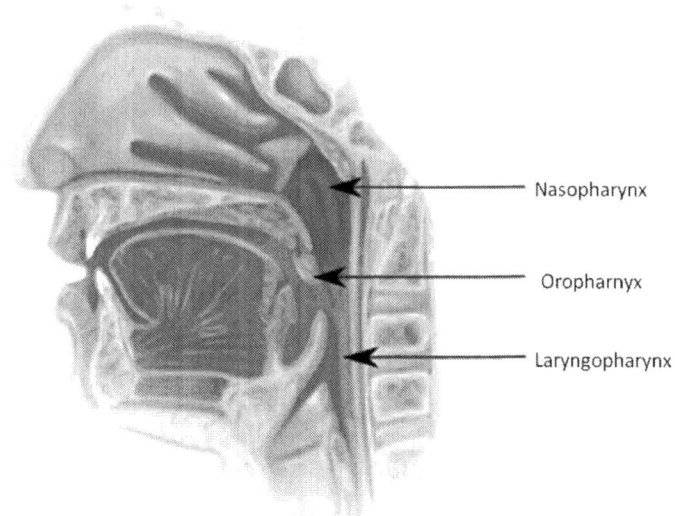

Figure 19.5. Pharynx

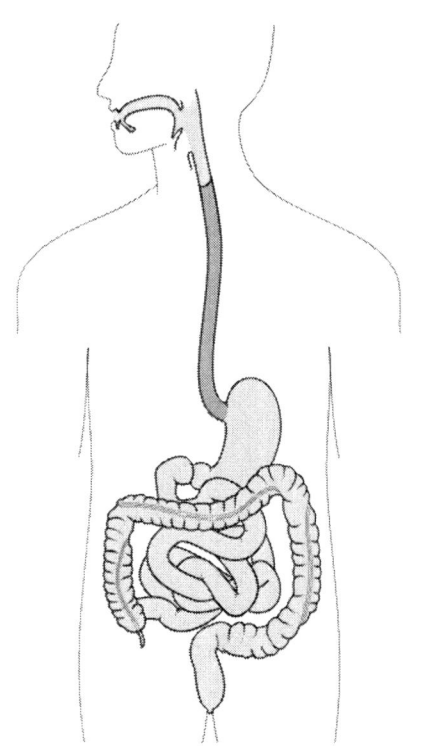

Figure 19.6. Esophagus

Stomach

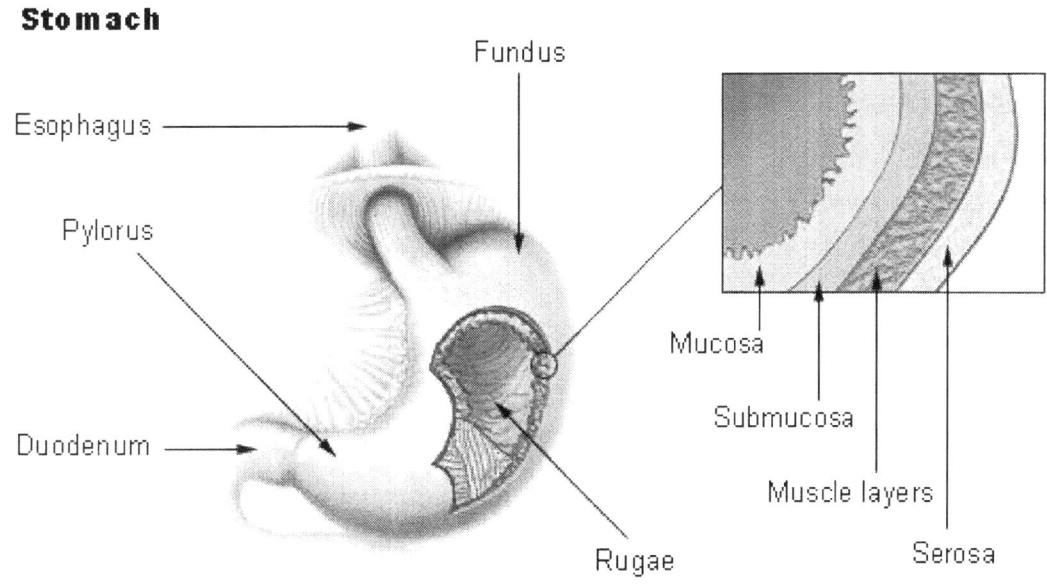

Figure 19.7. Stomach

Small Intestine

After the stomach, food enters the small intestine. The small intestine consists of three parts. The proximal section is the duodenum which is followed by the jejunum and ileum (fig. 19.8).

The small intestine is built for absorption with a large surface area. The inside of the small intestine consists of circular folds called plica circulares. The plicae also contain numerous finger-like projections known as villi. The villi contain blood vessels and a lymphatic system tubule called a lacteal. The intestines are lined with cilia containing epithelium. The epithelial membrane resembles a brush and is sometimes referred to as a brush border. Cells lining the intestines secrete enzymes and mucous. The membrane also contains intestinal crypts (crypts of Lieberkuhn) that undergo rapid mitosis. The crypts help the intestinal membrane to renew itself. Old cells are pushed out of the villi and are replaced by new cells.

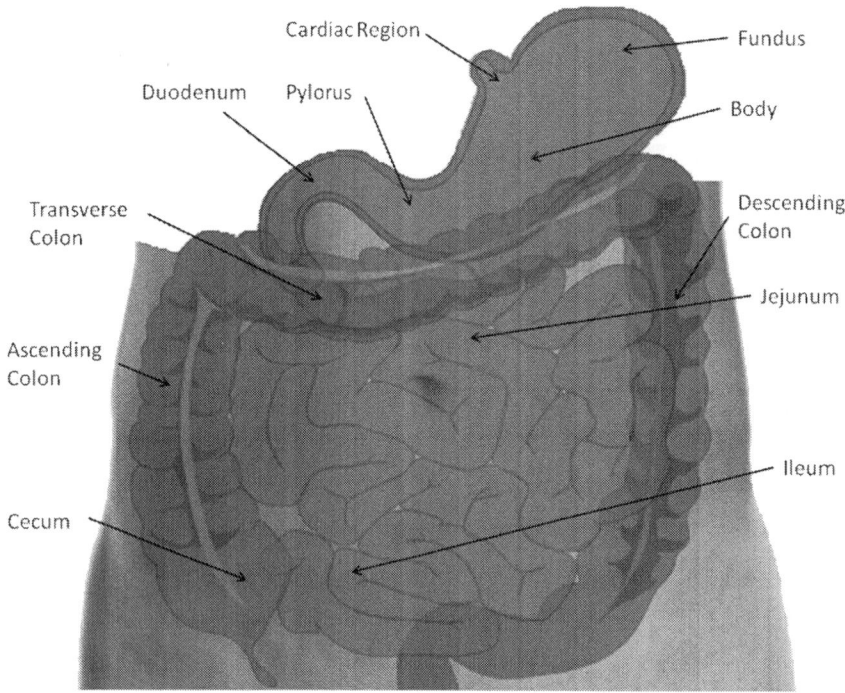

Figure 19.8. Duodenum and Intestines

Large Intestine

The last large section of the tube is the large intestine or colon (fig. 19.8). The large intestine begins at a pouch called the cecum. The junction between the ileum and cecum occurs at a smooth muscle sphincter in the cecum known as the ileocecal valve or sphincter.

The large intestine continues as the ascending colon which makes a turn at the liver called the hepatic flexure. It then continues as the transverse colon which makes a turn at the spleen called the splenic flexure. It then continues inferiorly as the descending colon until it makes an "S" shape called the sigmoid colon which connects to the rectum and anus.

The large intestine contains numerous mucous secreting glands. Along the outside of the colon are bands of smooth muscles called taenia coli that run longitudinally. There are also rings of smooth muscle that divide the colon into pouch-like structures called haustra.

By the time material has reached the large intestine, digestion and absorption are primarily complete. The colon secretes mucous that holds feces together. There are also resident bacteria known as intestinal flora. The bacteria help to break down undigested substances.

Substances move through the *small intestine* by slow peristaltic movements. In the *large intestine* substances move in what are called mass movements. Mass movements are stimulated by parasympathetic impulses, stretch of the colon walls and impulses from the enteric nervous system. The mass movements move large sections of the colon (up to 20 cm) resulting in the movement of feces toward the rectum.

Water and electrolytes are absorbed in the large intestine with the resulting material called feces. When feces enter the rectum it triggers the defecation reflex. Distention of the rectum sends sensory impulses to the conus medullaris of the spinal cord. The resultant motor impulses produce peristaltic contractions in the colon and cause the internal anal sphincter muscle to relax. Impulses also travel to the cerebrum where the urge to defecate is sensed. This allows for voluntary control of the external anal sphincter (this is a good thing). Relaxation of the external anal sphincter results in defecation.

Defecation can be produced voluntarily by holding the breath and bearing down. This increases the intraabdominal pressure and moves feces toward the rectum triggering the defecation reflex.

So that brings us to "the end" of our journey through the tube.

The Liver

We need to mention a couple of organs associated with the digestive system. These are the liver and pancreas. The liver is located in the right upper quadrant of the abdominal cavity close to the diaphragm (fig. 19.9). The liver consists of four lobes including right, left, quadrate, and caudate lobes. The right and left lobes are separated by the falciform ligament (fig. 19.12).

The liver performs many functions and is considered a vital organ. Its functions include detoxifying the blood, producing bile, metabolism of carbohydrates, fats and proteins, storing iron, blood and vitamins, recycling red blood cells and producing plasma proteins.

Since we are covering the digestive system we need to elaborate a bit more on bile. Bile is a yellowish-green substance secreted by the liver and stored in the gallbladder. Bile contains bile salts that are formed from cholesterol. Bile works to break down fat by emulsification and eliminates products from the breakdown of red blood cells.

Bile works something like soap on a greasy pan. Let's say that I just cooked a good ole country breakfast complete with some fatty bacon (yum). When I proceed to clean up the pan I squirt a few drops of dish soap on the pan with some water and low and behold, the grease just breaks up! The same kind of thing happens with bile. It breaks some of the bonds holding the fat molecules together so the enzymes can get in there and do their digestive work.

Bile is made in the liver and stored in the gallbladder. The gallbladder is about 3-4 inches long. In some cases bile can precipitate and form gallstones. The gallbladder can become inflamed in a condition known as cholecystitis.

The Pancreas

The pancreas has a dual endocrine and exocrine role (fig. 19.10). We investigated the endocrine role in the endocrine system chapter. The exocrine portion consists of glands that secrete substances into ducts. The smaller ducts merge with the larger pancreatic duct. The pancreatic duct merges with the common bile duct at an area in the duodenum known as the hepatopancreatic ampulla. The hepatopancreatic ampulla is encircled by smooth muscle forming the hepatopancreatic sphincter.

The exocrine glands secrete digestive enzymes (we'll cover these later). The endocrine cells are called alpha and beta cells. The alpha cells secrete glucagon and the beta cells secrete insulin.

The pancreas consists of a body, head and a tail (just like your pet dog except the tail can't wag). It is located in the curve of the duodenum.

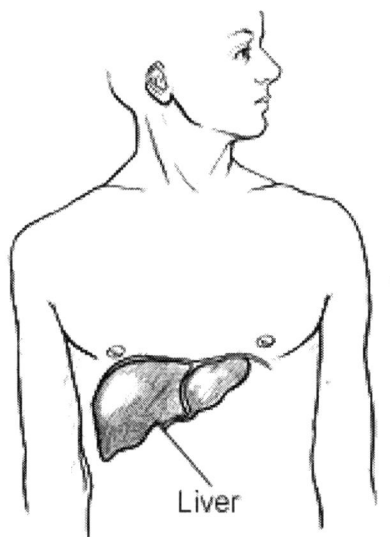

Figure 19.9. The liver is located in the right upper quadrant.

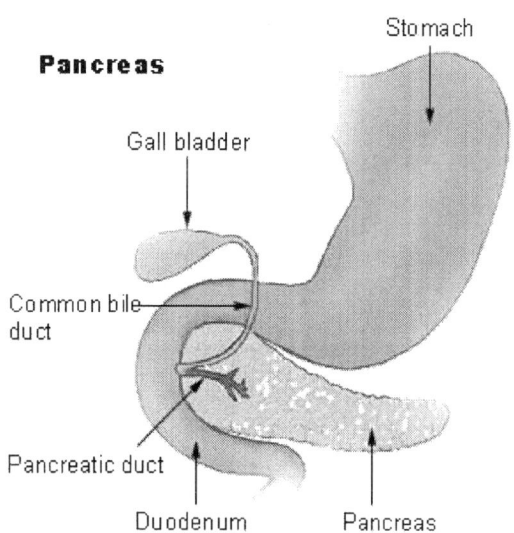

Figure 19.10. Pancreas

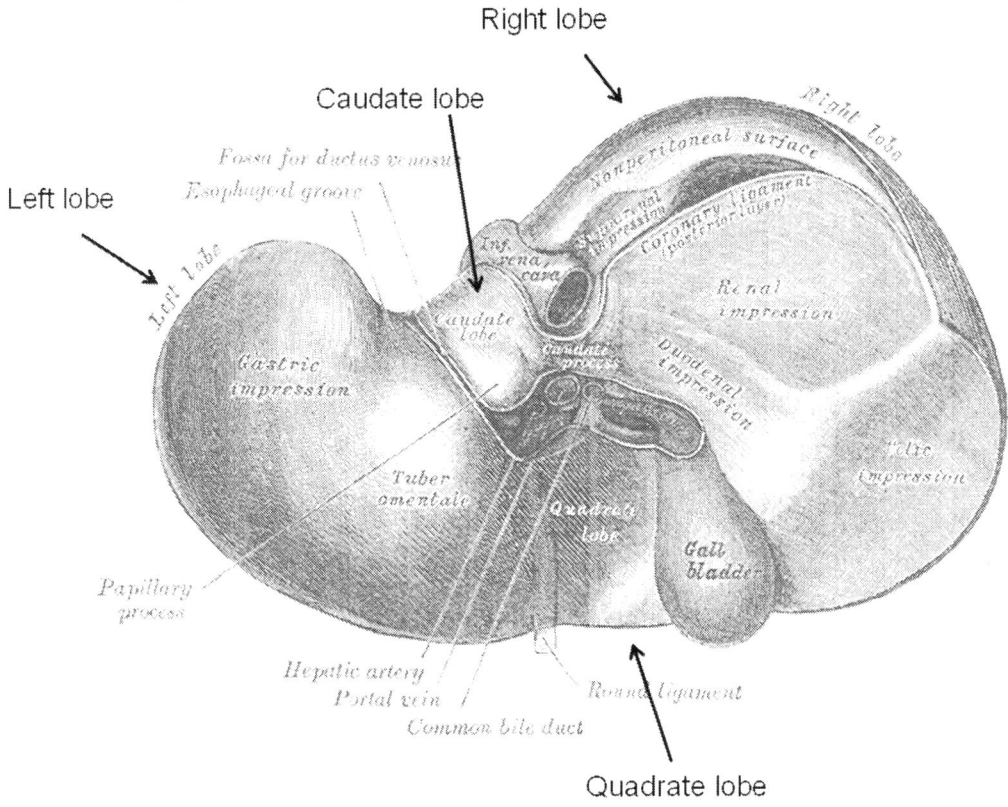

Figure 19.12. Liver

It is important to know the major digestive enzymes, where they are secreted and what they do. So here is a chart that summarizes the enzymes:

Enzyme	What it Digests	Where it's Secreted
Salivary Amylase	Carbohydrates	Salivary Glands
Pancreatic Amylase	Carbohydrates	Pancreas
Maltase	Carbohydrates	Small Intestine
Pepsin	Proteins	Stomach
Trypsin	Proteins	Pancreas
Chymotrypsin	Proteins	Pancreas
Nucleases	Proteins	Small Intestine
Gastric Lipase	Fats	Stomach
Pancreatic Lipase	Fats	Pancrease
Intestinal Lipase	Fats	Small Intestine

Feedback Systems

There are a few feedback systems that we need to cover. These help to control the secretions of the digestive system.

Phases of Gastric Secretions

Big Picture: three phases of gastric secretions

1. *Thinking about food.*
2. *Eating food.*
3. *Food exits stomach.*

There are three phases of gastric secretions. Let's say that I am going to one of those delicious all you can eat buffets (I used to go to these in my student days). I am driving to the buffet and thinking about all of the delicious (and cheap) food. My stomach begins to gurgle with excitement. This is the first stage of digestion called the cephalic stage. Just thinking about, or smelling food can cause my stomach to begin to secrete digestive juice (fig. 19.11).

The next phase occurs when I have filled a plate (or two) of food and begin eating it. The food slides down my esophagus and into my stomach kicking off what is called the gastric phase of digestion. Since digestion is under parasympathetic control my parasympathetic nervous system kicks in (fig. 19.12).

The last phase occurs when food leaves my stomach and moves into my duodenum. Here the gastric secretions are inhibited (fig. 19.13).

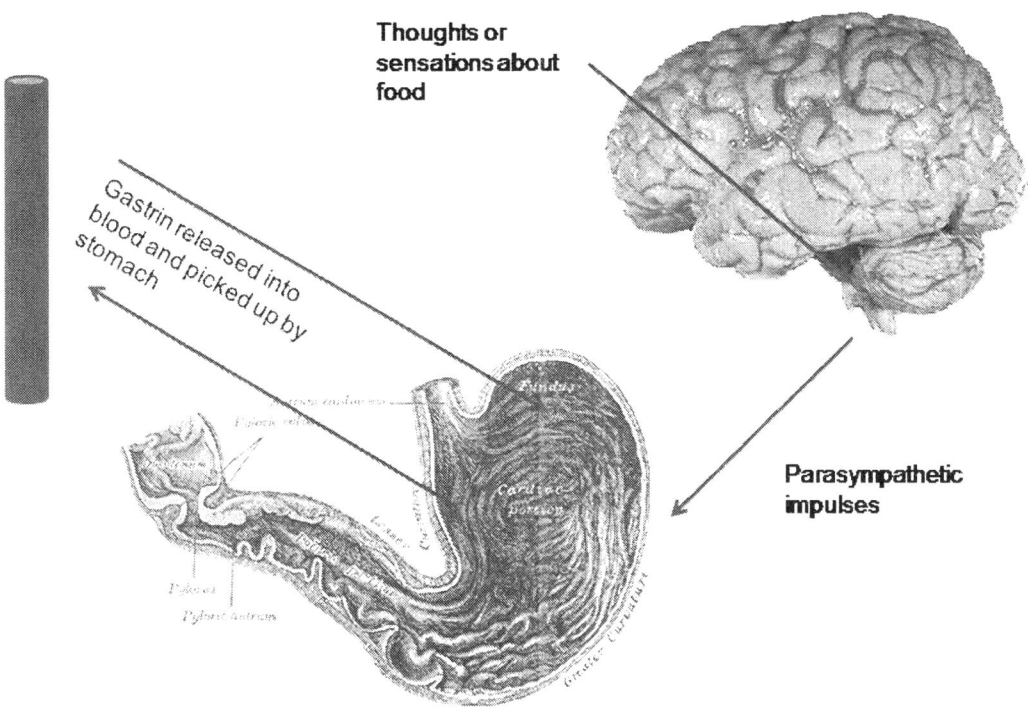

Figure 19.11. The cephalic phase of gastric secretion. Thoughts and smells cause the parasympathetic nervous system to send impulses by way of the vagus nerve to the stomach causing the release of the hormone gastrin. Gastrin is picked up by the stomach and stimulates gastric secretions.

Digestive 318

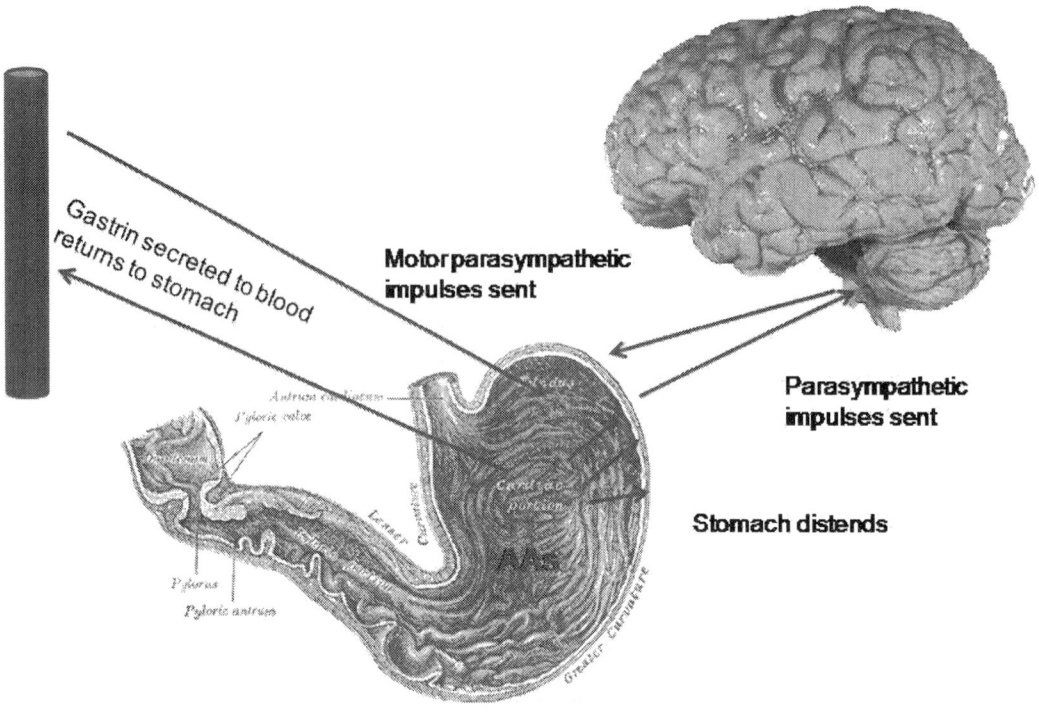

Figure 19.12. The gastric phase of gastric secretion. Distention of the stomach and the presence of amino acids causes the parasympathetic nervous system to stimulate gastric secretions.

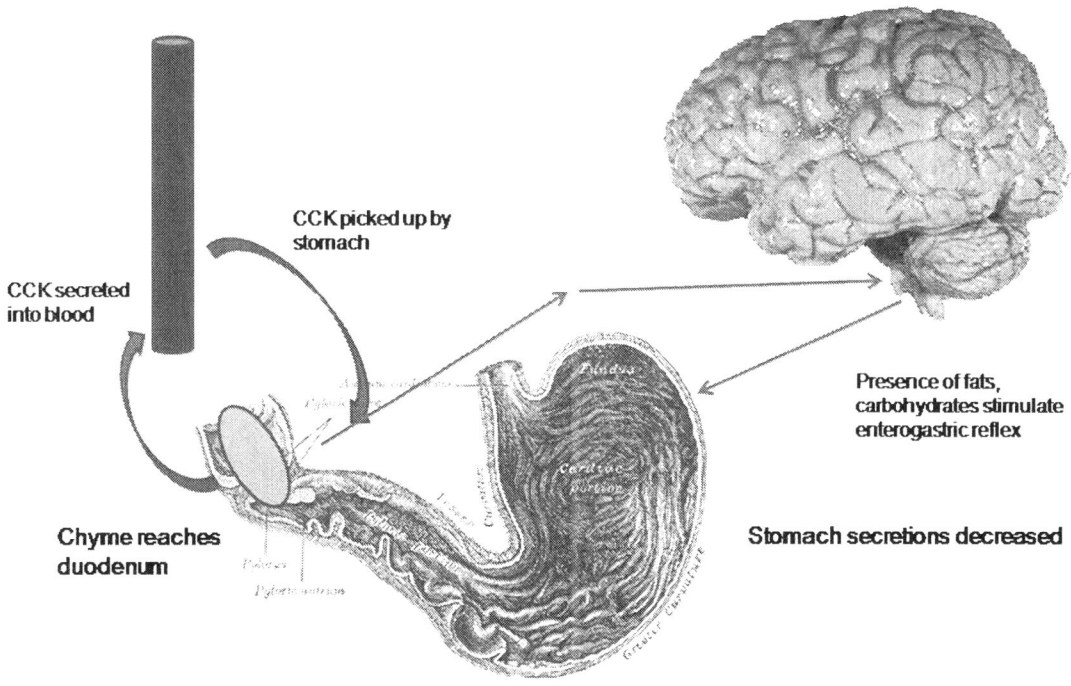

Figure 19.13. Intestinal phase of gastric secretion. Chyme moves into the duodenum causing the release of CCK and secretin that inhibit gastric secretions.

Once food in the form of chyme hits the duodenum it needs to be more fully digested. That's where more enzymes from the pancreas and bile from the gallbladder come into the picture. When that greasy bacon I just ate makes its way to the duodenum it responds by secreting a hormone called:

Cholecystokinin (CCK)

CCK is picked up by the gallbladder causing it to release bile. Bile helps to break some of the bonds of the greasy bacon fat so that the enzymes can more fully digest it.

Movement of chyme into the duodenum also stimulates the release of another hormone called:

Secretin

Secretin inhibits stomach secretion and causes the pancreas to secrete its digestive stuff (see chart). Secretin also causes the pancreas to secrete good ole bicarbonate ions which help to make the chyme more alkaline. Digestive enzymes work better in alkaline conditions (fig. 19.14).

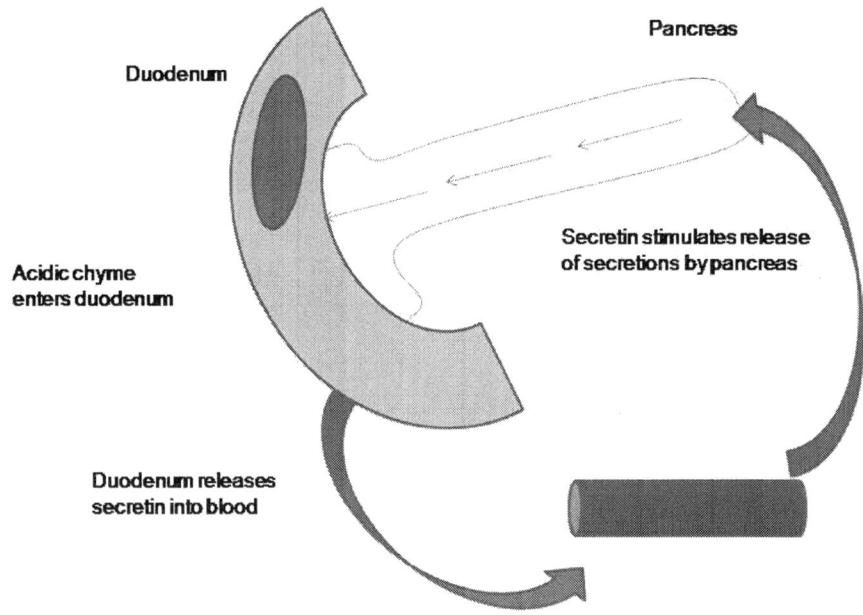

Figure 19.14. Pancreatic secretions are regulated by the release of secretin by the duodenum. Secretin causes the release of bicarbonate ions that help to neutralize acidic chyme.

How Digested Substances Move into the Blood

Big Picture: Absorption

Carbs and proteins are absorbed directly into the blood while fats move into lymph.

At this point everything is pretty much broken down into small molecules of carbohydrates, fats and proteins. These molecules now need to be absorbed so they can be used by the body.

The key to learning about absorption is to understand that carbs and proteins move from the small intestine into the blood in a similar way while fats move from the small intestine to the blood by way of the lymphatic system.

The big picture with regard to carb and protein transport is that they both move via transport proteins (remember symporters in the kidney section) and diffusion directly into capillaries in the intestine.

Fats diffuse into the intestinal cells and then are reassembled into triglycerides. The triglycerides are then packaged into small packages called chylomicrons. The chylomicrons move into the lymphatic system by way of small tublular structures located in the intestinal villi (fingerlike projections). The chylomicrons are eventually delivered to the circulatory system (fig. 19.15, 19.16).

The liver converts lipids by combining them with proteins to form lipoproteins. Lipoproteins are named for the amount of protein and lipid within them. Very low density lipoproteins (VLDL) contain about 92% lipid and 8% protein. Low density lipoproteins (LDL) contain about 75% lipid and 25% protein. High density lipoproteins (HDL) contain about 55% lipid and 45% protein.

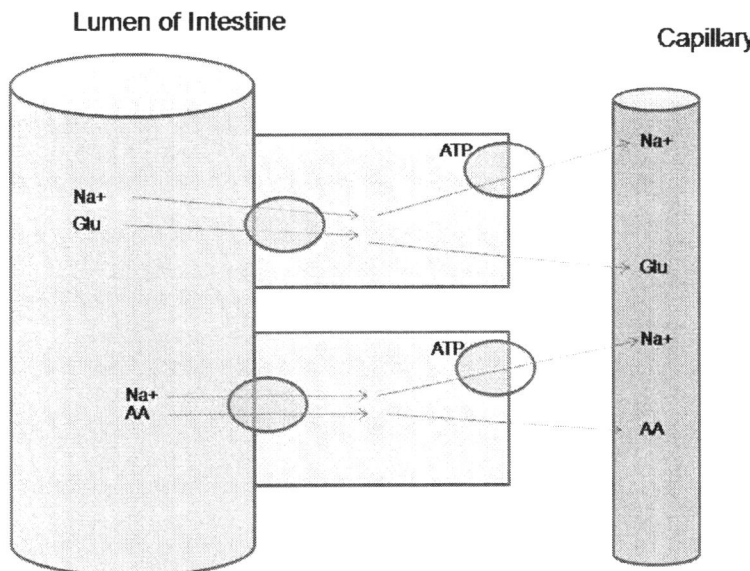

Figure 19.15. Absorption of glucose (Glu), sodium and amino acids (AA) in the intestinal wall. Transport proteins use the sodium gradient to symport glucose and amino acids into the cells. Sodium is removed by active transport while glucose and amino acids are passively transported into the blood capillaries.

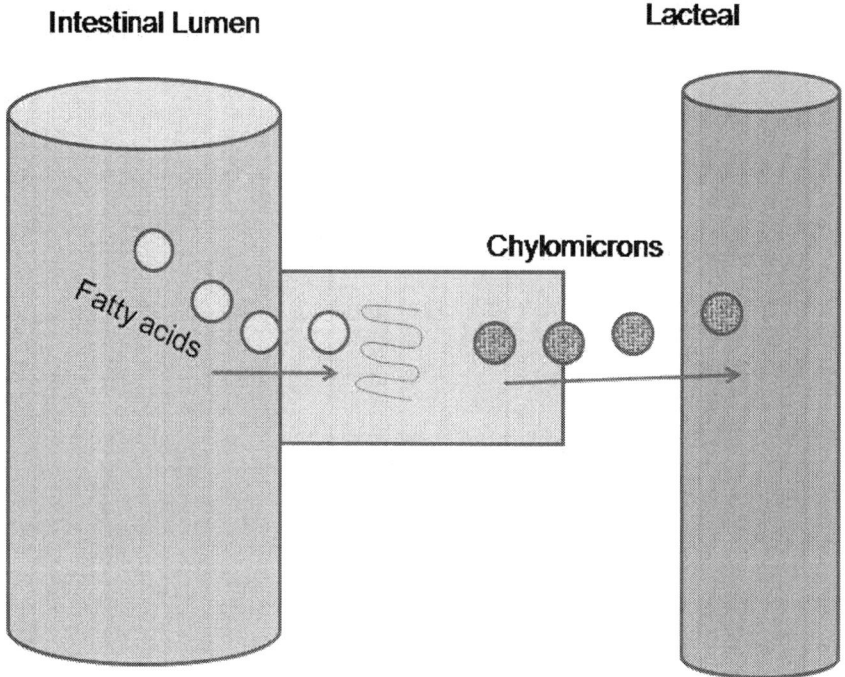

Figure 19.16. Lipid molecules diffuse into epithelial cells lining the lumen of the small intestine. Once inside lipids are repackaged by smooth endoplasmic reticulum into chylomicrons. Chylomicrons diffuse out the cells and into the lacteals which are part of the lymphatic system.

Review Questions

1. Which of the following is not part of the alimentary canal:
   a. Stomach
   b. Ileum
   c. Liver
   d. Duodenum

2. The alimentary canal contains ____ smooth muscle layers:
   a. 1
   b. 2
   c. 3
   d. 4

3. Which of the following papillae contain taste buds:
   a. Vallate
   b. Circumvallate
   c. Fungiform
   d. Filiform

4. Which of the following salivary glands secrete salivary amylase:
   a. Parotid
   b. Submandibular
   c. Sublingual
   d. All of the above

5. How many secondary teeth are there in the adult human:
   a. 24
   b. 26
   c. 30
   d. 32

6. Which of the following is not a stomach structure:
   a. Rugae
   b. Fundus
   c. Pylorus
   d. Omentum

7. Pepsinogen is secreted by which stomach cells:
   a. Chief
   b. Parietal
   c. Columnar
   d. Serous

8. Which of the following is not a stomach secretion:
   a. Pepsinogen
   b. Hydrochloric acid
   c. Vit B12
   d. Intrinsic factor

9. Where is the vomiting center located:
   a. Cerebrum
   b. Medulla oblongata
   c. Sympathetic nervous system
   d. Pons

10. Where are the crypts of Lieberkuhn located:
    a. Stomach
    b. Small intestine
    c. Esophagus
    d. Pancreas

11. Which best describes the structures known as haustra:
    a. Pouches in the small intestine
    b. Long smooth muscle segment in the large intestine
    c. Pouches in the large intestine
    d. Circular structures on the inner portion of the small intestine

12. Bile is formed in which of the following:
    a. Pancreas
    b. Duodenum
    c. Gallbladder
    d. Liver

13. During which phase of swallowing do the uvula and epiglottis move to close off the pharynx and larynx:
    a. First
    b. Second
    c. Third
    d. Fourth

14. Which of the following best describes the action of pepsin:
    a. Helps to digest carbohydrates
    b. Inhibits the release of gastrin
    c. Helps to digest fats
    d. Helps to digest proteins

15. Which best describes a mechanism for the release of gastrin:
    a. Release is stimulated by sympathetic nervous system
    b. Salivary amylase stimulates release of gastrin
    c. Acetylcholine secreted by parasympathetic nervous system stimulates release of gastrin
    d. Fatty chyme leaving the duodenum stimulates release of gastrin

16. Which of the following best describes the function of secretin:
a. Inhibits the release of gastric secretions
b. Stimulates the release of salivary amylase
c. Stimulates release of gastric secretions
d. Causes the release of bile

17. Which best describes the reason bicarbonate ions are secreted by the pancreas:
a. To help breakdown of carbohydrates
b. To help stimulate the release of bile
c. To control stomach secretions
d. To make the chyme more alkaline

18. Which best describes the function of bile:
a. Breaks down carbohydrates
b. Inhibits stomach secretions
c. Emulsifies fats
d. Stimulates release of pancreatic secretions

19. Fats are repackaged in the small intestine into structures called:
a. Chyme
b. Chylomicrons
c. Low density lipoproteins
d. Vesicles

20. Which of the following structures absorbs the most nutrients:
a. Stomach
b. Pancreas
c. Small intestine
d. Large intestine

21. Which of the following best describes the sodium/amino acid symporter in the small intestines:
a. Moves sodium and amino acids in the same direction powered by the sodium gradient
b. Moves sodium and amino acids in opposite directions powered by the sodium gradient
c. Moves sodium and amino acids in the same direction powered by the amino acid gradient
d. Moves sodium and amino acids in the opposite direction powered by the amino acid gradient

22. Which of the following substances is not absorbed in the large intestine:
a. Water
b. Electrolytes
c. Fats
d. All are absorbed

Chapter 20

Fun with Reproduction

Chapter 20

Fun with Reproduction

The reproductive system is often one of the last systems covered in a traditional A&P class. It shouldn't be too difficult since most students know at least a few of the parts. When you get to this system there is light at the end of the tunnel. So let's get the big picture.

The Big Picture: Reproduction

Both males and females contain packages of genetic information that are passed on to offspring.

The overall function of both male and female reproductive systems is to pass on genetic information to offspring (kiddos). The male produces half of the genetic information and packages it in sperm cells. These cells develop and travel through the male to the female. Likewise the female also produces half of the genetic information and packages it in an egg cell called an oocyte. The oocyte is cyclically produced and is either fertilized to complete its development or is not and subsequently discarded. Hormones work to control and support these processes.

The Male

Let's begin our anatomical journey by covering the male (fig. 20.1). First of all we can divide both male and female systems functionally into primary and secondary sex organs. The primary sex organs in the male are the testes while the primary organs in the female are the ovaries. All of the other organs are considered secondary.

The testes are located in the scrotum which hangs outside of the body (I always thought this was kind of a strange system). The scrotum helps to keep the testes at a slightly lower temperature because those sperm cells like things a bit cooler (they develop better). Inside the testes are structured like a series of tubes (seminiferous tubules). The immature sperm cells are located around the perimeter of the insides of the tubes and move to the middle of the tubes (lumen) when they mature. There are also cells surrounding the tubes that work to support the sperm cells and produce the famous male hormone testosterone.

The non-sperm cells in the testes include:

Leydig cells—secrete testosterone

Sertoli cells (sustentacular cells)—in the adult these cells secrete a hormone called inhibin (we'll cover this later) and substances that help sperm develop and mature.

Once sperm cells mature they move to a structure on the surface of the testes called the epididymis. These paired structures each have three portions consisting of a head, body and tail. The epididymis works to help sperm cells mature and they can spend up to three weeks in the tubule system within the epididymis. Sperm move through the epididymis to the vas deferens.

The vas deferens or ductus deferens is a muscular tubular structure that connects with the tail of the epididymis. The muscular layers help to propel sperm cells through the tube. Each vas deferens moves through the inguinal canal and travels through the abdominal cavity and over the top of the bladder to the seminal vesicle. As the vas deferens nears the seminal vesicle the tube widens into an area called the ampulla.

The seminal vesicles are located between the bladder and rectum. They contain epithelium that secretes an alkaline substance, fructose and prostaglandins.

The prostate is a walnut shaped gland just below the bladder. The prostate gland secretes an alkaline milky fluid that helps to nourish and mobilize sperm. The fluid also contains enzymes (hyaluronidase) and prostate specific antigen (PSA).

The urethra (prostatic urethra) passes through the prostate gland. The prostate also contains another set of paired ducts that connect the seminal vesicles to the urethra called the ejaculatory ducts.

The paired bulbourethral glands (Cowper's glands) are pea-shaped glands that secrete an alkaline substance and mucous to help protect and transport the sperm.

The urethra begins at the base of the urinary bladder and passes through the prostate gland and through the penis ending at an opening known as the urinary meatus of the penis. The urethra is lined with a mucous membrane. There are three parts to the male urethra. These include the portion traveling through the prostate (prostatic urethra), the portion extending from the base of the prostate gland to the penis (membranous urethra) and the portion running through the center of the penis (penile urethra).

The penis consists of three columns of tissue called erectile columns surrounded by fibrous coverings surrounded by skin. The two superior columns are called the corpus cavernosum and the lower column is called the corpus spongiosum. Each corpus cavernosum contains a deep artery and is surrounded by a fibrous covering called a tunica albuginea. The corpus spongiosum contains the urethra. The distal portion of the penis contains a slightly larger structure called the glans penis. The glans penis is covered by loose skin called the prepuce which is sometimes removed by circumcision.

And that's about it for an overview of the male anatomy.

So, it's on to the female.

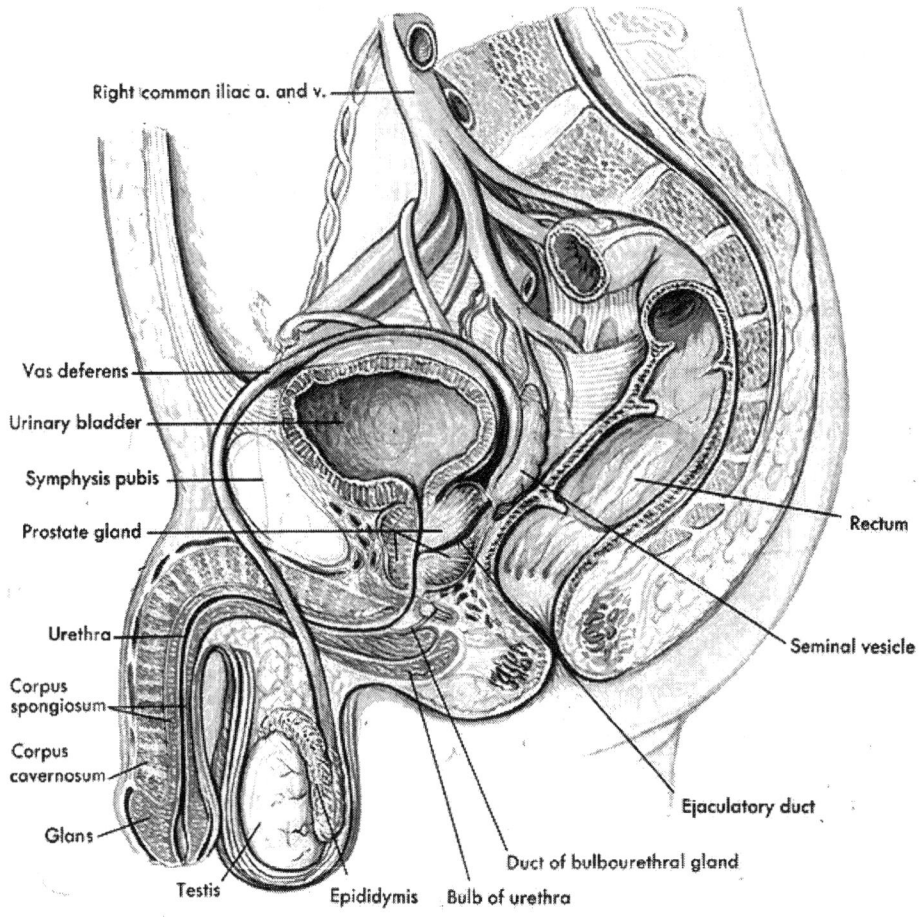

Male pelvic organs as viewed in a median sagittal section.

Figure 20.1. Male reproductive system

The Female

Females are bit more complex than males (no offense to males). Not only can they produce egg cells but can carry them to term and deliver a fully formed human baby. So let's take a look at the female reproductive system (fig. 20.2).

The primary sex organs of the female reproductive system are the ovaries. All of the other structures are considered secondary organs. Ovarian follicles are located inside of the ovaries. The ovarian follicle contains the egg cells known as oocytes. The oocytes are released at about half way through the menstrual cycle in what is known as ovulation.

The Fallopian tubes (aka uterine tubes) extend from the uterus and continue to near the ovaries but do not contact them. Hey, the vas deferens carries the sperm cells in the male and the Fallopian tubes carry the oocytes in the female.

The Fallopian tubes have three sections. These include the first third that extends from the isthmus of the uterus, the second third which ends in a widened area called the infundibulum and the final third which ends in finger-like projections called fimbrae.

The Fallopian tubes transport the oocyte to the uterus after fertilization and are the sites for fertilization by sperm cells. Most fertilized oocytes move to the uterus but occasionally they will deposit somewhere in the pelvic cavity causing what is known as an ectopic pregnancy.

The uterus is a pear shaped structure about three inches long and two inches in width (fig. 20.3). The uterus has two divisions including the body and cervix. The body ends anteriorly as a narrow region called the cervix and posteriorly as a rounded structure called the fundus.

The uterus has three layers. The inner layer is called the endometrium. The endometrium varies in thickness and is thinner just after menstruation and thicker at the end of the cycle. The endometrium has an extensive blood supply and contains mucous secreting cells. The mucous changes its consistency during various times of the menstrual cycle. It is normally thicker during most of the cycle and contains more water near the time of ovulation to help move sperm cells through.

The middle layer or myometrium is a thick smooth muscle layer. The smooth muscle is capable of producing very strong contractions during childbirth. The outer layer or perimetrium consists of a serous membrane.

The body of the uterus lies on top of the bladder in what is called an anteflexed position. Hey, just ask any pregnant female about having to pee frequently! The cervix of the uterus connects with the vagina at an upward right angle. This connection allows for pockets around the cervix called the anterior and posterior fornix that allow for pooling of semen to increase the chances of fertilization.

The uterus can lie in retroflexion in which the uterus tilts backward. Retroflexion can sometimes cause prolapse of the uterus. The uterus is held in place by a series of ligaments. These include two broad ligaments, two uterosacral ligaments (posterior and anterior), and two round ligaments.

The posterior ligament forms a pouch called the posterior cul de sac or rectouterine pouch (of Douglas). Likewise the anterior ligament also forms a pouch called the anterior cul de sac or vesicouterine pouch.

The vagina is located between the rectum and urethra. It is a tubular structure about three inches long that opens to the outside and extends superior and posterior to the cervix of the uterus. The vagina is primarily smooth muscle lined with an epithelial mucous membrane. The mucous membrane can form around the opening of the vagina. This structure is called a hymen. In some cases the opening to the vagina can be completely covered by the hymen (imperforate hymen). An imperforate hymen needs to be medically punctured to allow discharge of the menstrual flow.

The vulva consists of several externally located structures of the female reproductive system. These include the labia majora and minora, mons pubis, clitoris, vestibule, urinary meatus, greater and lesser vestibular glands.

The labia majora are skin covered structures consisting of primarily adipose and connective tissue. The outer surface of the labia majora contains hair while the inner surface does not. They also contain a mucous lining. They are analogous to the scrotum of the male. The labia minora are hairless structures located medially to the labia majora. The space between both labia minor is known as the vestibule.

The clitoris is an organ consisting of erectile tissue. It is located just superior and behind the labial junction. The clitoris contains two corpus cavernosum but no corpus spongiosum so it is similar in structure to the penis. The superior aspect of the clitoris contains a covering of tissue known as the prepuce.

Between the clitoris and opening to the vagina (vaginal orifice) is the urinary meatus which is the external opening of the urethra.

On the sides of the vagina are the greater vestibular glands or Bartholin's glands that open into the area between the labia minor and hymen. The lesser vestibular glands or Skene's glands are located near the urinary meatus.

The perineum is the area between the vagina and anus. The perineum helps to form the muscular floor of the pelvis and can be torn during vaginal childbirth. The perineum contains the urogenital triangle which is formed by drawing a line between the ischial tuberosities with the anterior point of the triangle just superior to the prepuce.

We also need to cover the mammary glands. The mammary glands or breasts are superficial to the pectoral muscles. Internally they consist of a series of lobes separated by connective tissue. The lobes subdivide into lobules containing secretory cells. The cells are arranged in clusters around a central duct. The smaller ducts combine to form larger ducts called lactiferous ducts for each lobe. The lactiferous ducts open to the outside at the nipple. The breasts also contain suspensory ligaments (of Cooper) that help to support them. Each breast contains a circular pigmented area called an areola. The areola contains sebaceous (oil secreting) glands to help protect the nipple.

The breast also contains adipose tissue and lymphatics that drain into the axillary region.

And that's the female anatomy.

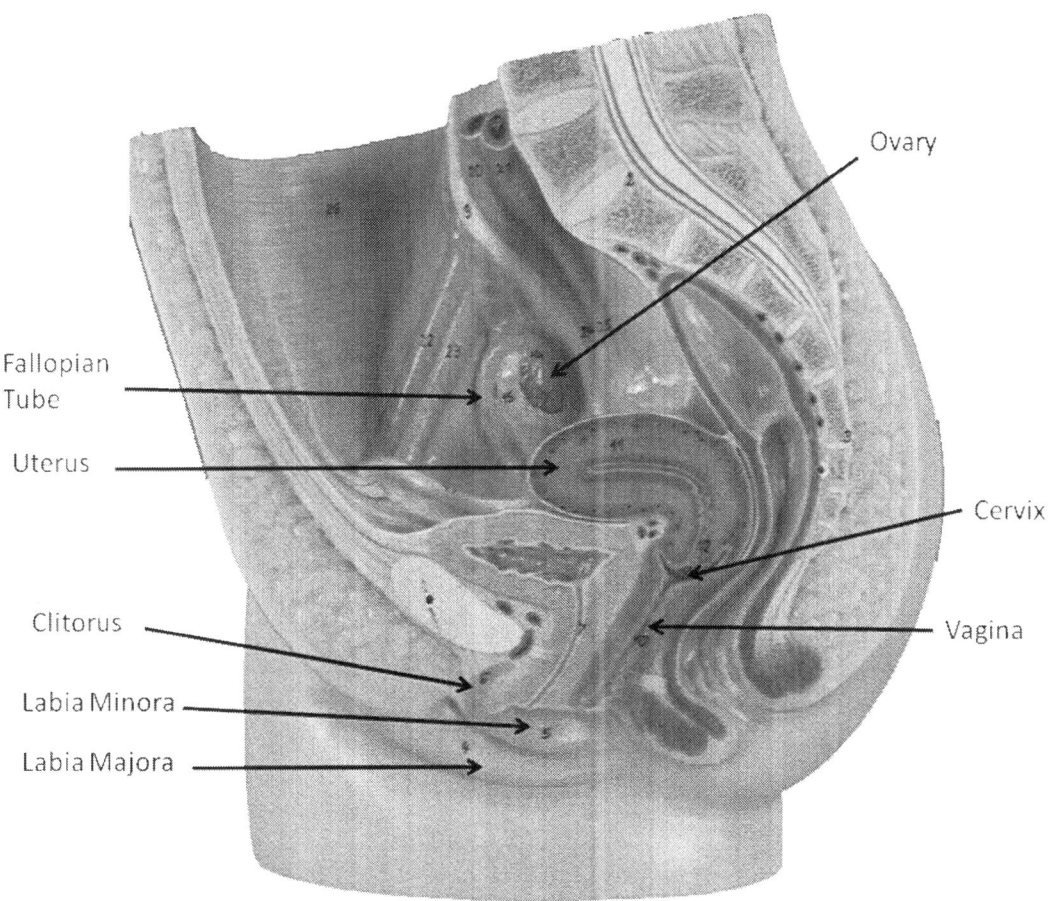

Figure 20.2. Female reproductive system

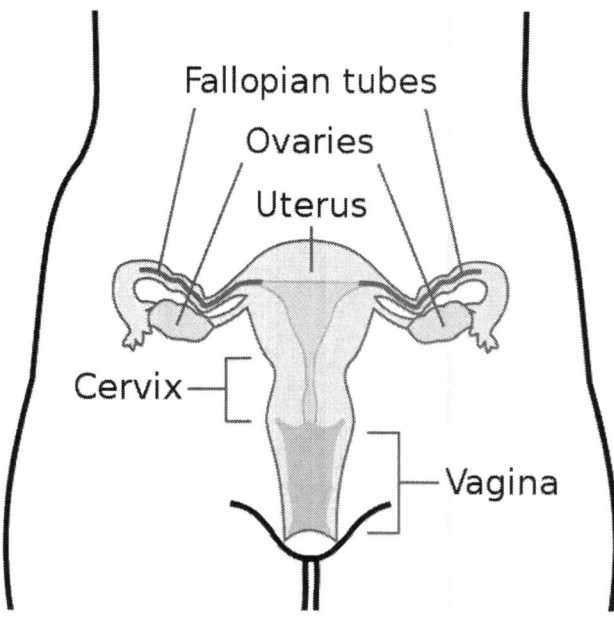

Figure 20.3. Uterus and Ovaries

Makin those Sperm Cells

Makin sperm cells is called spermatogenesis. Spermatogenesis begins with the undeveloped sex cells called spermatogonia. Spermatogonia reside in the testes and will begin to mature around the age of puberty. They continue to do so throughout an adult male's life.

The process begins with the secretion of hormones called gonadotropins from the anterior pituitary gland. These include follicle stimulating hormone (FSH) and leutenizing hormone (LH). Both are secreted in response to the releasing factor gonadotropin releasing hormone secreted by the hypothalamus. Hey, aren't these hormones female hormones? Well, not exactly, this is one of those similarities between males and females. These hormones are found in both sexes.

Leutenizing hormone targets the interstitial cells (Leydig cells) of the testes and promotes the secretion of testosterone. Follicle stimulating hormone (FSH) targets the sustentacular cells (Sertoli cells) of the testes and promotes their maturation and response to testosterone. Both FSH and testosterone work to facilitate the maturation of spermatogonia.

Spermatogonia begin to develop while the male is also developing inside of the mother's uterus. They don't completely develop in utero as their maturation is halted until puberty. They will divide and develop into primary spermatocytes. At this point they basically wait around while the male is enjoying a carefree childhood. Then when testosterone rears its head during puberty, they complete their development.

Once reaching puberty the primary spermatocytes undergo another type of cell division called meiosis (fig. 20.4). There are two stages to meiosis including meiosis I and meiosis II. Meiosis is similar to mitosis with one big difference. In mitosis you end up with two cells with the same number of chromosomes but in meiosis you end up with two cells with half the number of chromosomes.

The normal adult human has 46 chromosomes (called the diploid number of chromosomes).

Chromosomes form pairs that have the same but not necessarily identical genes. These are called homologous chromosomes. The pairs essentially split into two sections of homologous chromosomes with each new cell having 23 chromosomes (haploid number of chromosomes). The chromosomes may contain different variants of genes. For example one cell may contain a different variant for the gene for eye color than the other cell.

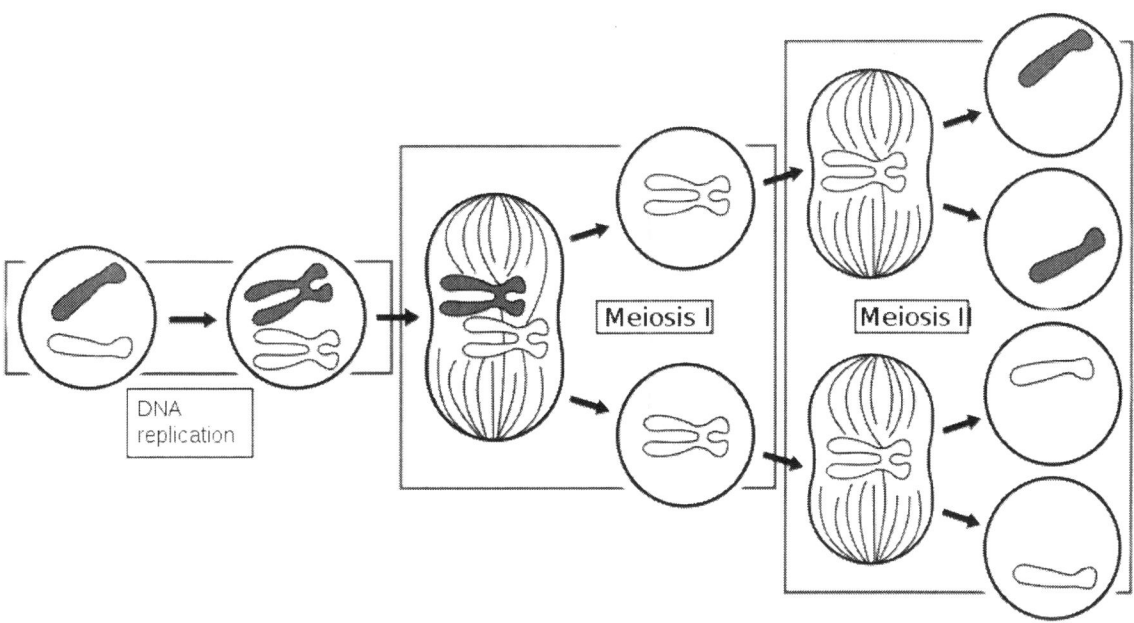

Figure 20.4. Meiosis

Sperm Cells

Once those sperm cells mature during puberty they are ready to go. Mature sperm cells contain three parts including a head, midpiece and tail (figure 20.5). The head contains the genetic material and has an enzyme containing structure called an acrosome on its outer surface. The acrosome contains enzymes such as hyalouronidase that help the sperm cell penetrate the egg cell of the female. The midpiece (engine) contains many mitochondria that produce a good deal of energy in the form of ATP to power the long tail or flagellum. The tail contains the flagellum which is constructed of protein microtubules.

Fun with Reproduction 334

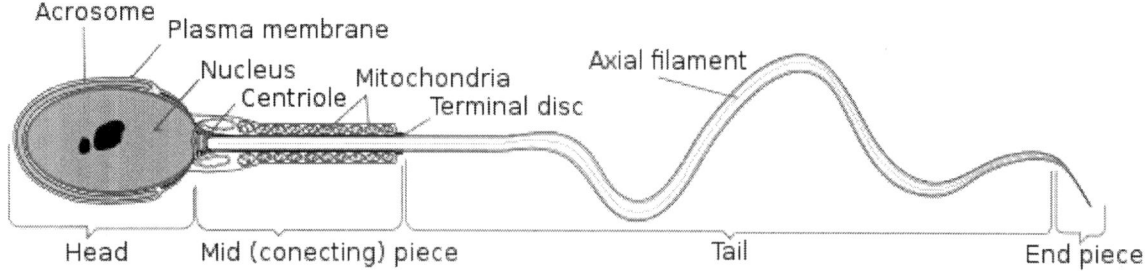

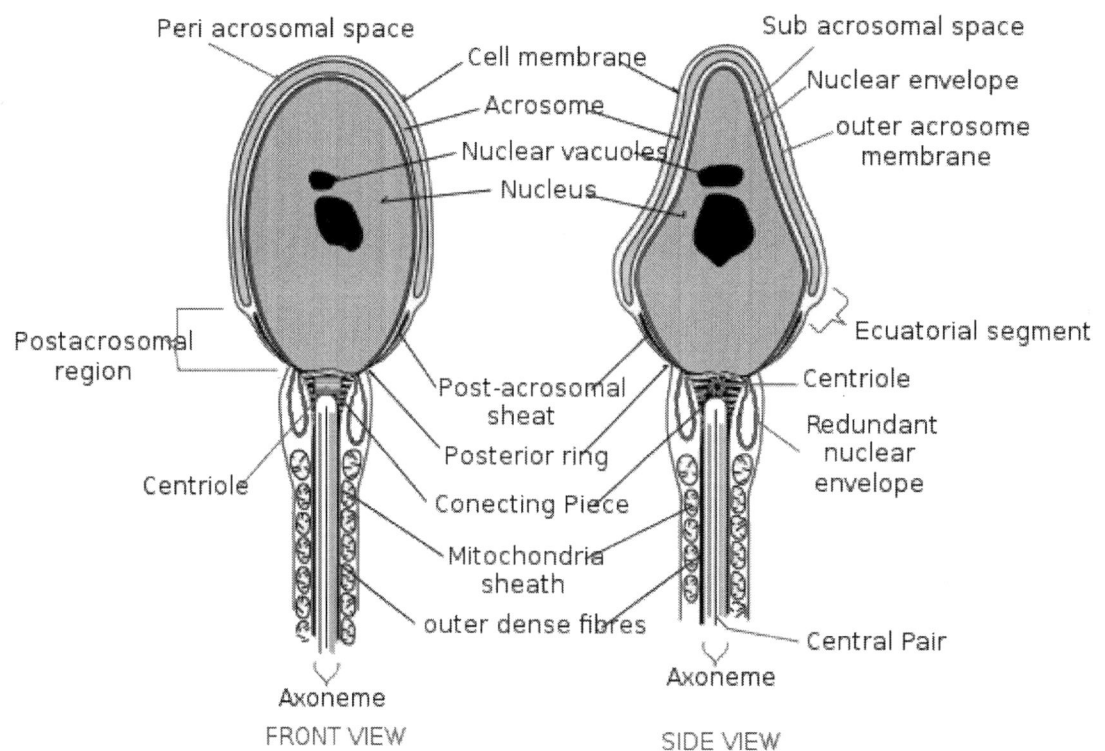

Figure 20.5. Sperm cell

Erection and Ejaculation

This is a complicated process involving both sympathetic and parasympathetic nervous systems. The sacral spinal cord sends parasympathetic impulses to the penis during sexual stimulation. The impulses result in the secretion of nitric oxide which causes vasodilation of the deep arteries in the erectile columns of the penis. Blood then fills the erectile columns which in turn close off the return pathway for blood by compression the dorsal vein of the penis. The penis then becomes erect.

Sexual stimulation then results in orgasm, emission and ejaculation. Emission is the movement of sperm and secretions from the seminal vesicle, prostate and bulbourethral gland into the urethra. Emission is under sympathetic control from the sacral spinal cord which results in smooth muscle contractions throughout the reproductive tract. Skeletal muscles at the base of the penis contract to cause ejaculation which is the forceful expelling of semen from the urethra. Following ejaculation sympathetic impulses cause vasoconstriction of the arteries of the penis, the penis again becomes flaccid and the male falls asleep.

Testosterone

Besides facilitating spermatogenesis, testosterone has other important functions in the male reproductive system. Testosterone is one of a group of hormones called androgens (male hormones).

Testosterone levels are higher during fetal development to help initial development of the male reproductive system and descent of the testes. Testosterone levels then fall during childhood until puberty where they again rise to essentially finish the job of maturation of the male reproductive system.

The actions of testosterone during puberty include the following:

Enlargement of the vocal cords and deepening of the voice.

Increased muscular growth.

Increased body hair on face, axilla and pubic areas.

Strengthening of bones.

Increased metabolism.

Maturation of the sex organs.

Testosterone is regulated by a feedback mechanism involving hormones from the hypothalamus and anterior pituitary gland. We saw that testosterone is secreted in response to LH secreted by the anterior pituitary gland. LH is secreted in response to gonadotropin releasing hormone from the hypothalamus. Blood concentration of testosterone is monitored by the hypothalamus which responds through negative feedback to control the secretion of gonadotropin releasing hormone. The testes also secrete a hormone called inhibin which feeds back to the hypothalamus exhibiting the same effect as testosterone.

Makin those Egg Cells

Oogenesis

Each ovary contains millions of sex cells called oocytes. The oocytes are encased in packages called follicles. At the premature stage the follicles are known as primordial follicles and each contains a primary oocyte. The primary oocytes begin meiosis but do not complete it until puberty. The development of oocytes is known as oogenesis.

As oogenesis continues at puberty the primary oocytes finish meiosis I which results in two cells each containing the haploid number of chromosomes (23). When the oocytes finish meiosis I they are called secondary ooctyes. Unlike spermatogenesis in the male the resultant cells consist of one secondary oocyte and a polar body. The polar body is not a viable cell but helps the secondary oocyte conserve resources to help make it as viable as possible. Development stops at this point unless fertilization occurs. Once the secondary oocyte is fertilized it completes meiosis and produces a second polar body. The fertilized cell is now called a zygote and has the diploid number of chromosomes (46).

The follicle plays an important role in oogenesis as well. The follicle matures under the influence of FSH. It first becomes a primary follicle and contains a region known as the zona pellucida. The zona pellucida contains glycoprotein that gradually separates the ooctye from the inner walls of the follicle. The follicle continues to mature into a secondary follicle which is characterized by the presence of a cavity called the antrum. The oocyte is pushed against the inner wall of the follicle at this stage. Finally the follicle reaches the end of maturation as it becomes a mature or Graffian follicle. The antrum is filled with fluid and the follicle moves to the surface of the ovary. Maturation of the follicle occurs in half of the menstrual cycle.

At about midway through the menstrual cycle the follicle pushes the oocyte out in what is called ovulation. This occurs in response to a surge of LH from the anterior pituitary gland. The oocyte moves toward the Fallopian tube. If it becomes fertilized it will eventually move to the uterus for implantation. If it is not fertilized it will degenerate.

Female Sex Hormones

During fetal development the hormones gonadotropin releasing hormone (GnRH), FSH and LH cause the initial development of the reproductive system as well as descent of the ovaries to their normal position in the pelvic cavity. Secretion of GnRH then decreases until puberty which occurs at about age 10. During puberty the levels of these hormones increases causing the secretion of estrogens and progesterone.

Estrogens are a group of molecules with estradiol as the most abundant. Estrogens are secreted by the ovaries as well as the adrenal glands, adipose tissue and the placenta (during pregnancy). Estrogens promote development of the secondary sex organs of the female. Actions of estrogen include:

Development of the breasts.

An increase in adipose tissue under the skin in specific areas of the body (thighs, buttocks, breasts).

There are also changes associated with the secretion of androgens during puberty including increased hair in the genital and axillary regions.

Estrogens provide negative feedback to the hypothalamus and anterior pituitary gland. For example a rise in estrogen levels works to inhibit the secretion of GnRH which in turn inhibits the secretion of FSH and LH.

Menstrual Cycle

Big Picture: Menstrual cycle

The menstrual cycle consists of three phases:

1. *Proliferative phase—uterus thickens.*
2. *Ovulation—leutenizing hormone causes egg to be released.*
3. *Luteal phase—inner layer of uterus sloughs off.*

The female menstrual cycle begins during puberty (between the ages of 10-13 years). It is characterized by changes in the endometrium of the uterus. The first menstrual cycle is called menarche. GnRH is secreted by the hypothalamus causing the secretion of FSH and LH. FSH causes maturation of the ovarian follicle and secretion of estrogens by the granulosa cells. LH also helps the follicle to mature and stimulates the production of estrogens.

Estrogens cause an increase in the thickening of the endometrium during the first phase of the menstrual cycle (proliferative phase) (fig. 20.6).

During the proliferative phase the follicle secretes estrogen that works to inhibit the release of LH. Instead of being released by the anterior pituitary, LH is stored. At about the 14th day LH is released (LH surge) causing the follicle to release the oocyte in what is called ovulation. The follicle then moves toward the Fallopian tube and is either fertilized or not.

Following ovulation, the follicle becomes what is known as a corpus luteum. The corpus luteum secretes large amounts of progesterone and estrogens during this second half of the cycle.

The estrogens and progesterone inhibit the release of LH and FSH from the anterior pituitary which in turn keeps other follicles from maturing. Progesterone also facilitates increased vascularization and thickening of the endometrium. If the oocyte is not fertilized the corpus luteum begins to degenerate near the end of the cycle at around the 24th day. What is left of the degenerated corpus luteum is called a corpus albicans. When the follicle goes from the corpus luteum to the corpus albicans stage the secretions of estrogen and progesterone diminish. This causes the thickened endometrium to slough off. The endometrium and accompanying blood constitute the menstrual flow which continues for about 3-5 days.

Menstruation continues throughout the female lifespan until the late 40s or early 50s where it begins to become irregular and eventually stops completely. This process marks the period of menopause. During this time the few remaining follicles no longer respond to FSH and LH. Since the follicles do not mature there is a subsequent drop in estrogens and progesterone. The consequences of low levels of these hormones include thinning of the vaginal, urethral and uterine linings, osteoporosis, and thinning of the skin.

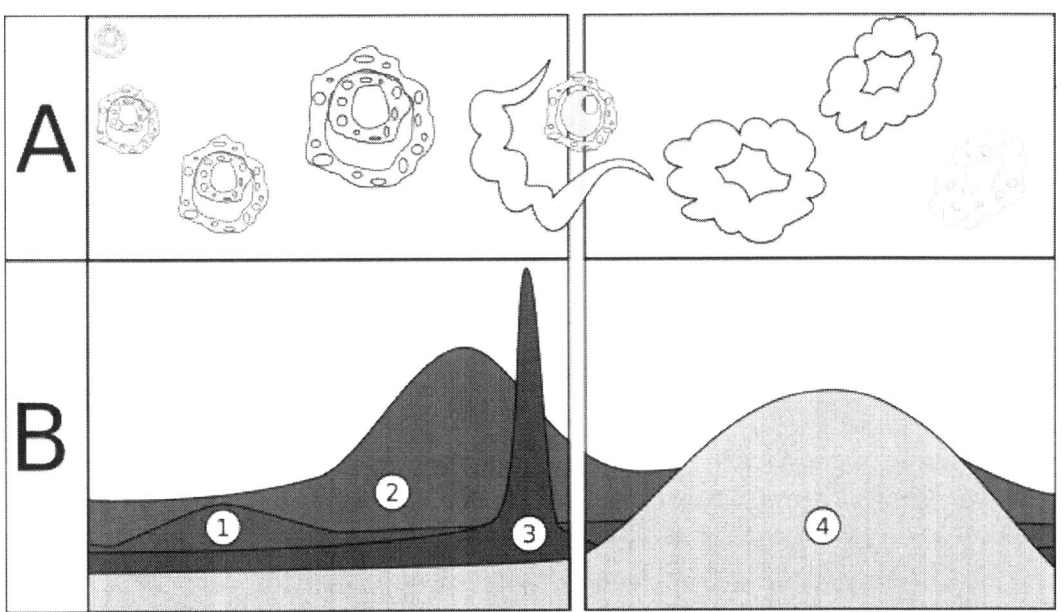

Figure 20.6. Menstrual Cycle

1 Follicle-Stimulating Hormone
2 Estrogens
3 Luteinizing Hormone
4 Progesterone
A Maturing follicle & corpus luteum
B Hormone levels

Fertilization

Sperm cells that reach the oocyte attempt to penetrate it with the help of the enzymes located in the acrosome. Once a sperm cell penetrates the oocyte it sheds its tail and the oocyte becomes unresponsive to other sperm. The nucleus of the sperm cell enters the oocyte and the oocyte undergoes meiosis II creating a second polar body. The genetic material from sperm and oocyte combine and the resultant cell is called a zygote. The zygote continues mitotic divisions to form a group of cells that migrates to the uterus. The Fallopian tube helps the migration along with its ciliated epithelial lining and smooth muscle contractions. The cells eventually implant in the wall of the uterus.

The layer surrounding the embryo secretes a hormone called human chorionic gonadotropin which helps to maintain the corpus luteum throughout the pregnancy. This results in high levels of estrogens and progesterone. After the first trimester the placenta takes over the job of secreting these hormones.

Breast Milk Production

The hormone prolactin works to stimulate milk production after birth. The first milk to appear is called colostrum which contains some nutrients including proteins and antibodies. When the infant suckles the breast the sensory impulses travel to the hypothalamus which in turn causes the release of oxytocin from the posterior pituitary gland. Oxytocin causes milk ejection by stimulating contraction of the myoepithelial cells of the breasts.

You should now have a much better understanding of the workings of both males and females.

Review Questions

1. Which of the following cells secrete testosterone:
a. Sustentacular
b. Leydig
c. Seminiferous
d. Tubular

2. Which of the following is a correct statement:
a. Sperm move from the vas deferens to the epididymis
b. Sperm move from the seminal vesicle to the ejaculatory duct
c. Sperm move from the epididymis to the vas deferens
d. Sperm move from the prostate to the seminal vesicle

3. Which of the following structures secretes fructose:
a. Epididymis
b. Testes
c. Seminal vesicle
d. Prostate gland

4. Which of the following is correct regarding the erectile columns:
a. 1 corpora cavernosum and 2 corpora spongiosum
b. 1 corpora cavernosum and 1 copora spongiosum
c. 2 corpora cavernosum and 2 corpora spongiosum
d. 2 copora cavernosum and 1 corpora spongiosum

5. This mucous secreting gland is located at the base of the penis:
a. Prostate
b. Bulbourethral
c. Seminal vesicle
d. Epididymis

6. Which are considered primary sex organs in the female:
a. Vagina
b. Ovaries
c. Uterus
d. Fallopian tubes

7. Which of the following is the thickest layer of the uterus:
a. Myometrium
b. Endometrium
c. Ectometrium
d. Perimetrium

8. The space between the labia in the female is known as:
a. Labial space
b. Majoral space
c. Vestibule
d. Perineum

9. The widened area of the Fallopian tube is called:
a. Infundibulum
b. Fimbrae
c. Ampulla
d. Endometrium

10. The normal uterus is in this position:
a. Retroflexed
b. Anteflexed
c. Retroextended
d. Anteextended

11. Which of the following is the main difference between meiosis and mitosis:
a. The number of chromosomes are doubled in meiosis
b. Meiosis is much slower than mitosis
c. Mitosis produces better cells
d. The number of chromosomes are halved in meiosis

12. Which best describes the function of the acrosome:
a. Helps to propel the sperm
b. Helps to nourish the sperm
c. Helps the sperm gain access to the oocyte
d. Helps to develop the sperm

13. Which of the following is not a function of testosterone:
a. Thickens vocal cords
b. Increases muscular growth
c. Causes testes to descend
d. Decreases metabolism

14. Which of the following best describes the function of inhibin:
a. Inhibits the secretion of gonatotropin releasing hormone
b. Stops growth and metabolism
c. Inhibits secretion of estrogen
d. Inhibits parasympathetic impuses

15. A mature ovarian follicle is known as:
a. Corpus albicans
b. Oocyte
c. Graffian follicle
d. Leuteal follicle

16. Which hormone causes ovulation:
a. Follicle stimulating hormone
b. Leutenizing hormone
c. Estrogen
d. Progesterone

17. How long is the proliferative phase of the ovarian cycle:
a. 10 days
b. 12 days
c. 14 days
d. 16 days

18. What happens when the corpus leuteum degenerates:
a. It secretes estrogen
b. It secretes progesterone
c. It secretes leutenizing hormone
d. It ceases to secrete estrogen and progesterone

19. The first breast milk to appear is called:
a. Prenatal milk
b. Colostrum
c. Estrostum
d. Prolactin

20. This hormone works to maintain the corpus leuteum:
a. Estrogen
b. Human chorionic gonadotropin
c. Gonadotropin releasing hormone
d. Progesterone

Answers to Review Questions

C1 Body Basics	C2 Taking the Creepiness out of Cells	C3 Monstrous Metabolism
1. B	1. B	1. D
2. B	2. A	2. A
3. A	3. C	3. A
4. A	4. C	4. B
5. D	5. A	5. C
6. D	6. C	6. C
7. D	7. A	7. D
8. A	8. D	8. C
9. B	9. C	9. B
10. C	10. C	10. C
11. B	11. C	
12. C	12. C	
13. C	13. B	
14. A	14. A	
15. B	15. B	
16. B	16. D	
17. B	17. A	
18. D	18. B	

C4 Don't be Terrified of Tissues	C5 Gimme Some Skin Man	C6 Bone, Bumps and Holes, Oh My!
1. C	1. B	1. D
2. B	2. D	2. D
3. D	3. B	3. D
4. B	4. B	4. A
5. B	5. A	5. B
6. A	6. D	6. B
7. A	7. A	7. A
8. C	8. D	8. C
9. C	9. C	9. B
10. B	10. A	10. A
11. D	11. C	11. D
12. D	12. C	12. B
		13. B
		14. C
		15. C
		16. D
		17. B
		18. A
		19. D
		20. B

Answers to Review Questions

C7 Managing Muscles	C8 What's a Joint Like this Doing in a Girl Like You	C9 Don't Get Nervous About the Nervous System
1. B	1. D	1. C
2. A	2. B	2. C
3. B	3. B	3. B
4. C	4. C	4. C
5. C	5. D	5. C
6. C	6. D	6. C
7. A	7. D	7. B
8. B	8. D	8. C
9. A	9. B	9. B
10. B	10. D	10. A
11. B		11. D
12. B		12. C
13. D		13. B
14. B		14. C
15. B		15. B
16. C		16. D
17. A		17. A
18. A		18. C
19. C		19. B
20. A		20. B
21. A		21. C
22. D		22. B
23. C		23. C
		24. B
		25. D
		26. C
		27. B
		28. A
		29. B
		30. B
		31. B
		32. A
		33. D
		34. D
		35. D
		36. B
		37. B

Answers to Review Questions

C10 Making Sense of the Sensory System

1. D
2. C
3. A
4. C
5. B
6. D
7. B
8. C
9. A
10. A
11. B
12. B
13. A
14. C
15. B
16. C

C11 Learning the Hell Out of Hormones

1. D
2. B
3. B
4. C
5. D
6. B
7. C
8. A
9. C
10. B

C12 I Want to Drink Your Blood, But I Want to Know what's In It First

1. A
2. B
3. A
4. B
5. A
6. A
7. B
8. C
9. B
10. C
11. C
12. C

C13 Emphatic Over the Lymphatic System

1. D
2. C
3. D
4. A
5. C
6. A
7. B

C14 Don't Be Immune to the Immune System

1. C
2. A
3. C
4. B
5. A
6. C
7. D
8. B

Answers to Review Questions

C15 Getting the Blood to all the Right Places	C16 In With The Good Air, Out With the Bad	C17 There's a lot to Making Pee Pee
1. C	1. B	1. D
2. C	2. A	2. C
3. B	3. D	3. D
4. B	4. B	4. A
5. A	5. A	5. C
6. C	6. C	6. A
7. B	7. B	7. C
8. D	8. D	8. A
9. B	9. A	9. D
10. A	10. D	10. D
11. C	11. C	11. C
12. C	12. B	12. D
13. B	13. C	13. C
14. C	14. A	14. A
15. C	15. C	15. B
16. C	16. B	16. B
17. C	17. B	17. A
18. C	18. C	18. C
19. B	19. A	19. D
20. C	20. D	20. B
21. C	21. A	21. C
22. B	22. D	22. D
23. A	23. C	23. C
24. B		
25. C		
26. A		
27. D		
28. C		
29. B		
30. A		

Answers to Review Questions

C18 You Already Know About Fluids and Electrolytes. How About Some Acid-Base Balance

1. D
2. C
3. D
4. A
5. D
6. C
7. D
8. D
9. A
10. D
11. B
12. C

C19 Taking the Indigestion Out of the Digestive System

1. C
2. B
3. C
4. A
5. D
6. D
7. A
8. C
9. B
10. B
11. C
12. D
13. B
14. D
15. C
16. A
17. D
18. C
19. B
20. C
21. A
22. C

C20 Fun With Reproduction

1. B
2. C
3. C
4. D
5. B
6. B
7. A
8. C
9. A
10. B
11. D
12. C
13. D
14. A
15. C
16. B
17. C
18. D
19. B
20. B

Appendix One

Detail of Selected Skeletal Structures

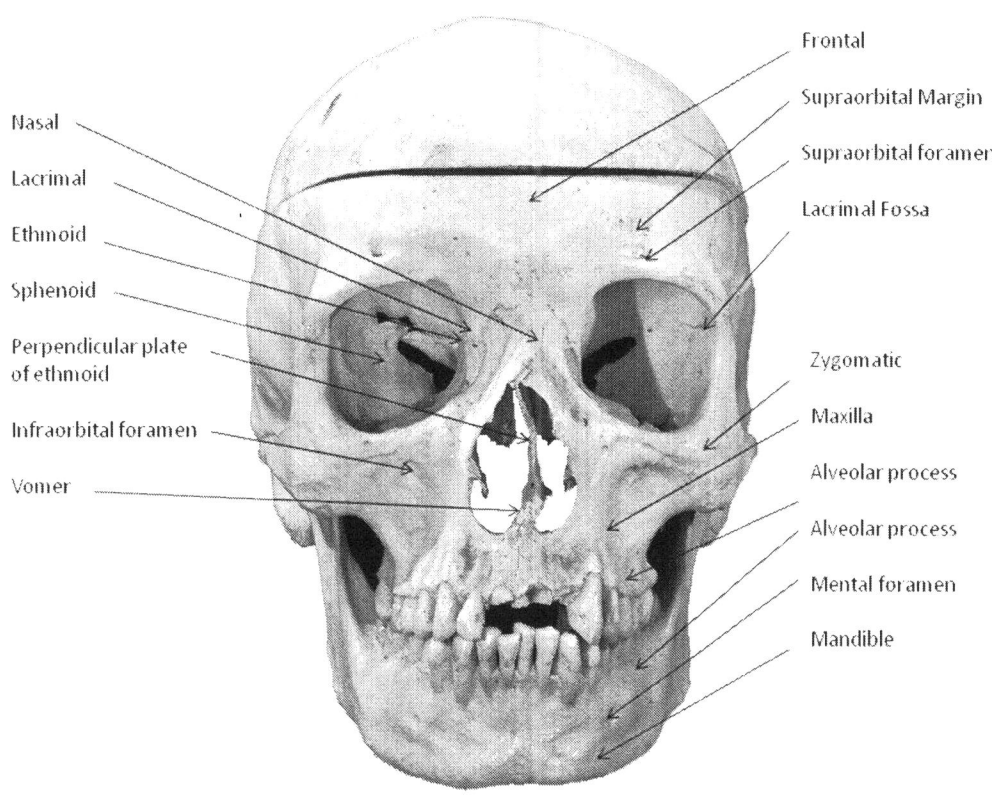

Fig. A1.1 Anterior skull detail.

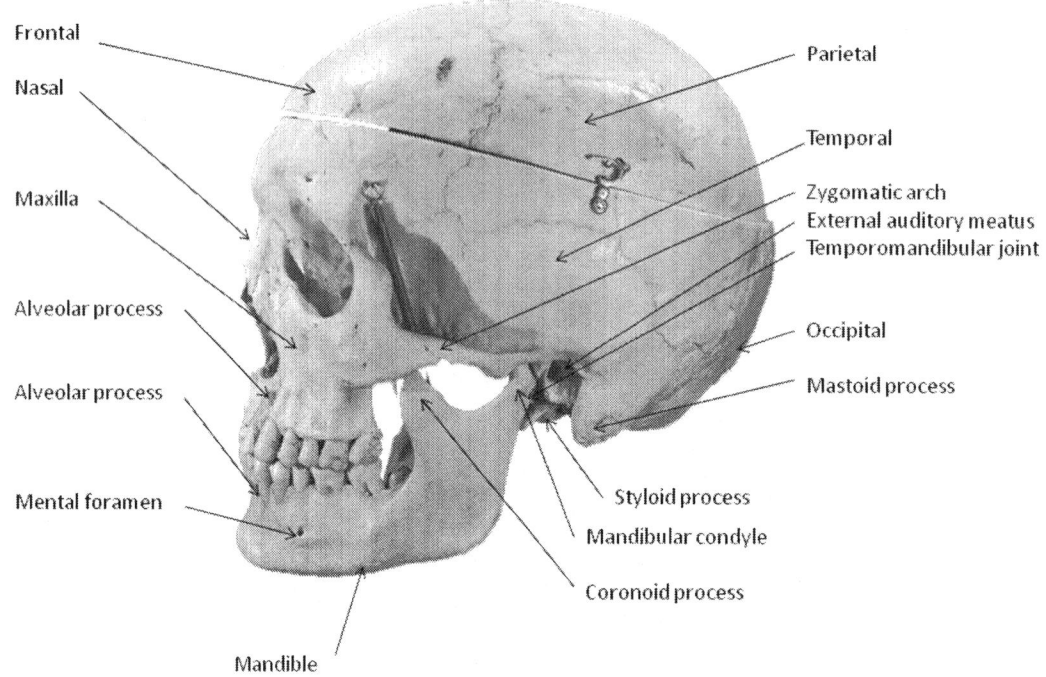

Fig. A1.2 Lateral Skull

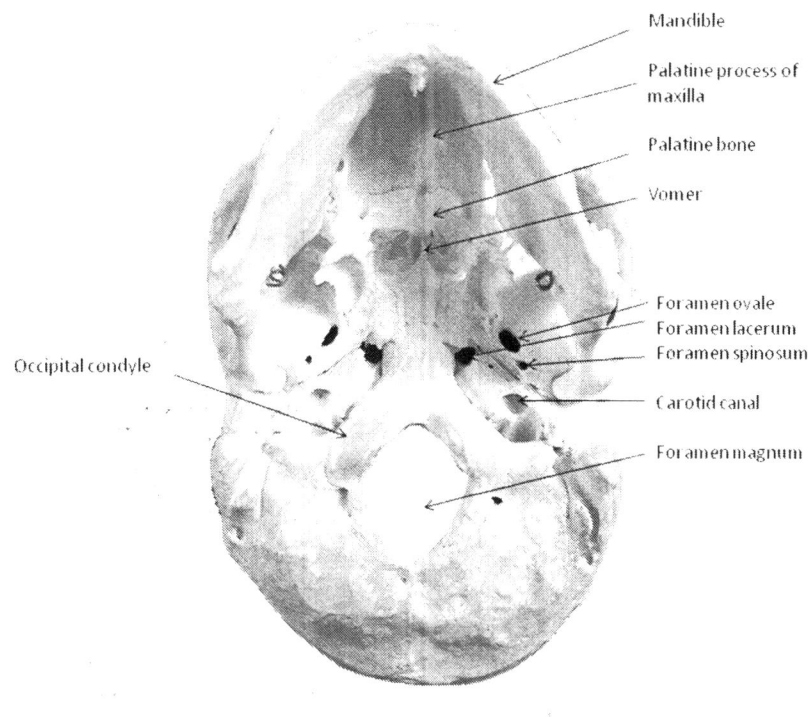

Fig. A1.3 Inferior skull

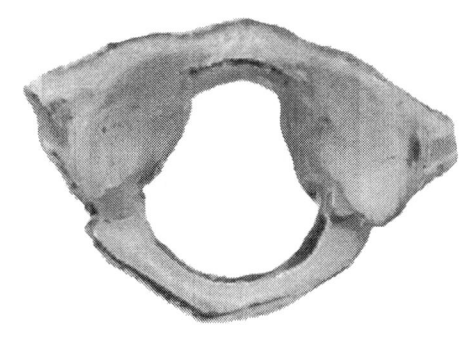

Fig. A1.4 Atlas (C1)

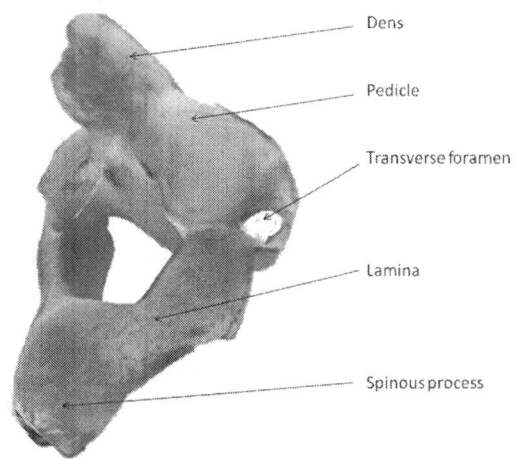

Fig. A 1.5 Axis (C2)

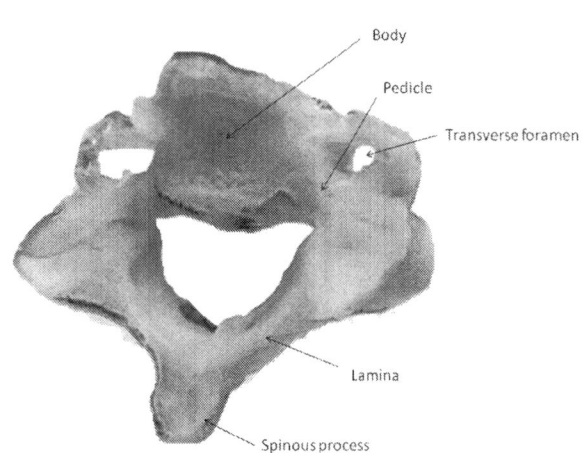

Fig. A1.6 Typical cervical vertebra (C3-C7)

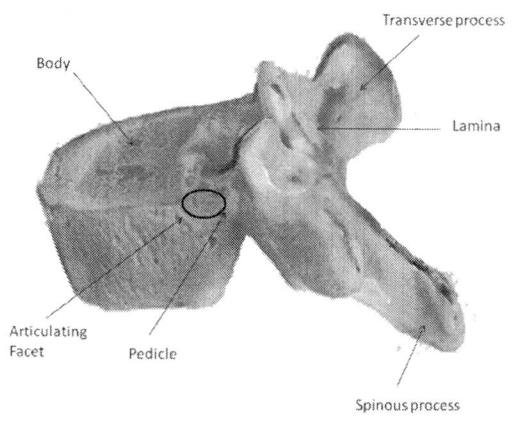

Fig. A1.7 Typical thoracic vertebra (T1-T12)

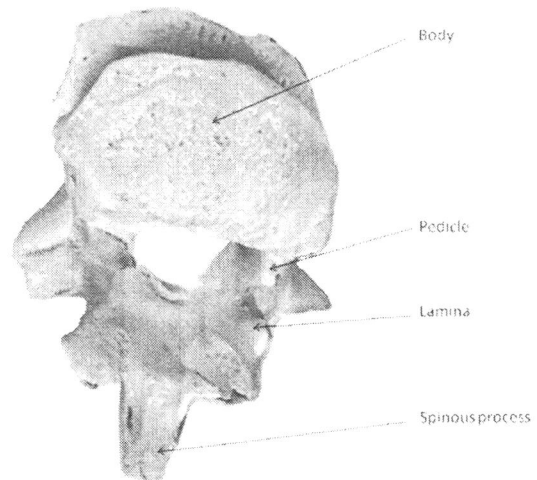

Fig. A1.8 Typical lumbar vertebra (L1-L5)

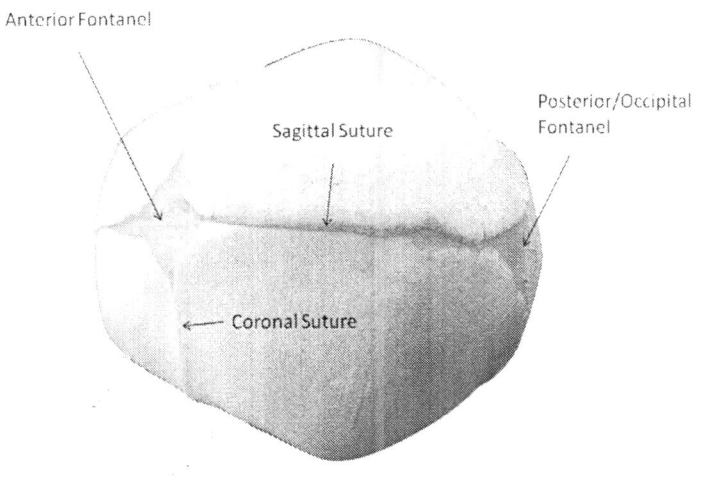

Fig. A1.9 Fetal skull superior projection

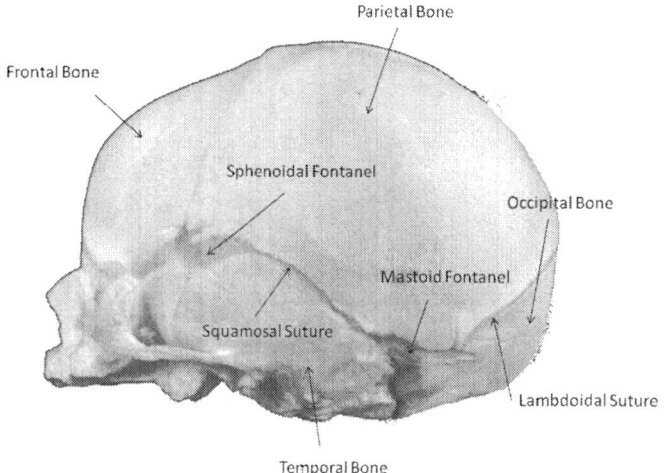

Fig. A1. 10 Fetal skull lateral projection.

Index

1, 3 diphosphoglycerate, *45*
7-dehydrocholesterol, *62*
a primary ossification center, *72*
Abdomen, *14*
Abdominal, *14, 116, 257*
abdominal aorta, *245, 246*
Abdominals, *264*
abdominopelvic, *12*
Abducens, *171, 172*
Abduction, *144, 145, 152*
Abductor digiti minimi, *126*
Abductor hallucis, *126*
abductors, *110*
about red blood cells, *219*
Accessory, *171, 172*
acetabulum, *90, 148*
Acetylcholine, *140, 182, 186, 187, 292, 324*
acetyl-coenzyme A, *45*
Achilles, *125*
acidic, *270, 296, 301, 302, 327*
Acromial, *14, 21*
Acromion, *14*
acromion process, *84, 97, 108*
actin, *128, 129, 139*
action potential, *129, 179, 180, 181, 248, 257*
active transport, *32, 33, 49, 288, 304*
Adam's apple, *214, 261, 274*
Addison's disease, *298*
Adduction, *144, 145*
adductor brevis, *121*
adductor femoris, *105*
Adductor hallucis, *126*
adductor longus, *121*
adductor magnus, *121*
adenine, *25*
adenohypophysis, *207, 208*
adenylate cyclase, *206*
ADH, *214, 216, 257, 288, 292, 297*
Adipose, *55, 58, 175*
ADP (adenosine diphosphate), *30*
adrenal gland, *227, 298*
adrenaline, *174, 249*
afferent arteriole, *278, 281, 282, 287, 288*
afferent pathway, *155*
Agglutination, *236*
agonist, *100*

aldosterone, *283, 287, 292, 297, 298, 300, 304*
Aldosterone, *215, 216, 283, 287, 297, 298, 304*
alkaline, *270, 271, 301, 302, 303, 308, 320, 325, 327, 328*
allergy, *237*
alveolar process, *79*
alveoli, *261, 262, 266, 273, 274, 275, 302*
amino acid, *29, 40, 41, 325*
amino acids, *29, 43, 51, 222, 301, 325*
ammonia, *283, 289*
Amphiarthrotic, *143*
anabolic, *43*
anal sphincter, *314*
anaphase, *30, 31*
anaphylactic, *237*
anatomical neck, *86, 146*
Anatomical Position, *12*
Anatomists, *14*
anatomy, *7, 8, 11, 12, 46, 69, 76, 78, 128, 155, 229, 242, 248, 261, 277, 328, 330, 366*
androgens, *335, 336*
anemia, *66, 220*
angiotensin II, *287*
annular ligament, *147*
antagonist, *100*
Antebrachial, *14*
Antebrachium, *14*
Antecubitus, *14*
Anterior, *12, 21, 22, 122, 123, 157, 159, 206, 208*
anterior cruciate ligament, *150*
anterior fontanel, *80*
anterior interventricular artery, *243*
antibodies, *221, 224, 227, 235, 236, 237, 238, 239, 338*
antibody mediated immunity, *235, 236*
antidiuretic hormone, *214, 297*
antigens, *224, 237*
antrum, *336*
aorta, *242, 243, 245, 246, 249, 256, 257, 262*
apneustic center, *272*
Apocrine glands, *64*
aponeurosis, *104, 112, 121, 123*
appendicular, *74, 82*
aqueous humor, *192*
arachnoid granulations, *168*

Arachnoid mater, *156, 185*
arcuate, *150, 277*
arcuate popliteal ligament, *150*
Arm, *14, 82, 86, 107, 364*
arrector pili, *63, 64*
arrector pili muscles, *63*
arrhythmias, *300*
arteries, *69, 232, 233, 243, 244, 245, 246, 249, 250, 252, 256, 266, 267, 291, 334, 335*
arterioles, *244, 278, 281, 282*
arthritis, *143, 150, 237*
Artificially acquired active immunity, *238, 239*
Artificially acquired passive immunity, *238, 239*
arytenoids, *261*
ascending colon, *313*
ascending limb, *287, 288, 292*
ascending spinal tract, *155*
aspirin, *303*
Astrocytes, *176, 187*
Atherosclerosis, *224, 257*
ATP, *30, 33, 40, 43, 45, 46, 49, 50, 51, 130, 333*
ATP (adenosine triphosphate), *30, 130*
atria, *241, 242, 244, 248, 251, 254, 256, 257*
atrial natriuretic hormone, *251*
atrial natriuretic peptide, *251, 297*
Atrial natriuretic peptide (ANP, *297*
atrioventricular bundle, *248*
atrioventricular node (AV node), *248*
Auditory, *171, 172*
Auricle, 197
autonomic, *159, 173, 185*
axial, *74, 82, 84, 96*
Axilla, *230, 233*
axon, *129, 140, 176, 177, 180, 181, 186*
baroreceptors, *249*
basilar membrane, *199*
basilic, *246*
Basophils, *221*
benign positional vertigo, *200*
bicarbonate, *222, 268, 269, 270, 283, 301, 302, 303, 320, 325*
bicarbonate ion, *268, 269, 270*
bicarbonate ions, *222, 284, 301, 302, 303, 320, 325*
biceps, *87, 100, 101, 103, 104, 105, 107, 110, 123, 146*
Bicuspid valve, *242, 256*
bile, *220, 289, 300, 314, 320, 325*

bilirubin, *220, 227*
biliverdin, *220, 227*
Bitter, *190, 201*
blood, *7, 11, 55, 56, 61, 63, 64, 65, 66, 70, 71, 72, 78, 79, 165, 168, 173, 176, 177, 182, 205, 206, 214, 215, 216, 219, 220, 221, 222, 223, 224, 225, 227, 229, 230, 231, 232, 233, 235, 236, 241, 242, 243, 244, 245, 246, 248, 249, 250, 251, 252, 253, 256, 257, 258, 261, 262, 265, 266, 268, 270, 271, 272, 275, 277, 278, 280, 281, 282, 283, 284, 287, 288, 289, 291, 292, 296, 297, 298, 300, 301, 302, 303, 304, 306, 312, 314, 322, 329, 334, 337*
Blood, *5, 56, 58, 63, 64, 175, 218, 219, 220, 222, 224, 225, 226, 232, 240, 241, 242, 244, 249, 250, 251, 253, 256, 257, 277, 280, 291, 322, 334, 335, 343, 344*
blood pressure, *249, 250, 251, 252, 282, 287, 297, 298*
blood vessel spasm, *222*
blood-brain barrier, *168, 176*
B-lymphocytes, *221*
Bohr effect, *271, 275*
bolus, *308*
Bone, *55, 58, 69, 71, 72, 73, 76, 77, 78, 82, 84, 86, 90, 96, 341, 364*
Bowman's capsule, *278*
Boyle's law, *262*
Brachial, *14, 256*
brachial plexus, *159, 183*
brachialis, *104, 107, 110*
brachiocephalic, *230, 246*
brachioradialis, *105, 107*
Brachium, *14*
brainstem, *155, 156, 158, 161, 165, 166, 168, 171, 183, 185, 272, 302*
breasts, *330, 336, 338*
bronchi, *261, 262, 272, 274*
bronchomediastinal trunks, *230*
brush border, *312*
buccinator, *112*
buffer, *301, 302, 303*
bulbospongiosus, *119*
bulbourethral glands, *328*
bundle branches, *248*
Bundle of His, *248*
Butt, *14*
Buttock, *14*

Calcaneus, *84, 93*
calcitonin, *214, 300*
Calcitrol, *73*
calcium, *62, 71, 73, 129, 181, 186, 214, 216, 222, 248, 283, 296, 300, 304*
Calf, *14*
calluses, *62*
calyces, *277*
cAMP, *206, 207*
Cancellous bone, *69*
capillaries, *54, 58, 168, 219, 229, 233, 244, 252, 253, 254, 262, 266, 278, 281, 288, 304, 322*
Capitate, *88*
capitulum, *87, 147*
carbaminohemoglobin, *268*
carbohydrates, *43, 222, 306, 314, 322, 324, 325*
carbon dioxide, *222, 261, 265, 266, 267, 268, 270, 271, 272, 275, 284, 301, 302, 303*
carbonic acid, *269, 301, 302*
carbonic anhydrase, *268, 275, 284*
cardiac accelerator nerves, *249*
cardiac cycle, *243, 244, 256*
cardiac impression, *262*
Cardiac output, *250*
cardiovascular system, *229, 241, 248, 249, 252*
carina, *262*
carotid canal, *78*
Carpal, *14*
Carpals, *82, 88*
Carpus, *14, 364*
Cartilage, *56, 58*
Cartilagenous, *143*
catabolic, *43*
cavity, *11, 71, 72, 74, 78, 79, 84, 96, 116, 119, 229, 241, 261, 262, 264, 274, 277, 278, 291, 308, 314, 327, 329, 336*
CCK, *320*
cell mediated immunity, *235*
cell membrane, *25, 31, 32, 33, 195, 206*
Cells, *5, 24, 25, 30, 32, 71, 219, 220, 274, 332, 333, 335, 341*
cellular metabolism, *43, 296*
cephalic, *246, 317*
Cephalic, *14, 256*
Cephalon, *14, 21*
cerebellum, *158, 161, 166, 167, 168, 171, 185*
cerebral aqueduct, *168, 171*

cerebral cortex, *161, 165, 272*
cerebral spinal fluid, *168, 296*
cerebrum, *161, 163, 165, 166, 167, 185, 314*
Cervical, *14, 21, 96, 233*
cervical plexus, *159*
cervical spine, *80*
Cervicis, *14*
cervix, *329*
Chemoreceptors, *189*
chemotaxis., *222*
Chest, *14*
Chief cells, *309*
Chin, *14*
chondrocytes, *73*
Chondrocytes, *72*
choroid coat, *192*
choroid plexi, *168, 171, 177*
chromosomes, *30, 31, 40, 332, 335, 339*
chylomicrons, *222, 229, 322*
chyme, *309, 320, 324, 325*
Chymotrypsin, *317*
ciliary body, *192*
cingulate gyrus, *167*
circulatory system, *25, 229, 233, 245, 322*
Circumduction, *145*
circumflex artery, *243*
cisterna chili, *230*
Citrate, *48*
class 1 lever, *99*
class 2 lever, *99*
class 3 lever, *99*
Clavicle, *82, 84*
clitoris, *330*
clones, *237*
clotting, *222, 223, 224, 235, 236*
coccygeus, *119*
Cochlea, *197, 201*
cochlear duct, *197*
collecting duct, *278*
colloid osmotic pressures, *254*
colon, *174, 313, 314*
color blind, *195*
Columnar, *54, 58, 324*
coma, *297, 298, 303*
common carotid, *246, 256*
Compact bone, *69, 70*
complement system, *235, 236*

compliance, *264*
concentric, *70, 101*
conchae, *261*
conduction, *64, 173, 180, 181, 186, 254*
Condyle, *74, 96*
condyles, *79, 90, 93, 106, 125, 150*
condyloid, *143*
cones, *195*
confusion, *65, 297, 298, 300, 304*
Contralateral, *14*
conus medullaris, *156, 314*
convection, *64*
convulsions, *298, 300*
Coracohumeral, *146*
coracoid process, *84, 104, 110, 112, 146*
cornea, *192*
corniculate, *261*
corns, *62*
coronal, *14, 21, 80, 96*
coronary arteries, *173, 243*
coronary sinus, *243*
coronary sulcus, *243*
coronoid process, *104, 147*
corpora quadrigemina, *155, 166*
corpus callosum, *161, 167*
corpus cavernosum, *328, 330*
corpus spongiosum, *328, 330*
corticospinal tract, *158, 183*
Cortisol, *215*
countercurrent, *288*
Coxal, *84, 90*
coxal bones, *90, 148*
cramping, *298*
cranial cavity, *12*
cranial nerves, *171, 172, 195*
cranium, *25, 74, 76, 77, 78, 79*
cricoid cartilage, *261*
cristae, *49*
Crus, *14*
crypts of Lieberkuhn, *312, 324*
CSF, *168, 171, 177, 186*
Cubital, *14*
cuboid, *93*
Cuboid, *84*
Cuboidal, *53, 58, 233*
Cuneaforms, *84*
cuneiform, *93, 106, 123, 125, 261*
Cushing's, *300*

cyclic adenylate monophosphate, *206*
Cyclic guanosine monophosphate (cGMP), *206*
Cytokines, *235*
cytoplasm, *25, 44, 62, 176, 206, 220, 296*
cytosine, *25*
cytoskeleton, *25*
cytotoxic, *236*
decussation, *158*
Deep, *14, 61, 121, 161, 163*
dehydrated, *297*
deltoid tuberosity, *86, 108*
dendrites, *177*
Dense, *55, 58, 291*
Deoxyhemoglobin, 219
deoxyribonucleic acid, *25*
depolarization, *129, 140, 179, 180, 186, 248, 254, 257*
depolarize, *179, 180, 248, 254*
dermal papillae, *62*
dermis, *55, 61, 62, 63, 64, 65, 66*
descending limb, *288*
descending spinal tracts, *157, 183*
diabetes, *237, 283, 289, 297, 298, 303*
Diacyglycerol (DAG), *206*
diapedesis, *222*
diaphoresis, *297*
diaphragm, *12, 119, 232, 242, 257, 262, 264, 272, 308, 314*
diarrhea, *297, 298, 300, 303*
Diarthrotic, *143*
diastole, *244, 250, 256*
diencephalon, *161, 165, 167, 168, 171*
differential, *222, 273*
diffusion, *32, 33, 40, 69, 288, 322*
digastricus, *105*
digestion, *215, 306, 308, 309, 313, 317*
digestive system, *229, 235, 306, 314, 317*
dissociate, *269, 301, 302*
Distal, *12, 21, 88, 291*
distal convoluted tubule, *278, 288, 292, 297*
dizziness, *65, 249, 297, 303*
DNA, *25, 28, 29, 30, 41, 206, 363*
Dopamine, *140, 182, 186*
dorsal, *11, 156, 158, 159, 183, 272, 334*
Dorsal interossei, *126*
Dorsiflexion, *144, 145, 152*
duodenum, *309, 312, 314, 315, 317, 320, 324*
Dura mater, *156, 185*

dynamic equilibrium, *172, 197*
dysrhythmias, *300*
ears, 12, 25, 155, 189, 199
eccentric, *101*
Eccrine sweat glands, *64*
ECG, *254, 257, 300*
edema, *254, 298*
efferent arteriole, *278, 281*
ejaculation, *174, 334*
ejaculatory ducts, *328*
Elbow, *14, 145, 147*
electrocardiogram, *254*
electron, *44, 46, 48, 50*
electron transport chain, *44, 46, 48, 50*
Elevation/Depression, *145*
Embolism, *224*
Embolus, *224*
emission, *334*
Empendymal cells, *177*
emulsification, *314*
endocardium, *242*
endochondral ossification, *72, 96*
endocrine system, *165, 205, 214, 314*
endomysium, *128*
enzyme, *29, 44, 50, 173, 206, 268, 308, 309, 333*
Eosinophils, *220*
epicardium, *242*
epidermal stem cells, *62*
epidermis, *58, 61, 62, 64, 65, 66*
epididymis, *327, 339*
epiglottis, *261*
epimysium, *128*
epiphyseal plates, *72, 73*
epiphyses, *69, 72*
epithalamus, *165*
Epithelial, *53*
epithelium, *53, 54, 58, 61, 229, 233, 242, 274, 278, 288, 291, 312, 327, 363*
Epitrochlear, *230*
equilibrium, *11, 32, 40, 178, 186, 199, 201, 301*
erythrocytes, *219*
esophagus, *30, 172, 308, 317, 367*
Estrogens, *336*
ethmoid, *76, 77, 261*
Ethmoid, *74, 77, 79, 96, 274*
Eustachian tube, *197, 201*
Eversion, *145, 152*

exhale, *261, 264, 265*
exocrine, *205, 314*
expiratory reserve volume, *265*
Extension, *144, 145, 152*
extensor carpi radialis, *105, 108*
extensor carpi ulnaris, *105, 108*
extensor digitorum brevis, *125*
extensor digitorum communis, *105*
extensor digitorum longus, *106, 123, 125*
extensor digitorum minimi, *105*
extensor retinaculum, *125*
external auditory meatus, *78, 96, 199*
External auditory meatus, *197*
External intercostals, *262, 274*
extrinsic), *223*
eyes, *155, 166, 189, 190, 237*
Facial, *171, 172, 186*
facilitated diffusion, *32*
factors, *73, 208, 223, 224, 236, 250*
FADH2, *44, 45, 46, 48, 49, 50*
Fallopian tubes, *329, 339*
fascicles, *128*
fasciculus cunneatus, *158*
fatty acids, *215, 301*
feces, *297, 313, 314*
female, *31, 90, 291, 327, 328, 329, 330, 332, 333, 336, 337, 339, 364, 367*
Femorus, *14*
Femur, *71, 84, 90, 96, 97, 364*
fetal circulation, *242*
fibrin, *65, 223, 224*
fibrinogen, *222, 224*
Fibrous, *143, 152*
Fibula, *84, 93, 97*
fibular head, *93, 150*
fibularis longus, *106*
fimbrae, *329*
First degree burns, *65*
first-order neuron, *158*
fissures, *161, 262*
flagella, *30*
flagellum, *30, 333*
Flat bones, *69, 71*
flexion, *100, 101, 105, 106, 107, 110, 113, 116, 122, 123, 125, 144, 152*
Flexion, *144, 145, 152*
flexor carpi radialis, *104, 107*

Flexor carpi radialis longus, *108*
flexor carpi ulnaris, *104, 107*
Flexor digiti minimi brevis, *126*
Flexor digitorum brevis, *126*
flexor digitorum longus, *125*
Flexor digitorum profundus, *108*
Flexor digitorum superficialis, *108*
flexor hallicus longus, *125*
Flexor hallucis brevis, *126*
fluid balance, *288, 296*
fluids and electrolytes, *296, 297*
FMN, *49*
FMNH2, *49*
follicle, *63, 64, 332, 335, 336, 339, 367*
Follicle stimulating hormone (FSH), *332*
Follicular cells, *214*
fontanels, *80*
Foot, *14, 93, 125*
Foramen, *74, 96*
foramen lucerum, *78*
foramen magnum, *78, 96, 156*
Forearm, *14, 87, 107, 364*
fourth ventricle, *168*
fovea centralis, *195*
Free nerve endings, *189*
Frontal, *74, 78, 79, 96, 185, 186, 187*
frontal lobe, *161, 165, 171, 172, 183*
Fructose, 1, 6 diphosphate, *45*
fulcrum, *99, 139*
Fumarate, *48*
Functional residual capacity, *265*
fundus, *308, 329*
funiculi, *156, 158, 159*
GABA, *182*
gallbladder, *172, 314, 320*
gastric juice, *309*
Gastric Lipase, *317*
gastric pits, *308*
gastric reflux, *308*
gastrocnemius, *106, 125*
glenoid, *84, 146*
Glenoid Labrum, *146*
glomerular capsule, *278, 281*
glomerulus, *278*
Glossopharyngeal, *171, 172*
Glucagon, *215, 216*
gluconate, *300, 304*

glucose, *11, 43, 44, 45, 46, 50, 51, 215, 280, 281, 282, 289, 300, 301*
Gluteal, *14*
gluteus maximus, *105, 120, 121*
gluteus medius, *105, 120, 121*
glycolysis, *43, 44, 45, 46, 50, 51*
GnRH, *336*
Golgi Tendon Organs, *189*
gomphosis, *143*
Gomphosis, *143, 152*
gonadotropin releasing hormone, *332, 335, 336*
G-protein, *206, 207*
gracilis, *105, 121, 156, 158, 183, 185*
granular, *176, 220*
Granulocytes, *220*
great cardiac vein, *243*
great saphenous vein, *247*
greater trochanter, *90, 96, 105, 121, 122*
greater tubercle, *86, 104, 110, 111, 146*
grey commissure, *156*
grey matter, *25, 156, 180*
Groin, *14*
Growth hormone, *73, 208, 216*
Growth Hormone Releasing Hormone, 208
guanine, *25*
gyri, *161*
hair bulb, *63*
hair cells, *196, 197, 199, 200*
hair follicles, *63*
Haldane effect, *272, 275*
Hamate, *88*
hamstrings, *101, 103, 123*
Hand, *14, 90, 107*
Haversian systems, *58, 70*
Head, *14, 97, 112, 113*
head of the radius, *87, 97*
heart, 12, 25, 56, 158, 165, 172, 173, 182, 215, 224, 241, 242, 243, 244, 245, 246, 248, 249, 250, 251, 252, 254, 256, 257, 262, 283, 297, 298, 366
helper, *236, 239*
Hematoma, *224*
hemisphere, *161*
Hemocytoblasts, *222*
hemoglobin, *219, 220, 227, 268, 271, 272, 275, 302*
hemoglobin saturation curve, *271*
hemostasis, *222, 227*

hepatic flexure, *313*
Hering-Breuer, *272*
hiatl hernia, *308*
homeostasis, *11, 173, 214*
hormones, *21, 63, 71, 73, 165, 205, 206, 207, 208, 214, 216, 297, 300, 309, 332, 335, 336, 337, 338*
humeralradial joint, *147*
humeralulnar joint, *147*
Humerus, *82, 86, 96, 97*
hyaluronidase, *328*
hydrogen, *44, 45, 49, 50, 269, 270, 271, 275, 283, 292, 301, 302, 303, 304*
hydrostatic pressure, *253, 254, 257, 258, 281, 282, 291*
hymen, *329, 330*
hyoid bone, *78, 82, 105*
hyperhydration, *298*
hyperkalemia, *300*
hyperkyphosis, *81*
hyperlorosis, *81*
hypernatremia, *298*
hyperpolarization, *180*
hypocalcemia, *300*
Hypoglossal, *171*
hypokalemia, *300*
hypokyphosis, *81*
hypolordosis, *81*
hyponatremia, *298*
hypoparathyroidism, *300*
hypothalamus, *64, 165, 185, 207, 208, 214, 216, 288, 297, 332, 335, 336, 338*
hypotonic, *288, 292, 296, 298*
IgA, *237, 239*
IgD, *237, 239*
IgE, *237, 239*
IgG, *237, 239*
IgM, *237*
ileum, *312, 313*
iliac crest, *90, 104, 105, 110, 121*
iliacus, *120*
iliofemoral ligament, *148*
iliotibial band, *121*
ilium, *90, 96, 105, 120, 121, 148*
immune system, *61, 229, 231, 232, 235, 236, 237, 238*
immunoglobulins, *237*

imperforate hymen, *330*
Incus, *197*
Infarction, *224*
inferior, *12, 14, 15, 21, 78, 79, 82, 84, 106, 111, 112, 121, 150, 165, 166, 167, 195, 242, 246, 261, 262, 278, 308*
inflammation, *150, 205, 221, 227, 235, 236, 237*
Inflammation, *235, 236, 239*
Infraclavicular, *230*
infraorbital foramen, *79*
infraspinatus, *104, 110, 111*
infraspinous fossa, *84, 104*
infundibulum, *165, 329*
Inguina, *14*
Inguinal, *14, 230, 257, 291*
Inhalation, *262, 366*
inhibin, *327, 339*
Inositol triphosphate (IP3), *206*
insipidus, *298*
Inspiratory capacity, *265, 274*
inspiratory reserve volume, *265, 274*
Insulin, *215, 216*
interatrial septum, *241*
intercalated disc, *56, 248*
interlobar, *277*
interlobular, *277*
Internal intercostals, *264*
internal urinary sphincter, *278*
interosseous membrane, *123, 125*
interphase, *30*
interstitial colloid osmotic pressure (ICOP)., *253*
interstitial fluid pressure, *253, 257*
interthalamic adhesion, *165*
intestinal flora, *313*
Intestinal Lipase, *317*
intoxication, *298*
intramembranous ossification, *71*
intravenous, *297, 300, 303*
intrinsic, *223, 224, 309*
Inversion, *145*
ion, *181, 270, 271, 275, 303, 304*
Ipsilateral, *14, 22*
iron, *49, 50, 220, 314*
Irregular bones, *69*
ischial tuberosity, *90, 105, 106, 123*
ischiocavernosus, *119*
ischiofemoral ligament, *148*

ischium, *90, 96, 105, 148*
Isocitrate, *48*
isokinetic, *100*
isometric, *100*
isotonic, *100, 288, 292*
jejunum, *312*
joints, *79, 90, 99, 107, 125, 143, 145, 146, 152, 158, 189*
jugular foramen, *78*
jugular trunks, *230*
jugulars, *246, 256*
juxtaglomerular apparatus, *287, 292, 297*
keratin, *61, 62, 63*
keratinocytes, *61, 62*
Ketoglutarate (alpha ketoglutarate), *48*
kidney, *62, 221, 245, 251, 277, 278, 280, 282, 283, 287, 288, 289, 297, 298, 300, 301, 303, 304, 322*
Kidneys, *277*
knee, *14, 21, 84, 101, 103, 106, 107, 122, 123, 125, 143, 145, 150, 182, 230*
KREBS cycle, *44, 45, 46, 50, 51*
labia majora, *330*
labia minora, *330*
laceration, *65*
lacrimal fossa, *78*
lacteal, *312*
lactiferous ducts, *330*
lacunae, *56, 58, 70, 71, 72*
large intestine, *313, 314, 324, 325*
larynx, *82, 112, 261, 262, 324*
Lateral, *12, 14, 125, 144, 145, 152, 157, 185, 201*
lateral collateral ligament, *150*
lateral epicondyle, *86, 105, 108, 147*
lateral funiculus, *158, 183*
lateral malleolus, *93*
lateral ventricles, *168*
latissimus dorsi, *104*
latissumus dorsi, *110, 139*
left coronary, *243, 256*
left marginal artery, *243*
lens, *191, 192*
lesser trochanter, *90, 120*
lesser tubercle, *86, 111, 146*
leukocytes, *220, 221*
leukocytosis, *222*
Leutenizing hormone, *332, 340*

levator ani, *119*
levers, *99*
Leydig cells, *327, 332*
ligaments, *55, 69, 78, 145, 146, 147, 148, 150, 155, 159, 329, 330*
lightheadedness, *303*
limbic system, *165, 167, 190, 272*
linea aspera, *90, 105, 106, 121, 122, 123*
lipid soluble, *32, 33, 168, 206, 216*
lipoproteins, *222, 252, 322, 325*
liver, *62, 64, 219, 220, 233, 303, 313, 314, 322*
Long bones, *69*
loop of Henle, *278*
Loose, *55, 58, 152*
loose connective tissue, *61*
Low Back, *14*
Lumbar, *14*
lumbar plexus, *159*
lumbar spine, *80*
lumbosacral plexus, *159*
Lumbricales, *126*
Lumbus, *14*
Lunate, *88*
lunula, *64*
lymph fluid, *229, 230*
lymph nodes, *58, 229, 230, 233*
lymphatic ducts, *230*
lymphatic system, *55, 229, 231, 233, 312, 322*
lymphatic trunks, *230*
lymphatic vessels, *230, 306*
lymphocytes, *221, 230, 231, 232, 235, 236*
Lymphocytes, *221, 232*
macrophages, *220, 230, 235*
macula densa cells, *287*
magnesium, *222, 283, 300*
Malate, *48*
male, *63, 90, 195, 278, 291, 327, 328, 329, 330, 332, 335, 364*
malignant melanoma, *62*
Malleus, *197*
Maltase, *317*
mamillary bodies, *165*
mammary glands, *330*
mandible, *78, 79, 97, 105, 113*
Mandible, *76, 79*
mandibular foramen, *79*
Manual, *14*
Manus, *14*

masseter, *105, 113, 172*
mastoid process, *78, 105, 113*
mastoiditis, *78*
matrix, *49, 53, 55, 56, 69, 71, 72, 73, 219*
maxilla, *79, 97, 274*
Maxilla, *76, 79, 274*
maxillary sinus, *79*
mean arterial pressure, *250, 282*
Mechanoreceptors, *189*
Medial, *12, 21, 22, 201, 256*
medial collateral ligament, *150*
medial epicondyle, *87, 104, 147*
medial malleolus, *93, 97*
medulla oblongata, *156, 158, 159, 165, 183, 249, 272*
medullary respiratory center, *272*
megakaryocytes, *222*
meiosis, *31, 332, 335, 338, 339*
melanin, *62, 63*
melanocyte, *62*
melatonin, *165*
Melatonin, *215, 216*
membranous urethra, *278, 328*
menarche, *336*
meninges, *77, 156, 159, 185, 365*
menstrual cycle, *329, 336*
Mental, *14, 304*
mental foramen, *79*
mental protuberance, *79*
Mentum, *14*
Merkel's discs, *189, 201*
metabolic, *43, 206, 296, 300, 301, 303*
Metabolism, *5, 42, 43, 341*
Metacarpals, *82, 90, 364*
metaphase, *30, 31*
metatarsals, *93, 106*
Metatarsals, *84, 93*
microfilaments, *30*
Microglia, *177*
microscope, *53*
micturation, *289, 290*
midbrain, *165, 166, 168, 183*
mitochondrion, *30, 44, 46, 49*
mitosis, *30, 40, 71, 312, 332, 339*
monocytes, *221*
motor neuron, *182*
mRNA, *28, 29*

Mucosa, *231, 306*
Mucosa Associated Lymphoid Tissue (MALT), *231*
Muscle, *53, 56, 58, 99, 104, 128, 129, 175, 190, 201, 306*
muscle spindles, *189*
muscles, *7, 25, 43, 44, 51, 53, 56, 64, 78, 82, 84, 86, 87, 90, 99, 101, 103, 107, 108, 110, 111, 112, 113, 115, 116, 117, 119, 120, 121, 122, 123, 125, 128, 129, 130, 150, 155, 157, 158, 159, 166, 171, 172, 174, 175, 183, 186, 189, 190, 195, 199, 201, 229, 242, 262, 264, 272, 300, 303, 308, 313, 330, 335, 364*
myelin, *156, 176, 177, 180, 181, 186*
mylohyoideus, *105*
myocardium, *242*
myometrium, *329*
myosin, *128, 129, 140*
NADH2, *44, 45, 46, 48, 49, 50*
nails, *64, 66*
nasal, *77, 78, 79, 197, 261, 274*
nasal septum, *77, 261*
nasopharynx, *261*
Naturally acquired active immunity, *237, 239*
Naturally acquired passive immunity, *238, 239*
navicular, *93, 125*
Navicular, *84*
negative feedback, *11, 21, 335, 336*
nephon loop, *288*
nephron, *278, 287, 288, 291, 292, 297*
nephron loop, *278*
nervous system, *56, 61, 63, 129, 130, 155, 156, 159, 161, 168, 171, 173, 176, 177, 182, 183, 185, 186, 189, 190, 195, 215, 248, 249, 251, 252, 257, 270, 272, 278, 282, 291, 298, 300, 314, 317, 324*
net osmotic pressure (NOP), *253*
neurohypophysis, 207, 208
neuron, *140, 158, 165, 176, 177, 180, 181, 182, 183, 186, 365*
neurons, *53, 56, 158, 166, 176, 177, 178, 181, 182, 183, 186, 187, 272*
neurotransmitter, *129, 140, 177, 181, 186, 187*
Neutrophils, *220*
night blindness, *195*
nipple, *330*
nitric oxide, *287, 334*

nitrogen containing base, *25, 41*
nociceptors, *189*
nodes, *248*
Nodes of Ranvier, *180*
non-self, *235, 236*
Non-specific defense, *235*
norepinephrine, *249*
Norepinephrine, *140, 182, 186, 215, 216, 292*
Nucleases, *317*
nucleotide, 25, 41
nucleus, *25, 28, 29, 30, 41, 49, 158, 206, 220, 227, 338*
oblique, *14, 105, 139, 150, 172, 195, 201, 262*
obturator foramen, *90*
Occipital, *74, 78, 96, 104, 185, 187*
occipital condyles, *78*
occipital lobe, *161, 172, 192*
Oculomotor, *171*
of homologous chromosomes, *333*
Olecranal, *14*
Olecranon, *14*
olecranon process, *87, 104*
olectranon process, *87*
Olfactory, *171, 172*
olfactory nerve, *77*
Oligodendrocytes, *177, 187*
oocyte, *327, 329, 335, 336, 337, 338, 339*
oogenesis, *335*
opsins, *195*
Optic, *171, 172, 201*
optic nerve, *76, 192, 195*
oral, *308*
orbit, *49, 77, 78, 79*
organ of Corti, *197, 199*
organelle, *28, 30, 44*
osmosis, *32, 33, 40, 251, 253, 283, 287, 296, 297, 298, 304*
ossicles, *197, 199*
Osteoarthritis, *150*
Osteoblasts, 71
Osteocytes, *71*
osteoporosis., *73*
otolithic membrane, *199*
outer tunic, *192*
Oval window, *197, 201*
ovarian follicle, *329, 336, 339*
ovaries, *73, 327, 329, 336*
ovulation, *329, 336, 340*

Oxaloacetate, *48*
oxidation, *49, 301*
oxygen, *44, 45, 50, 219, 220, 222, 229, 241, 261, 265, 266, 268, 271, 272, 275, 281*
Oxyhemoglobin, 219
oxytocin, *214, 338*
pacemaker, *173, 248, 257*
Pacinian corpuscles, *183, 189, 201*
Pain receptors, *189*
Palatine, *76, 79, 274*
palatine process, *79*
palmaris longus, *104, 107*
pancreas, *54, 216, 227, 314, 315, 320, 325*
Pancreatic Amylase, *317*
Pancreatic Lipase, *317*
papillary, *242*
Parafollicular cells, *214*
parahippocampal gyrus, *167*
parasagittal, *14, 15*
parasites, *220*
parasympathetic, *171, 173, 174, 186, 249, 278, 291, 313, 317, 324, 334, 339*
parathyroid glands, *214*
parathyroid hormone (PTH), *214*
Parietal, *74, 96, 185, 186, 187, 304, 309, 324*
parietal lobes, *161*
parietal pleura, *262*
Parotid, *308, 324*
partial pressure, *265, 266*
Patella, *14, 84*
patellar ligament, *122, 150*
pathogen, *235, 236, 237, 239*
pectineus, *121*
Pectoral, *14, 21, 230, 233*
pectoral girdle, *84*
pectoralis major, *104, 110*
pectoralis minor, *104, 112*
Pectoralis minors, *262*
Pectorus, 14
Pedal, *14*
peduncles, *166*
penile urethra, *278, 328*
penis, *278, 328, 330, 334, 335, 339*
Pepsin, *309, 317*
pericardial fluid, *242*
pericardium, *242*
perichondrium, *72*
perikaryon, *176*

perimetrium, *329*
perimysium, *128*
perineum, *330*
Peripheral resistance, *252, 257*
Peripheral Resistance, *250, 252*
peristaltic, *278, 313, 314*
peritoneum, *277*
Pernicious, *220*
peroneus brevis, *123*
peroneus longus, *123, 125*
peroneus tertius, *123*
perpendicular plate, *77*
Pes, 14
pH, *272, 281, 289, 296, 301, 302, 303*
phagocytosis, *177, 220, 236*
Phalanges, *82, 84, 90, 93*
pharynx, *82, 261, 308, 324*
phosphate, *25, 30, 40, 41, 43, 45, 48, 49, 50, 301*
phosphocreatine, *43*
Phosphophenolpyruvate, *45*
phosphorylation, *206*
phosphoryllate, *43, 45, 51*
phosphoryllation, *49, 50, 300*
photoreceptors, *195, 201*
Photoreceptors, *189*
physiology, 7, 8, 11, 21, 129, 178, 180, 199, 214, 248, 265, 271, 277, 280, 297, 298, 301, 302
Pia mater, *156, 185, 186*
Pierre-Simon Laplace, *273*
pineal body, *165*
pinna, *197, 199*
Pisiform, *88*
pituitary gland, *73, 76, 165, 185, 207, 288, 297, 332, 335, 336*
plane, *14, 15, 21, 80, 117*
Plantar interossei, *126*
Plantarflexion, *144, 145*
Plasma, *222, 253, 304*
plasma protein, *220*
plasma proteins, *220, 227, 235, 236, 253, 275, 281, 302, 314*
platelet plug formation, *222*
Platelets, *222*
plica circulares, *312*
pneumotaxic center, *272*
polar body, *31, 335, 338*

polyribosome, *29*
pons, *155, 165, 166, 183, 272*
pontine respiratory group, *272*
Popliteal, 14, 21, 230
positional terms, *12*
positive feedback, *11, 214*
post-central gyrus, *183*
Posterior, *12, 123, 125, 159, 206, 256*
posterior cruciate ligament, *150*
posterior interventricular, *243*
potassium, *178, 179, 180, 186, 222, 248, 283, 292, 296, 298, 300, 304*
P-R interval, *254*
precentral gyrus, *183*
pregnant, *329*
prepuce, *328, 330*
pre-synaptic neuron, *177*
progesterone, *336, 337, 338, 340*
prolactin, *214, 216, 338*
proliferative phase, *336, 340*
Pronation, *145, 152*
pronator teres, *104, 107*
prophase, *30, 31*
proprioception, *158, 167*
Proprioceptors, *189*
Prostaglandins, *205*
prostate, *278, 327, 328, 335, 339*
prostate specific antigen (PSA), *328*
prostatic urethra, *278, 328*
protein filaments, *25, 30, 56, 128*
protein kinase, *206*
Prothrombin, *224*
Protraction/Retraction, *145*
Proximal, *12, 22, 71, 88*
proximal convoluted tubule, *278, 291, 300*
Pseudostratified columnar, *54, 274*
psoas, *103, 120*
psoriasis, *62*
pterygoids, *113*
puberty, *64, 332, 333, 335, 336*
pubic arch, *90*
pubic tubercle, *90*
pubis, *90, 96, 105, 116, 148, 330*
pubofemoral ligament, *148*
pulmonary arteries, *244, 267*
pulmonary trunk, *242, 256*
Pulmonary valve, *242*

pulmonary veins, *244, 266*
Pulmonary veins, *242*
pulp, *232*
pupil, *173, 192*
Purkinjie fibers, *249*
P-wave, *254*
pyramids, *166, 277*
pyruvate, *44, 45, 50*
Pyruvate, *45*
QRS complex, *254*
Q-T interval, *254*
quadrants, *15*
Quadratus plantus, *126*
quadriceps, *103, 122, 150*
radial collaterals, *147*
radial notch, *87*
radial tuberosity, *87, 104, 107*
Radius, *82, 87, 96*
RBCs, *219, 220, 224*
receptor, *155, 158, 182, 183, 201, 206*
receptors, *155, 183, 189, 190, 195, 196, 201, 205, 206, 214, 249, 272, 283*
rectum, *174, 313, 314, 327, 329*
rectus, *105, 106, 116, 122, 172, 195, 201*
rectus abdominis, *105*
rectus abdominus, *116*
rectus femoris, *106*
rectus femorus, *122*
red marrow, *71*
reflex, *182, 183, 187, 272, 314*
Renal artery, *277, 291*
renal columns, *277*
renal cortex, *277*
renal hilus, *277*
renal papilla, *277, 278*
Renin, *215, 287, 292*
renin-angiotensin system, *257, 287, 297*
repolarize, *248*
reproductive system, *31, 327, 329, 330, 335, 336*
residual volume, *265*
respiratory acidosis, *270, 302, 303*
respiratory alkalosis, *270, 271, 303*
respiratory distress syndrome (RDS), *273*
respiratory system, *30, 235, 261, 274, 284, 301, 303*
Resting Membrane Potential, *178*
Reticular, *55, 58, 185, 233*

reticular formation, *166*
reticulospinal tracts, *159*
retina, *192, 195, 201*
retroperitoneal, *277*
Rh, *224, 225*
rhodopsin, *195*
rhomboideus, *104*
ribose sugar, *25*
ribosome, *28, 29*
ribs, *80, 81, 82, 104, 105, 110, 112, 113, 262, 264*
right coronary artery, *243*
right lymphatic duct, *230*
right marginal, *243*
rods, *25, 128, 195, 201*
Rotation, *144, 145*
rotator cuff, *101, 110, 111*
Round window, *197, 201*
rubrospinal tracts, *159*
Ruffini corpuscles, *189, 201*
rugae, *308*
sagittal, *14, 15, 21, 80*
salivary amylase, *308, 324*
Salivary Amylase, *317*
saltatory, *180, 181, 186*
Salty, *190, 201*
sarcomeres, *128*
sartorius, *105, 122*
scala, *197*
scala media, *197*
scala tympani, *197*
scala vestibuli, *197*
Scaphoid, *88*
Scapula, *82, 84, 97*
sclera, *192*
scoliosis, *81*
scrotum, *327, 330*
sebaceous gland, *64*
sebaceous glands, *63*
Second degree burns, *65*
second messenger, *205, 206, 207*
secondary ossification centers, *72*
second-order neuron, *158*
Secretin, *320*
segmental, *158, 277*
semicircular canals, *197, 200*
Semicircular canals, *197, 201*
semimembranosus, *106, 123*

seminal vesicles, *327, 328*
seminiferous tubules, *327*
semitendinosus, *106, 123*
senses, *64, 189, 190, 197, 199, 201, 288, 314*
sensory receptors, *61, 63, 66, 155, 183, 189*
sensory systems, *189*
septum pellucidum, *168*
Serosa, *306*
serotonin, *222*
Serotonin, *140, 182*
Serous fluid, *242*
serratus anterior, *104, 112*
Sertoli cells, *327, 332*
sex, *30, 63, 327, 329, 332, 335, 336, 339*
Shin, *14, 93*
Short bones, *69*
shoulder, 14, 21, 84, 101, 103, 104, 107, 108, 111, 143, 146, 148, 152
Shoulder, *14, 82, 108, 145, 146, 152*
sigmoid colon, *313*
Simon-Laplace Law, *273*
sinoatrial node (SA node), *248*
Sinus, *74*
skeleton, *25, 69, 74, 76, 82, 84, 96, 300, 364*
skin, 14, 17, 25, 32, 54, 58, 61, 62, 64, 65, 66, 155, 158, 159, 166, 183, 189, 235, 328, 330, 336, 337
small cardiac vein, *243*
small intestine, *312, 313, 322, 324, 325*
smell, *77, 161, 165, 172, 187, 189, 190, 201*
Smooth muscle, *56*
sodium, *178, 179, 180, 186, 222, 248, 251, 257, 282, 283, 284, 287, 288, 292, 296, 297, 298, 301, 303, 304, 325*
soleus, *106, 125*
somatic senses, *189*
Somatostatin, 208, 216
sound waves, *155, 199*
Sour, *190*
special senses, *189*
Specific defense, *235*
sperm, *30, 31, 327, 328, 329, 332, 333, 334, 338, 339*
spermatogenesis, *332, 335*
sphenoid, *76, 77, 79, 80, 274*
Sphenoid, *74, 76, 79, 96, 274*
sphincter urethrae, *119*

spinal canal, *12, 159*
spinal cord, *12, 78, 80, 155, 156, 157, 158, 159, 161, 165, 166, 168, 171, 177, 181, 182, 183, 185, 314, 334, 335*
spinal curves, *81*
spinal tract, *155, 156, 185*
spine, 74, 80, 81, 84, 90, 103, 104, 105, 106, 108, 110, 113, 116, 122, 139, 143, 145, 150, 156, 158, 159
spinocerebellar, *158*
spinothalamic, *157, 158, 183, 185*
spirometer, *265*
spleen, *219, 227, 231, 232, 233, 313, 366*
splenic flexure, *313*
splenius capitus, *113*
sprains, *150, 190*
squamous, *53, 54, 58, 61, 233, 242, 274, 291*
Stapes, *197, 201*
static equilibrium, *197, 199, 201*
Stensen ducts, *308*
Sternal, *14*
Sternocleidomastoids, 262
sternum, 12, 82, 84, 97, 104, 110, 113, 232, 241
Sternum, *14, 82, 97*
steroid, *206*
stomach, *11, 12, 14, 30, 172, 220, 229, 235, 297, 308, 309, 312, 317, 320, 324, 325, 367*
Stomach, *14, 21, 174, 233, 306, 312, 317, 324, 325*
strains, *150, 190*
Stratum basale, *61, 66*
Stratum corneum, *61, 66*
Stratum granulosum, *61, 66*
Stratum lucidum, *61, 66*
Stratum spinosum, *61*
Stretch receptors, *189, 272*
striated, *53, 56*
stroke volume, *250, 251*
styloid process, *78, 87, 105*
Styloid process, *74*
subclavian, *230, 246, 256*
subclavian trunks, *230*
subcutaneous layer, *61*
Sublingual, *308, 324*
Submucosa, *306*
substantia nigra, *166*
subthalamus, *165*

Succinate, *48*
Succinyl, *48*
sulci, *161*
Superficial, *14, 22*
superficial transverses perinea, *119*
superior, *12, 14, 15, 21, 78, 82, 90, 97, 105, 110, 111, 115, 121, 122, 161, 165, 166, 167, 172, 185, 195, 232, 242, 246, 256, 261, 262, 308, 328, 329, 330*
Supination, *145, 152*
suppressor, *236*
supraorbital margin, *78*
supraspinatus, *104, 110, 111*
supraspinous fossa, *84, 104*
suprorbital foramen, *78*
Sura, *14*
surface area, *219, 252, 309, 312*
surfactant, *272, 273*
suture, *80, 96, 143*
Suture, *74, 96, 143*
sutures, *79, 80*
sweat glands, *63, 64, 66, 175*
sweating, *175, 297, 298*
Sweet, *190, 201*
sympathetic, *63, 173, 174, 175, 186, 249, 251, 252, 257, 278, 282, 291, 324, 334, 335*
Synarthrotic, *143*
syndesmosis, *143*
Syndesmosis, *143*
Synovial, *143, 152*
systole, *244, 250, 254, 256*
T wave, *254*
T3 (triiodothyronine), *214*
T4 (tetraiodothyronine), *214*
taeniae coli, *313*
Talus, *84, 93*
target tissues, *205*
taste, *155, 171, 172, 186, 189, 190, 201, 324*
Taste receptors, *190*
tectorial membrane, *199*
teeth, *79, 112, 308, 324*
tegmentum, *166*
telophase, *30, 31*
Temporal, *74, 78, 96, 185, 187*
temporal lobes, *161*
temporal process, *78, 79*
temporalis, *113, 172*
temporomandiblular joint, *113*

temporomandibular joint (TMJ), *79*
tendons, *55, 69, 108, 125*
tensor fascia latae, *105, 120, 121*
teres major, *110*
Terminal hairs, *63*
testes, *73, 327, 332, 335, 339*
testosterone, *327, 332, 335, 339*
thalamus, *158, 165, 168, 171, 183*
The intertubercular groove, *86*
Thermoreceptors, *189*
thermostat, *11*
Thigh, *14, 90, 120, 122, 123*
Third degree burns, *65*
third ventricle, *168*
third-order neuron, *158*
thirst center, *297*
Thoracic, *14, 21, 233*
thoracic duct, *230*
thoracic spine, *80, 108*
Thorax, *14*
threshold, *179, 180*
thrombin, *224*
thrombocytes, *222*
thromboplastin, *224*
Thrombus, *224*
Thymosins, *215*
thymus, *227, 231, 232, 236, 366*
thyroid cartilage, *214, 261*
thyroid gland, *73, 214*
Thyroxine, *73*
Tibia, *84, 93, 96, 97*
tibial tuberosity, *93, 106, 122, 150*
tibialis anterior, *103, 106, 123*
tidal volume, *265, 274*
tissue, *53, 54, 55, 56, 58, 61, 62, 69, 71, 116, 121, 128, 139, 143, 146, 150, 152, 156, 168, 175, 176, 189, 205, 214, 219, 222, 224, 229, 230, 231, 232, 233, 235, 241, 242, 248, 252, 253, 257, 266, 278, 291, 300, 306, 328, 330, 336*
T-lymphocytes, *221, 232*
tongue, *82, 112, 115, 172, 186, 190, 308*
Total lung capacity, *265*
trachea, *261, 262, 274, 308*
transcription, *28, 29, 206*
transferrin, *220*
Transitional, *54, 58, 291*
translation, *28, 29, 150*

transverse, *14, 15, 21, 80, 81, 105, 116, 313*
Transverse humeral, *146*
Trapezium, *88*
trapezius, *104, 108, 172, 186*
Trapezoid, *88*
TRH, *214*
triceps brachii, *104, 107*
Tricuspid valve, *242, 256*
Triquetrum, *88*
tRNA, *28, 29, 41*
Trochanter, *73, 96*
trochlea, *87, 147*
Trochlear, *171, 172*
trochlear notch, *87, 147*
troponin-tropomyosin, *129, 139*
Trypsin, *317*
Tubercle, *73, 96*
Tuberosity, *73, 96*
tunica externa, *244*
tunica interna, *244*
tunica media, *244, 256*
tunics, *191, 196*
Tympanic membrane, *197, 201*
type A, *224, 237*
type AB, *224, 225*
type B, *224*
type O, *224, 225*
Ulna, *82, 87, 97*
ulnar collaterals, *147*
Ulnar/Radial Deviation, *145*
Umami, *190*
uogenital diaphragm, *119*
ureters, *277, 278, 280*
urethra, *277, 278, 280, 291, 328, 329, 330, 335*
urinary bladder, *54, 58, 277, 278, 328*
urinary system, *214, 251, 277, 291, 298*
urinating, *297*
urine, *251, 277, 278, 280, 281, 282, 287, 288, 289, 291, 298, 300, 301*
urogenital, *119, 330*
uterosacral ligaments, *329*
uterus, *329, 332, 336, 338, 339*
uvula, *261*
vaccines, *238*
vagina, *329, 330*

Vagus, *171, 172, 186, 257*
vagus nerve, *249, 272*
valves, *229, 242, 244, 256*
vas deferens, *327, 329, 339*
vasoconstrict, *222, 282, 288*
vasodilation, *235, 237, 252, 292*
vastus intermedius, *106, 122*
vastus lateralis, *106, 122*
vastus medialis, *106, 122*
veins, *69, 230, 232, 233, 242, 243, 244, 245, 246, 247, 256, 266, 267*
Vellus hairs, *63*
vena cava, *242, 246, 256*
vena cavae, *242, 246*
ventral, *11, 156, 159, 183, 272*
ventricles, *168, 171, 177, 186, 241, 242, 243, 244, 248, 249, 256, 257*
venules, *244*
vertebrae, *71, 80, 81, 82, 104, 159*
vestibule, *199, 330*
Vestibule, *197, 201, 339*
Vestibulocochlear nerve, *197*
villi, *186, 312, 322*
visceral pleural membrane, *262*
Vital capacity, *265, 274*
vitamin A, *195*
vitamin D, *62, 73, 281, 300*
vitreous humor, *192*
vocal cords, *261, 262, 274, 335, 339*
Volksmann's canals, *70*
vomer, *261*
Vomer, *76*
vomiting, *165, 297, 298, 300, 303, 304, 324*
WBC, *215, 222*
white matter, *156, 157, 161, 166, 177, 180, 185*
Wrist, *14, 88, 108, 145*
xiphoid process, *82, 105, 116*
yellow marrow, *71*
Zone of calcification, *73*
Zone of proliferation, *73*
Zone of resting cartilage, *73*
zygomatic, *78, 79, 97, 105, 112*
Zygomatic, *76, 79, 97*
zygomatic process, *78, 79*
zygomaticus, *112*
zygote, *335, 338*

Image Credits

Figures

Chapter 1
1.2 From: http://commons.wikimedia.org/wiki/File:Scheme_body_cavities-en.svg
1.3 From: http://commons.wikimedia.org/wiki/Image:Human_body_features.png
1.4 Modified from: http://commons.wikimedia.org/wiki/File:Human_body_features.svg
1.5 Author
1.6 http://commons.wikimedia.org/wiki/File:Line-drawing_of_a_human_man.svg
Courtesy of NASA
Original image modified by Dr. Bruce Forciea

1.7 From: http://commons.wikimedia.org/wiki/File:NormalerKorper_mit_full_bust.PNG
1.8 From: http://commons.wikimedia.org/wiki/File:Surface_projections_of_the_organs_of_the_trunk.png
1.9 Author
1.10 Author

Chapter 2
2.1 http://commons.wikimedia.org/wiki/Image:Animal_cell_structure_en.svg
2.2 Author
2.3 Author
2.4 From: http://commons.wikimedia.org/wiki/File:DNA
2.5 Author
2.6 Author
2.7 From: http://commons.wikimedia.org/wiki/Image:Scheme_facilitated_diffusion_in_cell_membrane-en.svg
2.8 Author
2.9 From: http://commons.wikimedia.org/wiki/File:Erythrozyten_und_Osmotischer_Druck.svg
2.10 From: http://commons.wikimedia.org/wiki/Image:Scheme_sodium-potassium_pump-en.svg
2.11 From: http://commons.wikimedia.org/wiki/Image:Endocytosis_types.svg

Chapter 3
3.1 Author
3.2 Author
3.3 Author
3.4 Author
3.5 Author
3.6 Author
3.7 Author

Chapter 4
4.1 From: http://commons.wikimedia.org/wiki/File:Illu_epithelium
4.2 From: http://commons.wikimedia.org/wiki/File:Illu_connective_tissues_1.jpg
4.3 From: http://commons.wikimedia.org/wiki/File:Illu_connective_tissues_2.jpg
4.4 From: http://commons.wikimedia.org/wiki/Image:Gray73.png
4.5 From: http://commons.wikimedia.org/wiki/File:Gray292.png
4.6 From: http://commons.wikimedia.org/wiki/File:Illu_muscle_tissues.jpg
4.7 Author
4.8 From: By Yvan Lindekens (Own work) [GFDL (www.gnu.org/copyleft/fdl.html) or CC-BY-SA-3.0-2.5-2.0-1.0 (www.creativecommons.org/licenses/by-sa/3.0)], via Wikimedia Commons

Chapter 5
5.1 From: http://commons.wikimedia.org/wiki/Image:Skinlayers.png
5.2 From: http://en.wikipedia.org/wiki/Image:Normal_Epidermis_and_Dermis_with_Intradermal_Nevus_10x.JPG
5.3 From: http://commons.wikimedia.org/wiki/Image:Skin.jpg

Chapter 6
6.1 From: http://commons.wikimedia.org/wiki/Image:Illu_long_bone.jpg

6.2 From: http://commons.wikimedia.org/wiki/File:Illu_compact_spongy_bone.jpg
6.3 From: http://commons.wikimedia.org/wiki/Image:Osteocyte_2.jpg
6.4 From: http://commons.wikimedia.org/wiki/Image:Bone
6.5 From: http://commons.wikimedia.org/wiki/File:Human_skeleton
6.6 From: http://commons.wikimedia.org/wiki/Image:Human_skull_side_bones_numbered.svg
6.7 From: http://commons.wikimedia.org/wiki/Image:Human_skull_front_bones_numbered.svg
6.8 Author
6.9 Author
6.10 From: http://commons.wikimedia.org/wiki/Image:Gray151.png
6.11 From: http://commons.wikimedia.org/wiki/File:Sutura_squamozomastoidea.PNG
6.12 From: http://commons.wikimedia.org/wiki/Image:Dolichocephalie.jpg
6.13 From: http://en.wikipedia.org/wiki/Image:Gray_111_-_Vertebral_column-coloured.png
6.14 From: http://commons.wikimedia.org/wiki/Image:Spinal_column_curvature.png
6.15 From: http://commons.wikimedia.org/wiki/Image:Gray112.png
6.16 From: http://commons.wikimedia.org/wiki/Image:Gray186.png
6.17 Author
6.18 Author
6.19 Author
6.20 Author
6.21 Author
6.22 Author
6.23 From: http://commons.wikimedia.org/wiki/Image:Carpus
Author: Benutzer Zoph
6.24 From: http://commons.wikimedia.org/wiki/Image:Metacarpals
6.25 From: http://commons.wikimedia.org/wiki/Image:Scheme_human_hand_bones-numbers.svg
6.26 Author
6.27 Author
6.28 From: http://commons.wikimedia.org/wiki/Image:Gray241.png
6.29 Author
6.30 Author
6.31 Author
6.32 Author
6.33 Author

Chapter 7
7.1 Author
7.2 From: http://commons.wikimedia.org/wiki/Image:Muscles_anterior_labeled.png
7.3 From: http://commons.wikimedia.org/wiki/Image:Muscles_posterior.png
7.4 From: http://commons.wikimedia.org/wiki/Image:Arm
7.5 From: http://commons.wikimedia.org/wiki/Image:Forearm
7.6 From: http://commons.wikimedia.org/wiki/Image:Forearm
7.7 From: http://upload.wikimedia.org/wikipedia/commons/3/3e/Forearm
7.8 From: http://commons.wikimedia.org/wiki/Image:Arm
7.9 From: http://commons.wikimedia.org/wiki/Image:Arm
7.10 From: http://commons.wikimedia.org/wiki/File:Illu_head_neck_muscle.jpg
7.11 From: http://commons.wikimedia.org/wiki/Image:Gray378.png
7.12 From: http://commons.wikimedia.org/wiki/Image:Splenius.png
7.13 From: http://commons.wikimedia.org/wiki/Image:Iliostalis.png
7.14 From: http://commons.wikimedia.org/wiki/Image:Genioglossus.png
7.15 From: http://commons.wikimedia.org/wiki/Image:Grays_Anatomy_image392.png
7.16 From: http://commons.wikimedia.org/wiki/Image:Transversus_abdominis.png
7.17 From: http://commons.wikimedia.org/wiki/Image:Ischiocavernosus-female
7.18 From: http://commons.wikimedia.org/wiki/Image:Ischiocavernosus-male
7.19 From: http://commons.wikimedia.org/wiki/Image:Gluteus_muscles
7.20 From: http://commons.wikimedia.org/wiki/Image:Sartorius_muscle.png
7.21 From: http://commons.wikimedia.org/wiki/Image:Semitendinosus_muscle.PNG
7.22 From: http://upload.wikimedia.org/wikipedia/commons/6/64/Tibialis_anterior_2.png
7.23 From: http://commons.wikimedia.org/wiki/Image:Gray438-cropped.png

7.24 From: http://commons.wikimedia.org/wiki/Image:Abductor_digiti_minimi_(foot).png
7.25 From: http://commons.wikimedia.org/wiki/Image:Musculus_flexor_digiti_minimi_brevis_(foot).png
7.26 From: http://commons.wikimedia.org/wiki/File:Skeletal_muscle.png
7.27 Author
7.28 Author
7.29 Author
7.30 Author
7.31 Author
7.32 Author
7.33 Author
7.34 Author
7.35 Author

Chapter 8
8.1 Author
8.2 From: http://commons.wikimedia.org/wiki/Image:Gray326.png
8.3 From: http://upload.wikimedia.org/wikipedia/commons/9/97/Gray329.png
8.4 From: http://commons.wikimedia.org/wiki/Image:Gray339.png
8.5 From: http://upload.wikimedia.org/wikipedia/commons/3/33/Gray340.png
8.6 From: http://commons.wikimedia.org/wiki/Image:Knee_diagram.png

Chapter 9
9.1 Author
9.2 Author
9.3 Author
9.4 From: http://commons.wikimedia.org/wiki/Image:Dermatoms.svg
9.5 From: http://commons.wikimedia.org/wiki/File:Brain_diagram_without_text.svg
Labelled by author
9.6 From: http://commons.wikimedia.org/wiki/Image:Cerebral_lobes.png
Labelled by Author
9.7 Modified by author from: http://commons.wikimedia.org/wiki/File:Mozok.gif
9.8 From: http://commons.wikimedia.org/wiki/File:Mozok.gif
Labelled by Author
9.9 From: http://commons.wikimedia.org/wiki/Image:Cerebellum_NIH.png
9.10 From: http://commons.wikimedia.org/wiki/Image:Brain_limbicsystem.jpg
9.11 From: http://commons.wikimedia.org/wiki/Image:Illu_meninges
9.12 Modified by author from: http://commons.wikimedia.org/wiki/Image:Gray750.png
9.13 Author
9.14 Author
9.15 From: http://commons.wikimedia.org/wiki/File:Complete_neuron
9.16 Author
9.17 Author
9.18 Author
9.19 Author
9.20 Author
9.21 Author
9.22 Modified by author from: http://commons.wikimedia.org/wiki/Image:Spinal_nerve.svg
Orignial authors: Mysid (original by Tristanb)

Chapter 10
10.1 Author
10.2 Author
10.3 Author
10.4 From: http://commons.wikimedia.org/wiki/File:Schematic_diagram_of_the_human_eye_en.svg
10.5 From: http://commons.wikimedia.org/wiki/File:Outer,_middle_and_inner_ear.jpg
10.6 Author
10.7 From: http://commons.wikimedia.org/wiki/File:Organ_of_corti.svg
10.8 From: http://commons.wikimedia.org/wiki/File:Hearing_mechanics_cropped.jpg

10.9 From: http://commons.wikimedia.org/wiki/File:Otolith_organ_of_vestibular_system.jpg
10.10 From: http://commons.wikimedia.org/wiki/File:Inner_ear%27s_cupula_transmitting_indication_of_acceleration.jpg

Chapter 11
11.1 From: http://commons.wikimedia.org/wiki/File:Illu_endocrine_system.jpg
11.2 Author
11.3 Author
11.4 Author
11.5 Author
11.6 Author
11.7 From: By Mikael Häggström (All used images are in public domain.) [Public domain], via Wikimedia Commons
11.8 From: http://commons.wikimedia.org/wiki/File:Illu_pituitary_pineal_glands.jpg
11.9 From: http://commons.wikimedia.org/wiki/File:Illu_adrenal_gland.jpg

Chapter 12
12.1 From: http://commons.wikimedia.org/wiki/File:Redbloodcells.jpg
12.2 From: http://commons.wikimedia.org/wiki/File:Neutrophil.jpg
12.3 From: http://commons.wikimedia.org/wiki/File:Eosinophil_1.png
12.4 From: http://commons.wikimedia.org/wiki/File:Basophil.jpg
12.5 From: http://commons.wikimedia.org/wiki/File:Monocyte.jpg
12.6 From: http://commons.wikimedia.org/wiki/File:Lymphocyte.jpg
12.7 From: http://commons.wikimedia.org/wiki/File:Coagulation_simple.svg
12.8 Author

Chapter 13
13.1 From: http://commons.wikimedia.org/wiki/File:Schematic_of_lymph_node_showing_lymph_sinuses.svg
13.2 From: http://commons.wikimedia.org/wiki/File:Illu_spleen
13.3 From: http://commons.wikimedia.org/wiki/File:Illu_thymus

Chapter 14
14.1 From: http://commons.wikimedia.org/wiki/File:Complement_pathway.png

Chapter 15
15.1 From: http://commons.wikimedia.org/wiki/File:Surface_anatomy
15.2 Derived from: http://commons.wikimedia.org/wiki/Image:Diagram_of_the_human_heart
15.3 From: http://commons.wikimedia.org/wiki/File:Human_healthy_pumping_heart
15.4 From: http://commons.wikimedia.org/wiki/File:Gray506.svg
15.5 From: http://commons.wikimedia.org/wiki/File:Circulatory_System_en.svg
15.6 Author
15.7 Author

Chapter 16
16.1 From: http://commons.wikimedia.org/wiki/File:Respiratory_system_complete_en.svg

16.2 From: http://commons.wikimedia.org/wiki/File:Gray961.png
16.3 From: http://commons.wikimedia.org/wiki/File:Inhalation
16.4 From: http://commons.wikimedia.org/wiki/File:Expiration_diagram.svg
16.5 Author
16.6 Author
16.7 Author
16.8 Author
16.9 Author
16.10 From: http://commons.wikimedia.org/wiki/File:Legameidrogeno-h2o.jpg

Chapter 17
17.0 From: http://commons.wikimedia.org/wiki/File:Urinary_system.gif
17.1 Author

17.2 Author
17.3 From: http://commons.wikimedia.org/wiki/File:Illu_bladder.jpg
17.4 Author
17.5 Author
17.6 Author
17.7 Author
17.8 Author
17.9 Author
17.10 Author
17.11 Author
17.12 Author
17.13 Author

Chapter 19
19.1 From: http://commons.wikimedia.org/wiki/File:Digestive_system_diagram_edit.svg
19.2 From: http://commons.wikimedia.org/wiki/File:Gray1024.png
19.3 From: http://commons.wikimedia.org/wiki/File:Cavities_evolution_1_of_5_ArtLibre_jnl.png
Labelled by Author
19.4 From: http://commons.wikimedia.org/wiki/File:Teeth_(PSF).png
19.5 From: http://commons.wikimedia.org/wiki/File:Parynx_simple_uk.svg
Labelled by Author
19.6 From: http://commons.wikimedia.org/wiki/File:Tractus_intestinalis_esophagus
19.7 From: http://commons.wikimedia.org/wiki/File:Illu_stomach2.jpg
19.8 From: http://commons.wikimedia.org/wiki/File:Intestine_and_stomach
Labelled by Author
19.9 From: http://commons.wikimedia.org/wiki/File:Liver2.png
19.10 From: http://commons.wikimedia.org/wiki/File:Illu_pancrease.jpg
19.11 Author
19.12 Author
19.13 Author
19.14 Author
19.15 Author
19.16 Author

Chapter 20
20.1 From: Derivative work from: http://commons.wikimedia.org/wiki/File:Male_anatomy.png
20.2 Author
20.3 From: http://commons.wikimedia.org/wiki/File:Scheme_female
20.4 From: http://commons.wikimedia.org/wiki/File:MajorEventsInMeiosis_variant.svg
20.5 From: http://commons.wikimedia.org/wiki/File:Complete_diagram_of_a_human_spermatozoa.svg
20.6 From: http://commons.wikimedia.org/wiki/File:Hormons_level_-_follicle

Appendix 1
A1.1 Derivative work from: By Didier Descouens (Own work) [GFDL (www.gnu.org/copyleft/fdl.html) or CC-BY-SA-3.0-2.5-2.0-1.0 (www.creativecommons.org/licenses/by-sa/3.0)], via Wikimedia Commons. Background removed and labeled by author.
A1.2 Derivative work from: By Didier Descouens (Own work) [GFDL (www.gnu.org/copyleft/fdl.html) or CC-BY-SA-3.0-2.5-2.0-1.0 (www.creativecommons.org/licenses/by-sa/3.0)], via Wikimedia Commons. Background removed and labeled by author.
A1.3-A1.10 Author

Made in the USA
Lexington, KY
05 February 2013